# CHEMICAL PHYSICS OF PYROLYSIS, COMBUSTION AND OXIDATION

# Chemical Physics of Pyrolysis, Combustion and Oxidation

A. A. Berlin
I. A. Novakov
N. A. Khalturinskiy
and
G. E. Zaikov
Editors

Nova Science Publishers, Inc.
*New York*

**Editorial Production Manager:** Donna Dennis
**Senior Production Editor:** Susan Boriotti
**Office Manager:** Annette Hellinger
**Graphics:** Andrea Charles, Magdalena Nuñez and Annemarie VanDeWater
**Editorial Production:** Marius Andronie, Maya Columbus, Keti Datunashvili, Andrew Kallio, Vladimir Klestov, Lorna Loperfido, Joann Overton and Rusudan Razmadze
**Circulation:** Ave Maria Gonzalez, Vera Popovic, Luis Aviles, Alexandra Columbus, Raymond Davis, Cathy DeGregory, Melissa Diaz, Marlene Nuñez, Jeannie Pappas, Lauren Perl and Katie Sutherland

*Library of Congress Cataloging-in-Publication Data*
*Available Upon Request*

ISBN: 1-59454-163-9.


400 Oser Ave, Suite 1600
Hauppauge, New York 11788-3619
Tele. 631-231-7269 Fax 631-231-8175
e-mail: Novascience@earthlink.net
Web Site: http://www.novapublishers.com



*Printed in the United States of America*

# CONTENTS

# LIST OF CONTRIBUTORS

Adamova L.V.
Alexeeva O.V.
Artemenko S.E.
Aseeva R.M.
Bakhtina G.D.
Begretov M.M.
Berlin A.A.
Besshaposhnikova V.I.
Bondarenko S.N.
Borukaev T.A.
Bui Dinh Thanh
Bychkova E.V.
Bystrov S.G.
Davydov E.Ya.
Didik A.A.
Donskoi A.A.
Fridrikh ).N.
Gajdalin A.N.
Gaponova I.S.
Garashchenko A.N.
Grishina O.A.
Hohlova T.V.
Kablov V.F.
Kadykova J.A.
Khalturinskiy N.A.
Kochnov A.B.
Kodolov V.I.
Kolesnikova N.N.
Konstantiniva M.L.
Kozlov G.V.
Kulikova T.V.
Kuperman A.M.
Kuznetsov A.P.

Lanska B.
Lipanov A.M.
Lomakin S.M.
Makarov G.G.
Nikitina T.G.
Novakov I.A.
Olifirenko V.N.
Panova L.G.
Panova L.G.
Pariiskii G.B.
Ponomarenko A.A.
Pustoshnyi V.P.
Razumovskii S.D.
Rusinova E.V.
Sashkina M.A.
Sebenda J.
Serkov B.B.
Shaov A.Kh.
Sheptalin R.A.
Shlyapnikov Yu.A.
Shuklin S.G.
Solodinov V.I.
Tuzhikov O.I.
Tuzhikov O.O.
Vasiljeva O.G.
Vshivkov S.A.
Zaikov G.E.

# List of Institutes

Basic Research-High Educational Center of Chemical Physics and Mesoscopy, Udmurd Scientific Center, Ural Division, Academy of Sciences, Izhevsk, Udmurdia, Russia

D.I.Mendeleev Russian Chemical-Technological University Ministry of High Education, Moscow

H.M.Berbekov Kabardino-Balkarian State University, Ministry of High Education, Nal'chik

Institute of Aviation Materials, Moscow

Institute of Macromolecular Chemistry Czech Academy of Sciences, Praque, Czech Republic

N. M. Emanuel Institute of Biochemical Physics of Russian Academy of Sciences, Moscow

N. N. Semenov Institute of Chemical Physics of Russian Academy of Sciences, Moscow

Saratov State Technical University, Engels Technological Institute, Ministry of High Education, Engels, Saratov District

State Fire Academy, Ministry of Civil Defence, Emergency Situation and Disaster Managment, Moscow

Ural State University, Ministry of High Education, Ekaterinburg

Volgograd State Technical University Ministry of High Education, Volgograd

Volzhsky Polytechnical Institute (Branch of Volgograd State University), Volzhsk-town, Volgograd District, Russia

# PREFACE

**Epigraphs**

1. If we can do this – we must.
V. Putin, the President of the Russian Federation

2. Nothing brings us to the aim faster than the lack of resources
A. Gerasimov, the analyst of the Independent Russian TV

Pyrolysis, combustion and oxidation reactions are intrinsically similar classes of chemical processes. Solid polymers do not combust themselves but gas products of their pyrolysis. Therefore, pyrolysis is the first stage of polymer combustion (the lower is the pyrolysis rate, the lower is the polymer combustibility). In 1939, this combustion mechanism was discovered by the leading scientists of the Institute of Chemical Physics of the Academy of Sciences of the USSR Yu. D. Khariton and Ya.B.Zeldovich and then was proved by many investigators of N.N. Semenov school (O.I. Leipunskii, D.A. Frank-Kamenetskii, A.F. Sokolik, N.Ya. Anin, A.F. Belyaev, A.G. Merzhanov, L.A. Gussak, A.H. Dremin, A.G. Merzhanov and others). On the other hand, combustion is a particular case of oxidation reactions. This precisely connection has forced the authors to make a collection devoted to these three classes of chemical reactions.

Some reviews and articles in the Collection are specially devoted to the problems of Pyrolysis and Flammability. Of special attention are also flame retardants based on nanocomposites and development of principles for creating flame retardants with nanostructures. Data on foamy coke structure control by introducing carbon, metal-containing nanostructures to flame retardant, foaming covers, shown in the Collection, are important for practical purposes.

Collected information concerns both artificial polymers of different classes and natural polymers. Hence, of special attention are factors affecting heat release.

Despite the fact that organochlorine compounds give dioxins (substances causing various diseases, including cancerous) when combusting, fire resistant and thermally stable materials derived from chlorinated polymers are still widely applied. The discussion of this problem is also included in the Collection as the humankind has not yet found the proper alternative to chlorinated polymers in the most important fields of fire retardant polymer application.

Multiple reviews in the Collection are devoted to the study and application of organophosphorous compounds as substances suppressing combustion. Various polymeric material including blocks, films, coatings, fibers and polymer-based composites were chosen as the objects for investigation.

The authors look forward for great interest of the readers – the representatives of pure and applied chemistry, to this Collection.

**Editors**

*Alexander A. Berlin,* Director of N.N. Semenov Institute of Chemical Physics (ICP), Russian Academy of Sciences (RAS), Moscow

*Ivan A. Novakov*, Rector of Volgograd State Technical University, Volgograd

*Nikolai A. Khalturinskiy*, Deputy Director of ICP RAS

*Gennady E. Zaikov*, Member of Directorium of N.M. Emanuel Institute of Biochemical Physics RAS, Moscow

In: Chemical Physics of Pyrolysis, Combustion and Oxidation ISBN 1-59454-163-9
Editors: A. A. Berlin et al., pp. 1-17 

*Chapter 1*

# PYROLYSIS AND FLAMMABILITY OF POLYURETHANE: ORGANOPHILIC CLAY NANOCOMPOSITES

***G. E. Zaikov, R. A. Sheptalin* and S. M. Lomakin***
Institute of Biochemical Physics of Russian Academy of Sciences,
4 Kosygin Street, Moscow, 119991, Russia,
e-mail: *chembio@sky.chph.ras.ru*,
tel: 7-095-9397191, fax: 7-095-1374101
* D.I Mendeleyev Russian Chemical-Technological University
9 Miusskaya Place, Moscow, Russia

## INTRODUCTION

Polymer layered silicate (clay) nanocomposites are materials with unique properties when compared with conventional filled polymers. Polymer nanocomposites, especially polymer-layered silicates, represent a new alternative to conventionally filled polymers.

In polymer industry there exists a need for new, more effective, and environmentally friendly flame resistance polymers. Recent data on the combustion of polymer nanocomposites indicate that they could be employed for this purpose [1].

There are several proposed mechanisms as to how the layered silicate affects the flame retardant properties of polymers [1]. The first is increased char layer that forms when nanocomposites are exposed to flame. This layer is though to inhibit oxygen transport to the flame front, as well as, gaseous-fuel transport from the polymer and therefore reduce the heat release rate of the burning surface. This may interrupt the burning cycle as radical species are needed to break polymer chains into fuel-fragments. The disordered nanocomposites also inhibit oxygen and combustible "fuel" species transfer by increasing the path length of these species to the flame front. There is also a high possibility of alumina-silicate solid phase catalysis of polymer decomposition, which can dramatically change the overall scheme of thermal degradation process kinetics.

Polyurethane foam (PU) is unique and most useful commercial polymeric material. In the present study a Polyurethane-Organophilic Montmorillonite Nanocomposite (PU-OM) was synthesized by known two-stage procedure [2]. Simulated Heat Release analysis was applied to provide the information on flame-resistance properties of PU-OM.

# EXPERIMENTAL

## Materials

The original sodium montmorillonite fraction having a cation-exchange capacity of 80 mEq/100 g was allocated from natural bentonite clay (Ural Region, Russia).

## Preparation of Organophilic Montmorillonite (OM)

Montmorillonite was gradually added to a solution of alkyltrimethylammoniumchloride (C16-C18) - Quartamin 60L, Kao Corporation S.A. and resultant suspension was stirred vigorously for 3 h. at 50°C. After filtration the product was placed in vacuum oven at 100°C for 24 hours.

## Synthesis of Propylene Oxide-OM (PO-OM)

OM and Propylene Oxide glycol (Laprol 5003-2B-10), provided by Nizjnemkamskneftkhim co., whose average molecular weight is about 5,000 (Reaction Scheme 1):

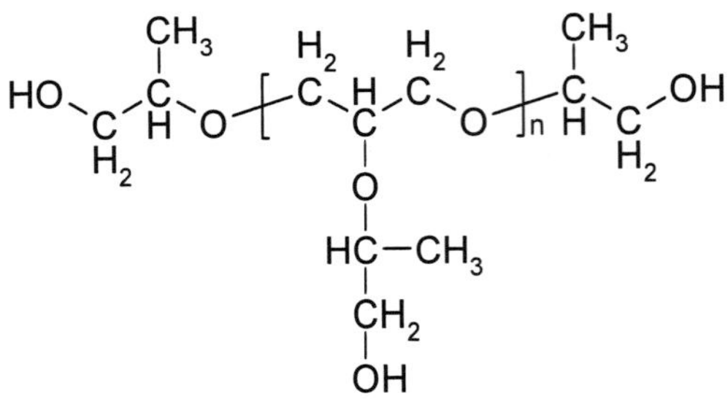

**Reaction Scheme 1.** Propylene Oxide glycol (Laprol 5003-2B-10).

Were mixed together, 1:1, with rapid stirring at 40°C for 1 hour. Then the colloidal PO-OM intercalated hybrid was obtained.

## Synthesis of Polyurethane-Organophilic Montmorillonite Nanocomposite (PU-OM

PU-OM was synthesized by polycondensation *in situ* of component A: hybrid polyol (PO-OM) and plain polyol (1:10), catalyst (triethylenediamine) -Tegostab 100, by

Goldschmidt, water and stabilizer and component B: toluene diisocyanate - Voranate T-80, by Dow Chemical co. and polyisocyanate, PITZ-B by NPO Korund co. (1:2). Component A and component B were mixtured with stirring. Then the mixture was poured out in the special form followed by its solidification and foaming during 15 minutes.

## XRD Characterization

X-ray diffraction (XRD) analysis of OM, PO-OM and PU-OM was performed on a Philips diffractometer using Cu $K_{\alpha}$ radiation, ($\lambda = 0.1540562$ nm).

The microstructure of PU-OM nanocomposite was characterized using XRD. The XRD patterns of OM and PU-OM, which reveal the intercalated structure of OM and delaminated structure of PU-OM nanocomposite, are shown in the Fig. 1. The interlayer (basal) spacing (001) for the OM is 3.5 nm.

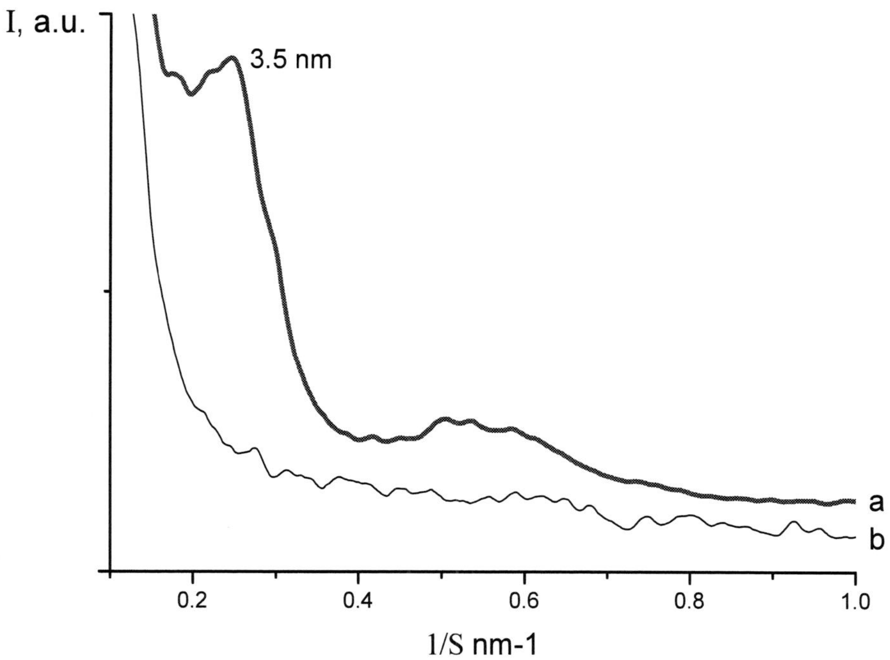


**Figure 1.** XRD patterns of OM and PU-OM delaminated nanocomposite: a – OM, b – PU-OM

## Pyrolysis

Extensive pyrolysis of PU and PU-OM/nanocomposite samples over a one minute period of time was carried out in a laboratory model pyrolizer in air atmosphere under 250 and 500°C. The oven temperature was monitored with a thermocouple and a stability of ±5°C was attainable. The products of pyrolysis were dissolved in hexane at 0°C temperature.

## Gas Chromatography/Mass Spectrometry (GC/MS) Analysis

GC/MS analysis of samples was performed using “Varian 3300” gas chromatograph connected with mass spectrometer detector (ion trap) “Finnigan MAT ITD 800”. A DB-5 fused capillary column (0.32 mm X 30 m) temperature programmed from $50^o$ to $270^o$ at $10^oC$ per minute was used in GS/MS analysis. Mass spectra detection were accomplished in the electron impact mode scanned from 40 to 650 Da with the energy of 70 eV. All mass spectra were searched against the Wiley275 mass spectral library.

## Combustion Tests

Simulated Heat Release analysis were performed on the spherical samples of foamed PU and PU-OM with OM clay fraction of 10% wt. on laboratory thermal analysis balance Mettler T300.

The samples were ignited in air at ambient conditions. The external heat flux was assumed equal to a zero.

Calculation of the heat of the combustion reaction of polymers was carried out using the principle of molar additivity of the heats of formation of the combustion products and reactants [PhysProps 1.6 by G & P Engineering Software Co.]. The concept derives from the fact that enthalpy (H) is a state function and therefore its change in any process is independent of the path from reactants to products. Thus, the overall enthalpy of a reaction is the simply the sum of the enthalpies of the component reactions. In practice, the heat of combustion of the reaction can be calculated by subtracting the heat of formation of the products from the heat of formation of the reactants (1).

$$\Delta h_c = \sum_i n_p \Delta h^o_{f,p} \pm \sum_j n_r \Delta h^o_{f,r} \qquad (1)$$

**Structural Formula 1.**

For polymeric reactants the molar heat of formation can be estimated from the tabulated molar contributions of the chemical groups which constitute the monomer or repeat unit.

# Results and Discussion

Thermochemical calculations of the gross heat of combustion from molar group additivity of the heats of formation of products and reactants achieves better accuracy than calculations based on oxygen consumption for the polymers examined in this study. This is not surprising since the group contributions to the heats of formation used in this study were originally determined from the gross heats of combustion of materials with known composition.

The net heats of combustion of PU polymers of known chemical composition were measured and calculated.

The thermochemical calculation based on complete combustion of PU (Reaction Scheme 2):

$$C_{50}H_{56}O_{11}N_7 + 71\,O_2 \longrightarrow 50\,CO_2 + 28H_2O + 7NO_2$$

**Reaction Scheme 2.** Complete combustion of PU

The heat of combustion of PU, as well as PU-OM, was calculated as 5.36 MJ/kg using the PhysProps 1.6 (2)

$$\Delta H_c = \sum \Delta H_{products} - \sum \Delta H_{reac\tan s} \qquad (2)$$

**Structural Formula 2.**

Bench-scale fire calorimeters have since been developed which use the oxygen consumption principle to determine the chemical heat release rate of burning materials. The oxygen consumption principle is based on the observation that combustion of a wide range of organic compounds and common polymers produces 13.1 ± 0.7 kJ of heat per gram of diatomic oxygen consumed independent of the chemical composition of the organic material. The gases evolved during polymer decomposition are usually unknown and do not burn to completion in real fires. Oxygen consumption is a means of measuring heat release without detailed knowledge of the fuel species. Oxygen consumption calorimetry measures the heat released by the burning of volatile polymer decomposition products, the net heat of complete combustion of which can be written (3)

$$\Delta h^o_{c,v} = \frac{\Delta h^o_{c,p} \pm \mu \Delta h^o_{c,\mu}}{1 \pm \mu} \qquad (3)$$

**Structural Formula 3.**

where $\Delta h^o_{c,v}$, $\Delta h^o_{c,p}$, $\Delta h^o_{c,\mu}$ are the heats of complete combustion for the volatiles, polymer and char respectively and μ is the char fraction. The heat of combustion of the volatile fraction can differ significantly from that of the polymer and the char, so polymer heats of combustion should not be used to calculate flaming combustion efficiency of materials.

The experimentally observed minimum irradiance level, below which no sustained burning occurred, was approximately 20 kW/m$^2$ [3]. The critical irradiance level derived from the experimental correlation was 12.6 kW/m$^2$. The difference between the minimum and critical irradiance for this type of material was explained earlier by Janssens [4]. At low irradiance levels, fuel volatiles are exhausted prior to the time that the lower flammability limit is reached in the gas phase. At higher irradiance levels, the minimum mass flux of volatiles to create flammable mixture is generated before fuel exhaustion. The ignition times for the polyurethane foam were short (between 2 - 6 seconds). In Cone Calorimeter tests the heat of combustion can be obtained as the ratio of heat release rate and mass loss rate, both measured in the Cone calorimeter. Our approach was based on Simulated Heat Release analysis, which operates with experimental Mass Loss Data and calculated Heat of Combustion.

An empirical correlation is needed to relate bench-scale data to full-scale results. It is normally dealt with by normalizing the results by the exposed surface area of the specimen. We have chosen the normalizing surface area as 1 $cm^3$ (4)

$$\dot{q}_{full-scale}(t) = \dot{q}''_{bench-scale} \times S(t) \tag{4}$$

**Structural Formula 4.**

Where $\dot{q}_{full-scale}$ is the total rate of heat release at any time *t;* $\dot{q}''_{bench-scale}$ is the bench-scale heat release rate; *S(t)*is the area of the full-scale specimen that is at any time ignited, covered with flames, but not yet burned out.

Figs. 2 and 3 compare the results obtained for PU bench-scale combustion tests. The spherical samples of PU were ignited and mass loss data were measured during flaming out.

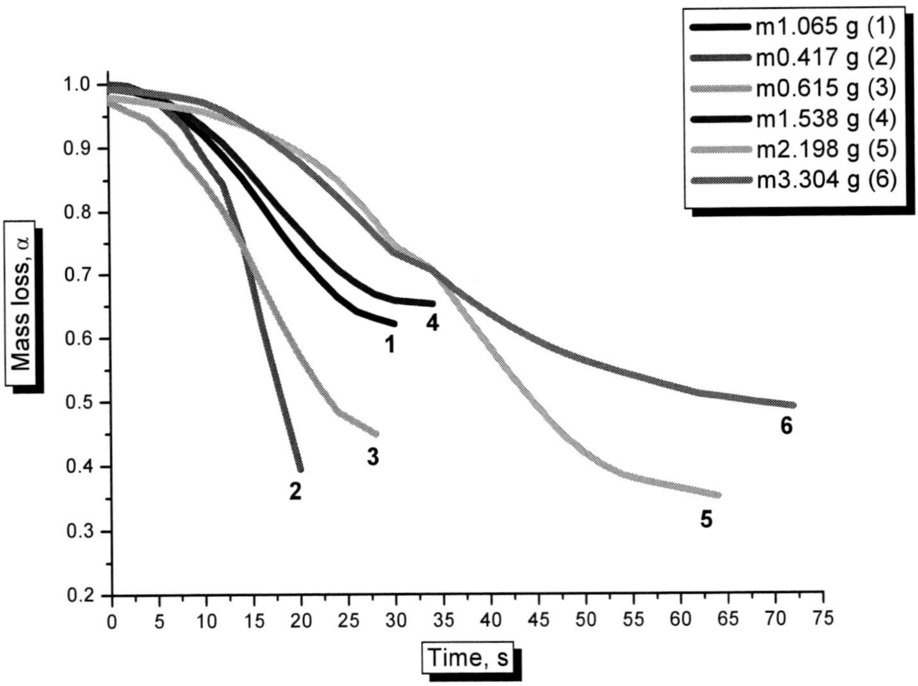


**Figure 2.** Bench-scale data of PU spheres combustion tests: Mass loss (conversion degree) *vs.* time.

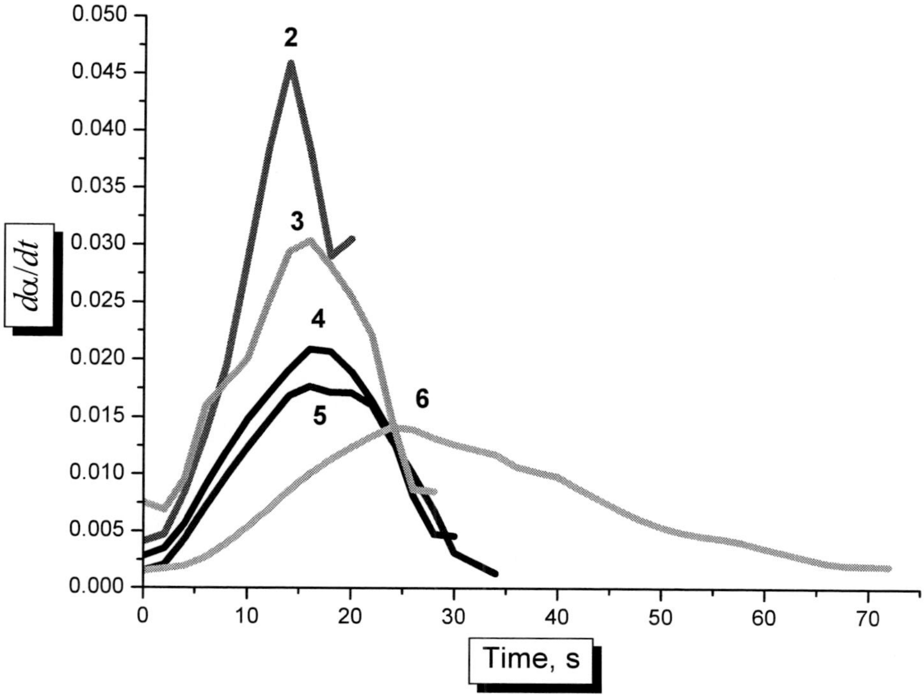


**Figure 3.** Bench-scale data of PU spheres combustion tests: Mass loss Rate (conversion degree) *vs.* time.

The similar results were obtained for PU-OM nanocomposite. The values of Maximum Mass Loss Rate at 1 cm$^2$ surface area o sample were obtain via approximation the Maximum Loss Rate (PU and PU-OM) vs. time to exponential equation (Fig. 4 and 5).

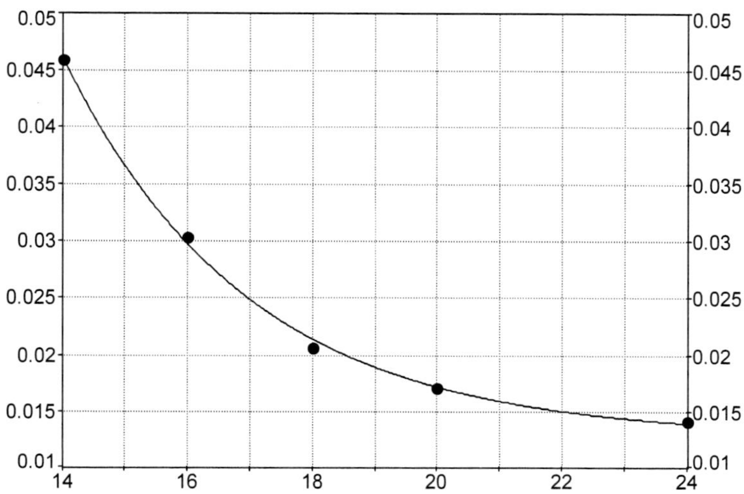


$t_{m=0}$ = 12.915 c, $d\alpha/dt$ = 0.0608

**Figure 4.** Approximation of Maximum Loss Rate (PU) vs. time to exponential equation $y=a+b\exp(-x/c)$.

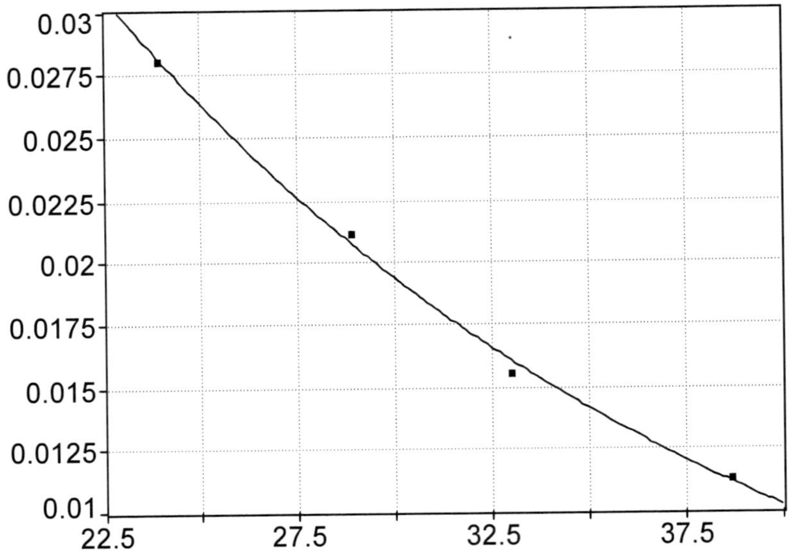


$d\alpha/dt = 0.035$, $t_{m=0} = 19.8$ c

**Figure 5.** Approximation of Maximum Loss Rate (PU – OM nanocomposite, 10%) vs. time to exponential equation *y=a+bexp(-x/c)*.

The values of Maximum Heat Release Rate were estimated using relation (5)

$$\dot{q}''_{bench-scale} = \dot{m}''_{bench-scale} \times \Delta H_{combustion} \quad (5)$$

**Structural Formula 5.**

Where $\dot{m}''_{bench-scale}$ is the bench-scale mass loss rate; $\Delta H_{combustion}$ is effective heat of combustion (kJ/kg). Effective Heats of Combustion for PU and PU-OM nanocomposite were calculated using the PhysProps 1.6 software by G & P Engineering Software Co.

Fig. 6 presents the Values of Maximum Heat of Release of PU and PU-OM nanocomposite.

These Data indicate about 40% improvement of PU-OM – 10% (nanocomposite) over the native PU Maximum Heat of Release Rate.

During the combustion test of the nanocomposite specimen, a carbon layer formed on its surface from the start, grew over time and resisted the heat. Formation of a carbonized layer on the surface of the polymer is a feature of all nanocomposites studied so far: the pattern illustrated decomposition, which can dramatically change the overall scheme of thermal degradation process kinetics by the way of solid phase catalysis or shifting the reaction mechanism from radical random-scission degradation to aromatization followed by carbonization.

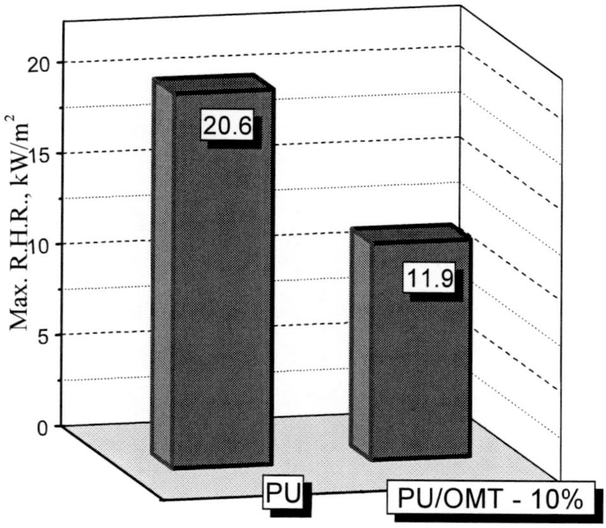


**Figure 6.** The values of Maximum Heat of Release of PU and PU-OM nanocomposite

## Ecological Issue of Isocyanates and Pyrolysis of Polyurethane Nanocomposite

Amines and isocyanates are used in the production of polyurethane (PU). Exposure to isocyanates is associated with respiratory disorders and may occur during production or processing of PU. During thermal degradation of PU, amines and aminoisocyanates are formed in addition to isocyanates. Some isocyanates and several aromatic amines are sensitizers and listed as carcinogens.

**Isocyanates** are a group of reactive compounds that are widely used in industry, mainly in the production of polyurethane. The industrial introduction of PU started in the middle of the twentieth century and subsequently, reports on isocyanate induced occupational diseases started to appear [5, 6]. Today, many people are still affected by isocyanate exposure, despite legislation and safety precautions [7, 8].

The most common isocyanates are the aromatic difunctional isocyanates toluene diisocyanate (TDI) and methylenediphenyl diisocyanate (MDI) together with polymeric MDI (pMDI), which is a mixture of MDI oligomer analogs. They accounts for more than 90 % of the total world production [9]. The major use of TDI is for the production of flexible PU foam, elastomers and coatings, while pMDI is used for the production of rigid PU foam and as binder resin for reconstituted wood products or foundry cores. MDI is mainly used for the manufacturing of PUR elastomers. The more expensive aliphatic diisocyanates are mainly used for the production of color stable PU for coatings and elastomers.

To reduce exposure levels for isocyanates with high vapor pressure, such as hexamethylene diisocyanate (HDI), they are often used as prepolymers or adducts such as biuret, allofanat and isocyanurate adducts.

## Occupational Exposure

The occupational exposure to isocyanates occurs mainly during the production or processing of PU. Numerous isocyanates are commercially available and technical grade qualities, which contain complex mixtures of isocyanate isomers and analogous, are often used in the production of PU. The different compounds that are released when processing of PU (e.g. thermal degradation) are even more complex, and many unknown isocyanates and other compounds, such as amines and aminoisocyanates are formed (Reaction Scheme 3).

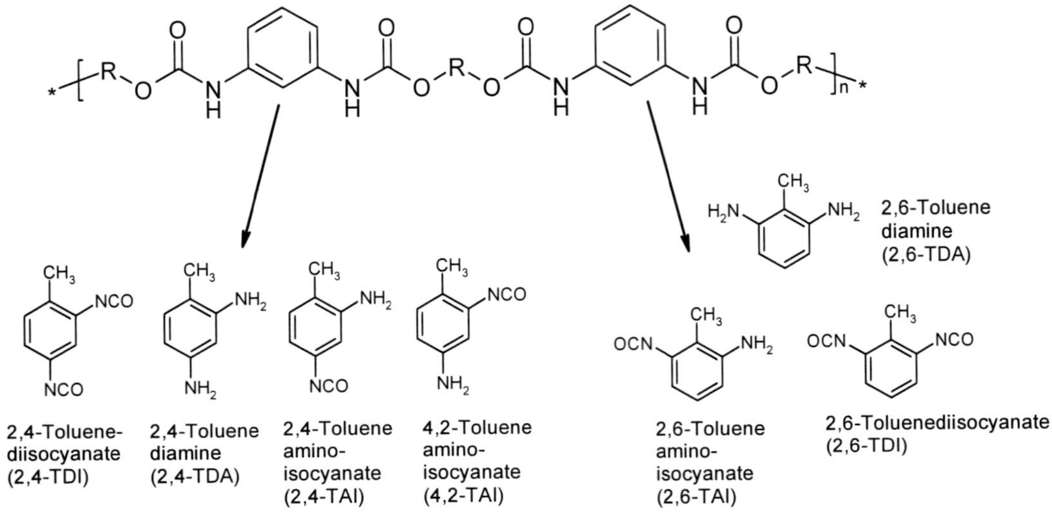


**Reaction Scheme 3.** Thermal degradation of TDI-based PU.

During the production of PU-foam, volatile isocyanates may be emitted to the atmosphere and the exothermic reaction will increase the emission. The exposure to TDI is mainly described during the manufacturing of flexible PU-foam [10, 11]. The exposure to MDI during production of rigid PUR foam has also been described Kaaria *et al.* [11]. Due to the low volatility of MDI, the air concentrations when compared with TDI are much lower, during production of flexible foam. However when isocyanates with low vapor pressure, such as MDI or prepolymerised HDI are used for example in spraying applications, high concentrations of isocyanates can be found in the working atmosphere [12, 13].

The urethane bonding in a PU polymer will start to dissociate at temperatures above 150-200°C [14]. During thermal degradation of PU, exposure to diisocyanates may occur in both the gas phase and the particle phase [15]. The exposure to thermal degradation products of PUR has been reported during work operations such as: processing of PU coated metal sheets in car repair shops [16] and flame lamination of PU with textiles [10].

## Health Effects

Exposure to isocyanates is irritating to the respiratory tract and some of the symptoms from occupationally exposed workers are cough, rhinitis and chest tightness [8]. Exposure to isocyanates is the most common cause of occupational asthma [17]. Even if the risk of developing asthma is greater at high exposure to isocyanates, asthma can occur at relatively low concentrations [18].

Methyl isocyanate (MIC) has been reported to be irritating on the respiratory tract as well as to the skin and eyes. MIC has been shown to be a reproductive toxicant, neurotoxicant and to give rise to systemic effects [19]. Most of the studies have been performed on the victims of the tragic disaster in Bhopal in India. The information regarding occupational exposure to MIC is insufficient. TDI and MDI are suspect carcinogens based on animal studies, but only TDI is classified as a possible carcinogen to humans. Aliphatic amines, such as Hexamethylene diamine (HDA) are classified as moderately toxic [20].

## GC–MS Pyrolysis

The additional aim of this study was to investigate the formation of the suspected carcinogen toluenediamine (TDA,) toluene diisocyanate (TDI) and toluene isocyanate (TI) as a degradation product of PU and PU-OM nanocomposite.

Figures 7 and 8 show the GC–MS analyses PU at 250 and 500°C in air. These peaks were assigned to the products of degradation by comparison of the mass spectrum with data from the Wiley275 spectral databases (Table 1). No detectable amounts of TDA and TDI could be found in extracts of both polyurethane compositions (PU and PU-OM 10% nanocomposite). By contrast, two isomers of toluene isocyanate TI: 1-isocyanato-2-methylbenzene (TI2), 1-isocyanato-4-methylbenzene (TI4) were found in extracts from both polyurethane compositions pyrolysis at 500°C.

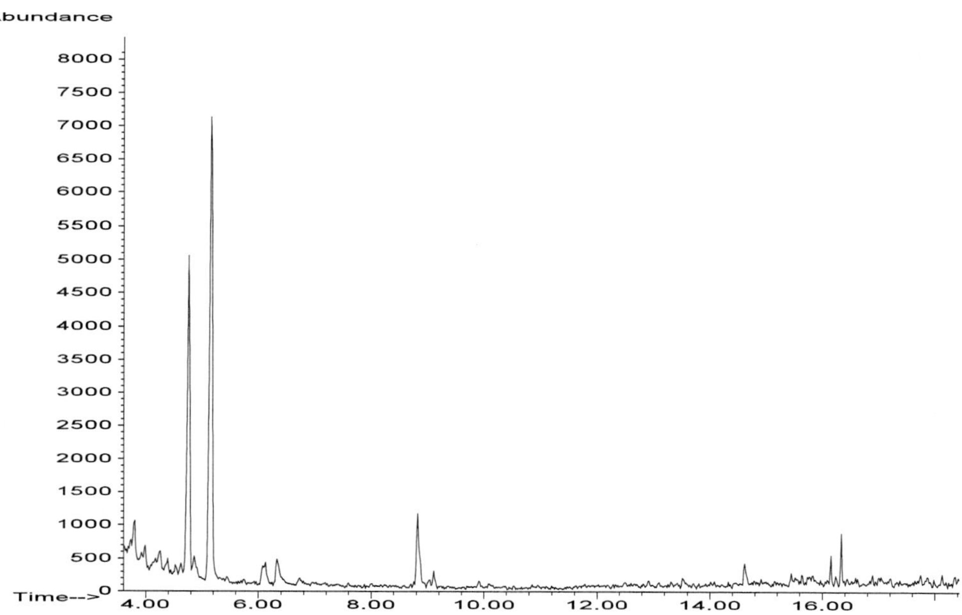


**Figure 7.** GC analysis of PU degradation products at 250°C

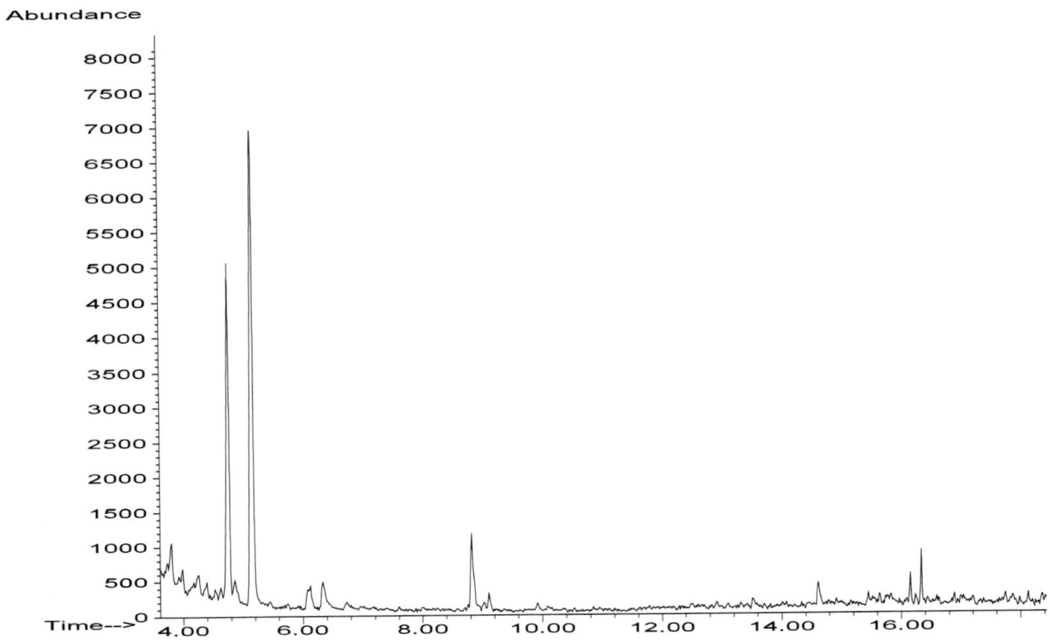


**Figure 8.** GC analysis of PU degradation products at 500°C

Figure 9 shows mass spectra of TI2 (**a**) and TI4 (**b**).

These findings confirm that the formation of TI2 and TI4 in and GC–MS analysis of polyurethane samples is 40% higher for PU pyrolysis than for PU-OM 10% (Table 1).

All degradation products could be attributed to 7 specific groups (Fig. 10):

## CONCLUSION

Analysis of the products degradation indicates that depolycondensation and reactions involving the urethane functional group are not so favorable for PU-OM nanocomposite than for plain PU, since there is a reorientation in the mechanism of thermal degradation. Nano-structure of PU-OM interferes with formation of toxic toluene isocyanates: TI2 and TI4.

Thus, the PU-OM nanocomposite samples have the lowered combustibility, as well as the lowered threshold of toxic degradation products formation in comparison with original PU.

**Table 1. Products of thermal degradation of PU compositions.**

| Products of thermal degradation of PU compositions | Retention time, min | 250°C | | 500°C | |
|---|---|---|---|---|---|
| | | PPU | PPU-OM 10% | PPU | PPU-OM 10% |
| 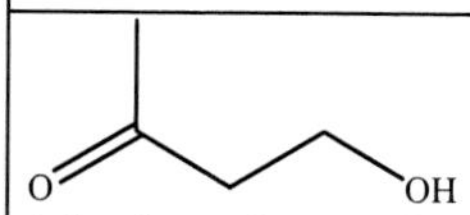 4-hydroxybutan-2-one | 4:15 | 2.3% | 4.9% | 1.2% | 3.2% |
| 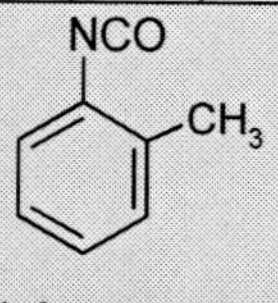 **1-isocyanato-2-methylbenzene** | **4:44** | **0.0%** | **0.0%** | **27.0%** | **18.2%** |
| 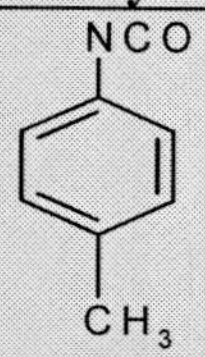 **1-isocyanato-4-methylbenzene** | **5:09** | **0.0%** | **0.0%** | **44.0%** | **27.2%** |
| 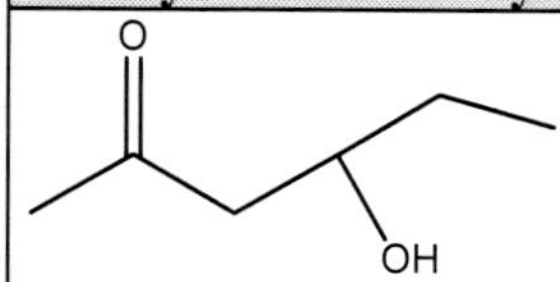 4-hydroxyhexan-2-one | 5:19 | 1.4% | 0.0% | 0.0% | 0.0% |
| 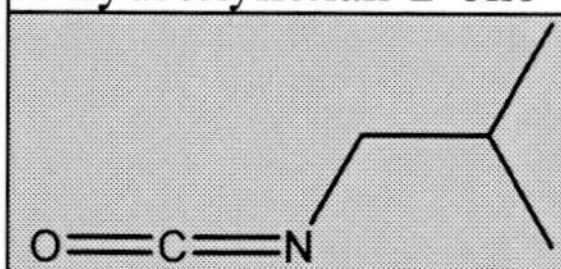 isobutyl isocyanate | **5:68** | **5.3%** | **1.9%** | **0.0%** | **0.0%** |
| 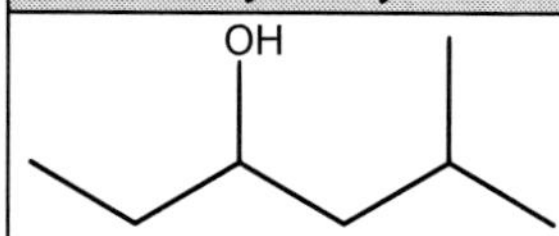 5-methyl-3-hexanol | 6:10 | 0.8% | 1.8% | 2.7% | 2.7% |
| Meta-xylene | 6:14 | 0.3% | 0.7% | 0.0% | 0.0% |
| dioxolan 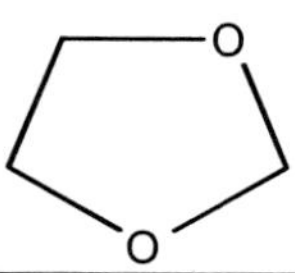 | 6:21 | 0.5% | 12.4% | 3.5% | 0.0% |
| Styrene | 7:02 | 0.8% | 0.0% | 0.0% | 0.0% |
| 2-acetoxy-1-propanol | 8:52 | 11.4% | 31.6% | 7.2% | 3.5% |
| 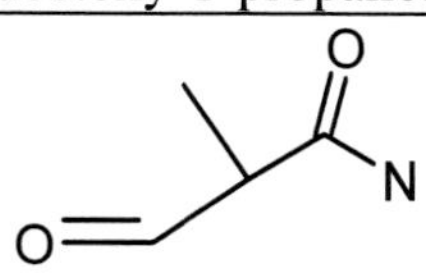 2-Methyl-3-oxo-propionamide | 9:06 | 0.0% | 0.0% | 1.4% | 1.7% |

**Table 1. Products of thermal degradation of PU compositions.**

| **Products of thermal degradation of PU compositions** | **Retention time, min** | **250°C** | | **500°C** | |
|---|---|---|---|---|---|
| Ethyl, methyl-benzene | 9:56 | 1.6% | 0.0% | 0.6% | 1.7% |
| 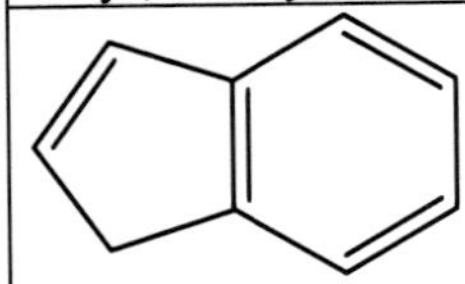 indene | 11:28 | 1.1% | 0.0% | 0.0% | 0.0% |
| Hexane-1-propoxy | 12:37 | 1.2% | 1.2% | 0.0% | 0.0% |
| 2-phenyl butadiene | 14:00 | 2.5% | 0.0% | 0.0% | 0.0% |
| Benzene, cyclopropyledene methyl | 14:06 | 2.2% | 0.0% | 0.0% | 0.0% |
| 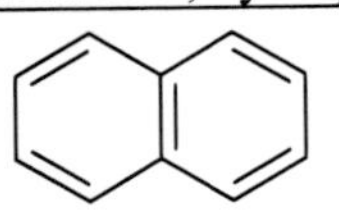 naphthalene | 14:37 | 9.1% | 4.3% | 1.7% | 5.3% |
| Aliphatics | 15:28 | 1.7% | 1.2% | 0.6% | 2.6% |
| | 15:32 | 0.7% | 0.0% | 0.4% | 1.0% |
| 7-octen-2-one | 15:50 | 0.9% | 1.7% | 0.0% | 0.7% |
| Naphthalene-1-methyl | 16:07 | 4.7% | 0.9% | 0.0% | 1.0% |
| Naphthalene-2-methyl | 16:10 | 6.4% | 3.5% | 1.4% | 1.1% |
| Aliphatics | 16:16 | 1.8% | 0.6% | 0.7% | 3.2% |
| Naphthalene-methyl, ethyl | 16:22 | 5.6% | 4.9% | 2.1% | 1.0% |
| Aliphatics | 16:26 | 3.0% | 2.4% | 0.3% | 3.3% |
| | 16:31 | 2.4% | 1.0% | 0.0% | 4.0% |
| 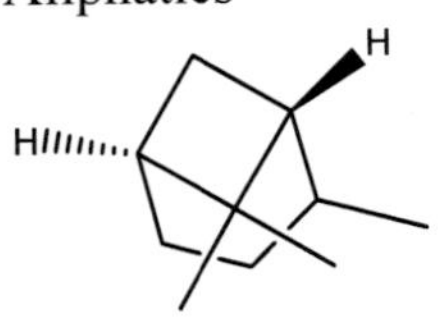 trans-pinane | 17:01 | 0.9% | 0.0% | 0.0% | 0.0% |
| biphenyl | 17:46 | 2.6% | 2.0% | 0.0% | 0.2% |
| 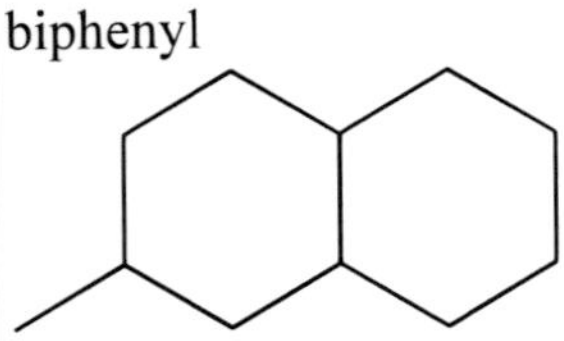 decahydro-2-methylnaphthalene | 18:09 | 3.3% | 1.9% | 0.4% | 0.6% |
| Aliphatics | 18:21 | 3.3% | 3.0% | 0.0% | 1.0% |
| | 18:42 | 2.6% | 1.4% | 0.0% | 1.7% |
| | 19:40 | 2.0% | 0.0% | 0.0% | 0.0% |
| **Non-identified products** | | **17.6%** | **16.7%** | **5.5%** | **15.1** |
| **Total Σ (prod.)** | | **100.0%** | **100.0%** | **100.0%** | **100.0%** |

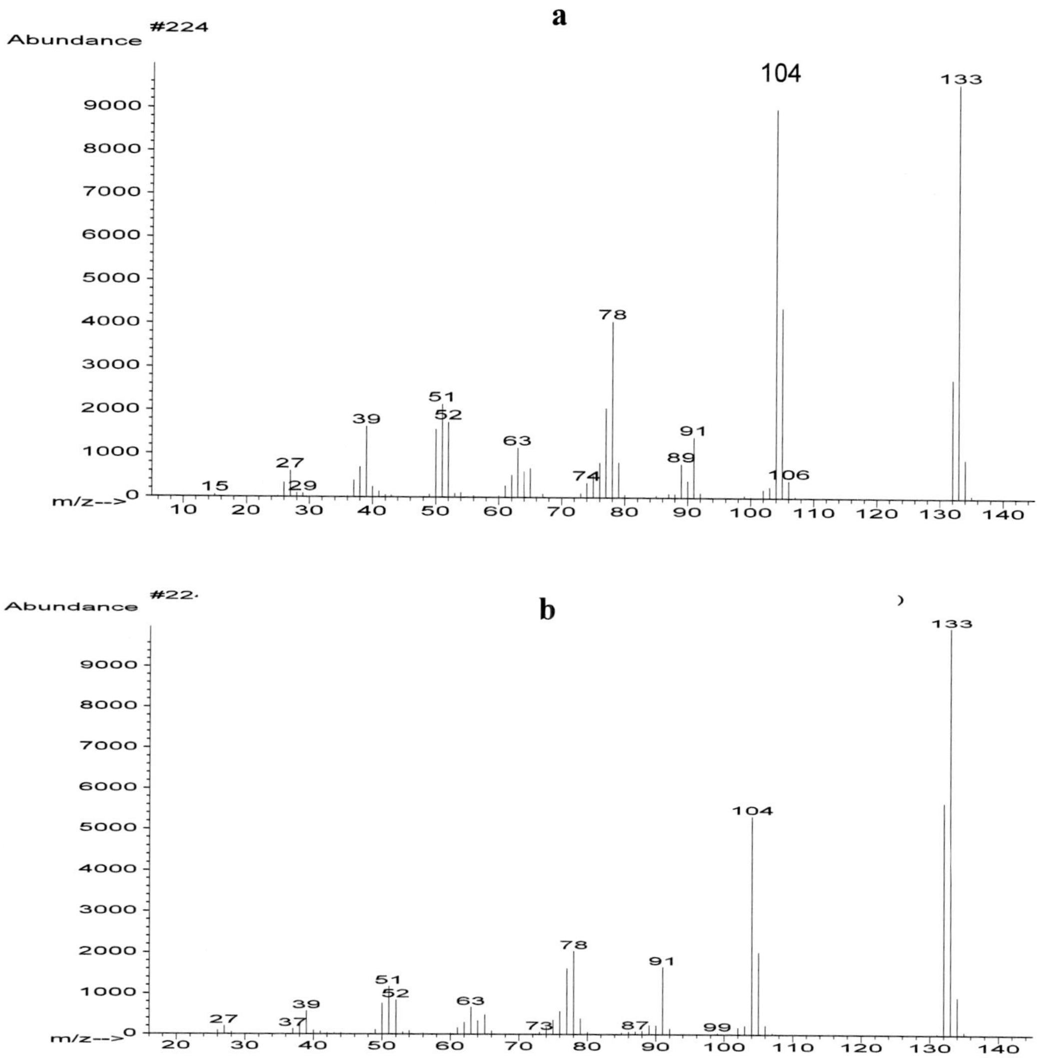


**Figure 9.** Mass spectra of TI2 (**a**) and TI4 (**b**).

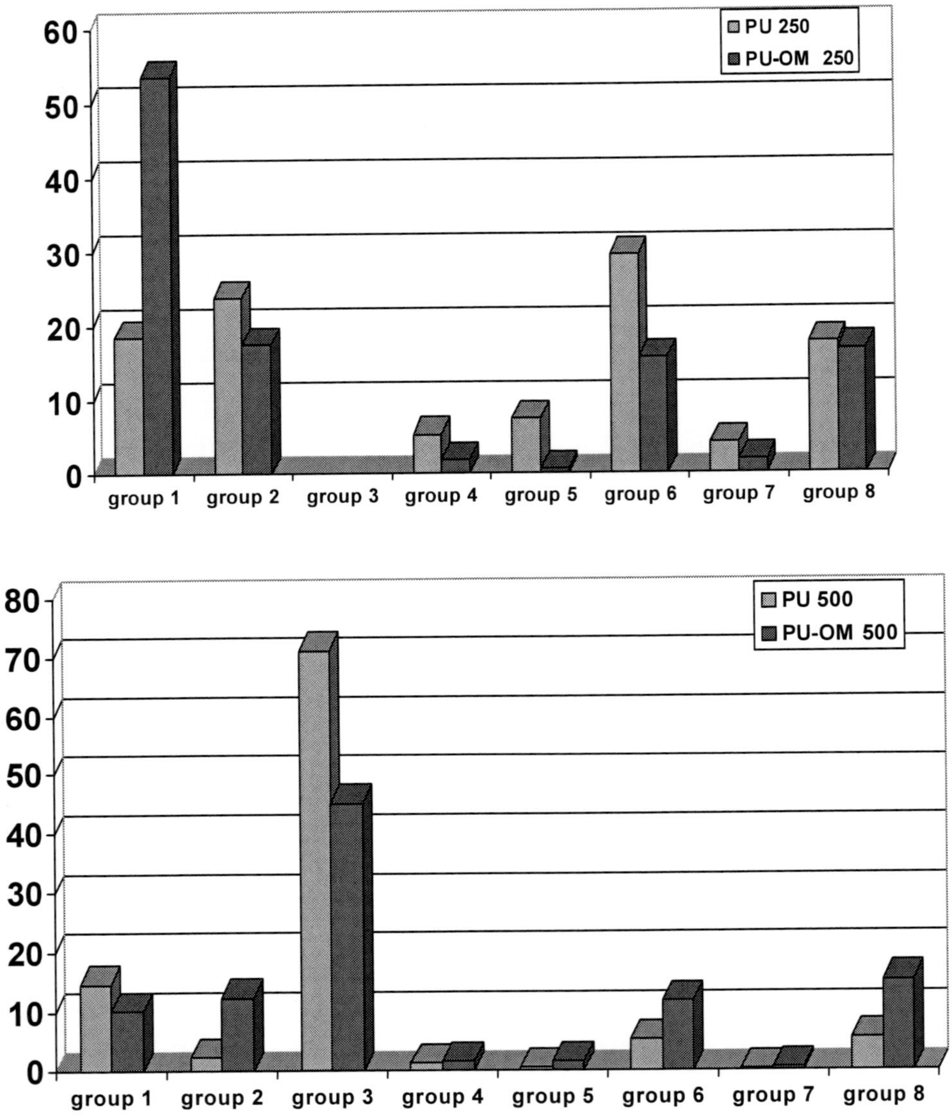


**Figure 10.** Diagram of the PU and PU-OM degradation products.

| | |
|---|---|
| Group 1 | 4-hydroxybutan-2-one, 4-hydroxyhexan-2-one, 5-methyl-3-hexanol, 2-acetoxy-1-propanol, hexane-1-propoxy, 7-octen-2-one, dioxolan. |
| Group 2 | Aliphatics. |
| Group 3 | 1-isocyanato-2-methylbenzene (TI2), 1-isocyanato-4-methylbenzene (TI4). |
| Group 4 | Isobutyl isocyanate, 2-methyl-3-oxo-propionamide. |
| Group 5 | Meta-xylene, styrene, Ethyl, methyl-benzene, 2-phenyl butadiene, benzene, cyclopropyledene methyl. |
| Group 6 | Indene, biphenyl, naphthalene, naphthalene-1-methyl, Naphthalene-2-methyl, naphthalene-methyl, ethyl. |
| Group 7 | Trans-pinane, decahydro-2-methylnaphthalene. |
| Group 8 | Non-identified products |

## REFERENCES

[1] J. W. Gilman, C. L. Jackson, A.B. Morgan, R. Harris Jr., *Chem. Mater.* 2000, **12,** 1866-1873

[2] Y. Hu, L Song, J. Xu, L Yang, Z.Chen, W. Fan, *Colloid Polym. Sci.*, 279, 819, (2001).

[3] O. Grexa, WOOD & FIRE SAFETY, *Proceedings of 3rd International Scientific Conference The High Tatras*, Hotel Patria SLOVAK REPUBLIC, 139, (1996).

[4] Janssen M, *Improved Method of Analysis for the LIFT Apparatus – Part 1: Ignition , in Proceedings of the Second Fire & Materials Conference*, Arlington VA, pp.37, (1993).

[5] Fuchs S., Valade P., *Arch. Maladies Prof.*, 12, 191-196, (1951).

[6] Swensson A., Holmquist C. E., Lundgren K. D.. Brit. *J. Industr. Med.*, 12, 50, (1955).

[7] Tarlo S. M., Liss G. M., Dias C., Banks D. E., *Am. J. Ind. Med.*, 32, 517-521, (1997).

[8] Ott M. G., Klees J E., Poche S. L., *Occup. Environ. Med.*, 57, 43-52, (2000).

[9] Ulrich H., *Chemistry and Technology of Isocyanates*. John Wiley & Sons Ltd, Chichester, England, p.344, (1996).

[10] Tinnerberg H., Spanne M., Dalene M., Skarping G., *Analyst*, 121, 1101-1106, (1996).

[11] Kaaria K., Hirvonen A., Norppa H., Piirila P., Vainio H., Rosenberg C., *Analyst*, 126, 1025-1031, (2001).

[12] Crespo J., Galan J., *Ann. Occup. Hyg.*, 43(6), 415-419, (1999).

[13] England E., Key-Schwartz R., Lesage J., Carlton G., Streicher R., Song R., *Appl. Occup. Env. Hyg.*, 15(6), 472-478, (2000).

[14] Gupta D. C., Wast D. V., Tapaswi M. A., Nigade B. N., *Macromolecular reports*, A31, 613-625, (1994).

[15] Streicher R. P., Kennedy E. R., Loiberau C. D., *Analyst*, 119, 89-97, (1994).

[16] Skarping G., Dalene M., Mathiasson L., *J. Chromatogr.*, 435, 453-468, (1988).

[17] Rosenman K.D., Reilly M.J., Kalinowski D.J., *J. Occup. Environ. Med.*, 39 (5), 415, (1997).

[18] Meredith S. K., Bugler J., Clark R. L., *J. Occup. Environ. Med.*, 57, 830-836, (2000).

[19] Varma D. R., Guest I., *J. Toxicol. Environ. Health.*, 40, 513, (1993).

[20] Dashiell O. L., Kennedy G. L.,. *J. Appl. Toxical.*, 4, 320-325, (1984).

In: Chemical Physics of Pyrolysis, Combustion and Oxidation ISBN 1-59454-163-9
Editors: A. A. Berlin et al., pp. 19-31 

*Chapter 2*

# Latest Achievements in the Field of Creation of Fire-Resistant Polymeric Materials with Use Phosphorusorganic Compounds

***A. Kh. Shaov, T. A. Borukaev and M. M. Begretov***
The Kabardino-Balkarian State University him. H.M.Berbekova,
360004, Nalchik, Chernishevskaya st. 173, KBR, Russia
Fax: (095)337-99-55; E-mail: ah_shaov@mail.ru

## Abstract

The basic directions of researches in the field of increase thermostability and fire resistance of polymeric materials are replacement of antioxidizers and fire resistance additions (FRA) from connections of bromine (and in general halogens) and ammonium three-hydrate on more effective and less cancerogenic substances. Growing cost of connections of antimony also has resulted in necessity of their replacement on phosphorcontaining antioxidizers and FRA which in a combination to connections of bromine, provide synergetic effect, that is obviously expressed in oxygenic polymers.

**Key words**: polymers; phosphor organics compounds; fire resistance additions.

The environmental problems of application brominations organic connections in quality fire resistance additions for polymers are considered in work [1].

The general tendency in the given area of a science are questions of compatibility of additives with polymers, their influence on coloring of materials, shock durability and adaptability to manufacture, and also development of target additives for concrete types of polymeric materials.

If to speak about the mechanism of action FRA it is necessary to note, that the most general moments of this question are well covered in the grant [2]. Historically usual ways of increase of fire resistance of polymeric materials considered [3] on an example of various types of high-molecular compounds.

From them numerous conferences [4-11] and exhibitions [12] are traditionally devoted to a problem of increase of fire resistance of polymeric materials and products. Mark [4], that one of such conferences has considered scientific and technological problems of creation of the fire-resistant materials used in arms. Believe, that speed of distribution of a flame depends on speed of gasification of a polymeric material, which is in turn defined by heat of the gasification making an interval for the most widespread polymers (from PE up to rigid phenoplasts and glass-filling polyesters) 1,39-3,74 kJ/g.

The contemporary industry demands new approaches and techniques on creation of fire-resistant polymers. So, firm T and Techn. Ltd [13] develops the new advanced program of definition long thermostability and stability to light-thermal aging of various polymers and compositions on their basis, in particular fire-resistant polyamide without halogencontaining fire resistance additions, the composites reinforced with fiber glass for electronics and electrical engineers where high thermostability is a determinative. By the given technique the thermal index of a material, including the big number of tests for dimensional stability of samples is defined under concrete conditions of operation and atmospheric influence, thermographymetry, DTA etc. analytical methods. Thus use laboratory extruder with the closed both closed computer control and adjustment of parameters of formation of samples with the subsequent thermostatic them within 1 year for more full end of chemical processes in a material, stabilization of structure of polymer in a finished article. Then define explosive durability, fire resistance, specific shock durability, dielectric properties which results are expressed as tables, diagrams, diagrams that quickly and reliably allows consumers to choose type of polymer or a composition for concrete area of use; for reception modern special thermo- and fire-resistant polymeric fibers the named characteristics are necessary for pawning [8] at the moment of their reception.

The choice of this or that polymer for needs of various industries is caused by a complex of physico-mechanical properties. But under equal conditions the preference will be given polymer or a composition on his basis, at which raised fire resistance. To a condition and prospect of development of the basic types superconstructional and heat-resistant thermoplasts, work [14] is devoted to this question.

One of widely and for a long time used fire resistance additions are elementary phosphorus and his compounds (organic and inorganic). They can be both low-molecular, and high-molecular. To this question frequently devote the whole conferences, for example, in work [15] results 2-nd conferences on fire-resistant polymers on which results of last achievements and perspective directions in this area were reported are generalized. Questions of creation FRA, development of test methods of polymeric materials (PM) on fire resistance and flammability, problems of toxicology chemical delay burning and products of combustion PM are considered. At creation fire-resistant PM wide application find halogen and phosphorus containing the organic compounds, foaming additives, depressants a smoke. Are developed fire-resistant PM, successfully used in manufacture of personal computers, wires, cables, and also at creation of fire-resistant fabrics, clothes, furniture, an accessories etc.

Frequently compounds of phosphorus in quality FRA add to polyolefines. So, compositions with the improved fire resistance and stability in water and water-acid environments on the basis of software contain [16] 10-250 % Mg-hydroxide as particles in the size 0,2-5 microns with a specific surface of 1-20 $m^2/g$, 0,5-10 % of salt of phosphate processed on a surface and amino alcohol formulas $(RO)_nP(O)(HO)[NH(R^1OH)]_{2-n}$ or metal salt of phosphate of the formula $(RO)_nP(O)(OM)_{3-n}$, where R - alkyl, alkenyl, arylalkyl $C_{10\text{-}30}$,

$R^1$- alkylen $C_{1-6}$, n =1-2, M - hydrogen or alkaline metal. Such compositions have shown combustibility on norm UL 94 V-O, durability at a stretching of 209 kG/sm$^2$, impact strength on Izode 9,5 kG·sm/sm and the raised stability to acid environments.

Sewed under action of heat or electronic radiation, the fire-resistant composition contains [17] weigh part (w.p.): A) 20 on 100 compositions sewed polyolefin (PO); B) the mix from 5 polyphenylenoxide, organic phosphate, di- or multifunctional compound, or polymer of formulas: $(R^1O)P(O)(R^2)OC_6H_{4-m}(X)_mOP(O)(R^2)_nOR^1$, $(R^5O)P(O)(R^4)OC_6H_{4-m}(X)_mR^3C_6H_{4r}OP(O)(R^4)_pOR^5$, $(R^6)(R^7)P(O)OC_6H_3OP(O)(R^6)(R^7)_2$, where $R^1$, $R^3$, $R^5$ - (it is independent) hydrocarbon group, $R_2$, $R_4$, $R_6$ and $R_7$ - hydrocarbon group (probably containing oxygroup), X - halogen, m and r = 0-4, n, p =1-30, stable bromine containing compound; C) sewing agent - organic peroxide, for example, 2,2-bis-(tertiary butylperoxy)-diizopropylbenzol (I). In quality PO use polymer polyolefin, for example, PE, polybuten; in quality polyfenilenoxide, for example, poly-(2,6-dimethyl-1,4-phenylen)oxide; as organic phosphate- an encore-resorcinodiphenylphosphate, hydroquinone or isopropylthree-phenylphosphat (II); in quality bromine containing compounds a product of the formula: $(Br)_qC_6H_4OROC_6H_4(Br)_s$, where R - not ramified alkylen $C_{1-10}$, q and s =1-5, for example, 1,2-bis-(2,4,6-three-brominephenoxy)-ethane (III), 1,2-bis-(2,3,4,5,6-pentabrominephenoxy)-ethane, or olygopolycarbonate on a basis tetra-brominediphenylpropane. In quality synergistic additives in a composition use oxide antimonies. Mark, that it is possible to enter into a composition elastoplastic, for example, a block - copolymer styrene - ethylene/butylene-styrene. A composition use for manufacturing electric isolation of metal conductors. The composition has durability at a stretching of 67 kG/sm$^2$ and lengthening of 100 %; after the first ignition does not light up, after the second - burns 2,4 seconds.

In work [18] use halogenated and unhalogenated fire resistance additions for plastic, and also a problem of toxicity, smoke-education, corrosion is analyzed. It is shown, that the additive three-oxide antimonies to brominated diphenyls in quality synergistic the agent successfully allows to use the given compounding for increase of fire resistance of various polymers (PE, software, polystyrene, ABS, epoxide resins, polyurethane's). It was successfully used unhalogenated additives, such as three-hydrate of aluminium, hydrate of magnesium with sylon a covering, additives on the basis of red phosphorus and connections of metals (oxides of zinc and iron, borate of zinc, the molybdenum, hydrogenated borates of zinc etc.). Efficiency of use of copolymers (for example, PETHP/PVCh) for increase of fire resistance instead of homopolymers (PVCh) is established at use as isolation of a cable and wires.

Way of reception FRA is offered [19] on the basis of the red phosphorus, having the improved stability and the safety, including the following stages: 1) heating of yellow phosphorus in the inert environment at temperature 250-600$^0$ C, raising his partial transformation (<70 %) in red phosphorus and formation of a fluid dispersion thin, basically of spherical particles of red phosphorus with the size 100 microns, dispergated units from such particles, dispergated in a matrix from the fused yellow phosphorus; 2) removal of not transformed yellow phosphorus from a fluid dispersion; 3) allocation of spherical particles of red phosphorus and their agglomerates, and particles basically of the spherical form, on their external surface is not present traces of destruction and sharp regional lines, and on a surface basically there are no the active centres capable to reaction with a moisture and oxygen that provides superficial stability of spherical particles, thus spherical they are formed, escape procedure of spraying, and 4) drawing of a uniform, homogeneous covering on an external surface of particles from one integumentary material chosen from thermosettings of pitches (PhPhR, MPhR, MLPhR etc.) And hydroxide of metals $Al(OH)_3$ or $Zn(OH)_2$, in quantity of 1-35 % (from weight of particles of red phosphorus). The fire-resistant polymeric composition with improved thermo-, water- and atmospheric resistance a basis on PE, software, copolymers ethylene (I) and acrylate resins, copolymers I and vinyl acetate, polybutene and others with the contents of 20-200 % hydrated inorganic filler $Al(OH)_3$, $Mg(OH)_2$, $Mg(CO_3)_2$ and their mixes and 0,1-30 % of spherical particles of red phosphorus is offered also.

The modified polymers and copolymers of propylene for fire-resistant polymeric compositions receive [20] at their heating at the presence of sources of free radicals (organic peroxides, azo compounds). Usually updating carry out in situ in the polymeric composition containing various fillers and the additives ($Sb_2O_3$, antimony Na, organic halogenated compounds, phosphorus compounds).

In many cases, for achievement of high parameters of fire resistance of polymeric materials, technologists apply synergistic mixes fire resistance additions. So, in work [21] it is marked, that for polyolefines the structure under name Melapur P on a basis synergistic mixes FRA, containing phosphorus and nitrogen is developed.

Compositions with the lowered combustibility contain [22] (w.p.) 100-200 polymers (polyolefins, polyesters, polyamids, synthetic rubber, polyethylene-terephthalate, styrol polymer etc.), 5-50 nitrogencontaining salts organic polyphosphorus oxyacides, 0-30 copolymers, consisting of polyethylene circuits and polyoxyalkylenes lateral circuits, 0-100 in addition carbonated compounds: phosphorus, a pine forest, sulphurcontainings, boronphosphates, 0-30 metallcontaining the compounds accelerating carbonated of burning surface (zincthiocarbamate, zinc oxide, zincborate etc.) 0-400 silicate filler, a carbonate, phenoplast, graphite, (hollows) glass beads etc. Fire resistance additions enter in polymers in quantity of 5-50 %.

For finding - out of an opportunity of recycling and repeated processing of polyethylene of high density [23] search of new modifiers which considerably allow to lower temperature of processing of polymer will be carried out, keeping thus thermooxidation stability a high level to give fire-resistant properties and maximum full to reuse a polymeric material. As those are offered organic derivative five-valent phosphorus, for example polyphosphonates. The data received as a result of realization of the complex comparative analysis speak about perspectivity of use phosphororganics polymers in quantity of 0,01-0,5 % as stabilizers and modifiers at recycling waste products of polyethylene of high density and use of a polymeric material for manufacture of products of constructional purpose.

Transparent thermoplastic compositions for manufacturing elastic packing or used as fire-resistant polymers, polymeric additives, for example for improvement of compatibility, contain [24] copolymers vinylaromatic monomers, vinylphosphonic acids and probably nonsaturated monomers which presence of parts allows to vary elasticity and a transparency, and 30-50 % from a composition of polymers (polyolefines, polyethers) with trailer amino groups and glass temperature $< 0^0$ C in quantities, sufficient for giving a composition of elasticity. Vinylphosphonic acids are connections of formula $CH_2{=}C(A)P(O)(OH)_2$, where A - aryl or halogenaryl. The nonsaturated monomers - connected dienes, vinylhalogens, olefin's, vinyl ethers carboxylic acids, aryl or alkyl ethers and amids (met) acroleic, phenylacroleic acids, nitriles nonsaturated carboxylic acids, simple alkyl vinyl ethers, anhydrides nonsaturated dicarboxylic acids.

Fire-resistant polymers on a basis polyolefines (or PVCh, nonsaturated polyester, polyurethanes, epoxy resins, polystyrol), processed at temperatures <$280^0$ C, contain [25] 1-40 % trihydrazine- and/or thriguanidinetriazine and/or salts of these compounds and a phosphorus or boric acid, and the maximal size 25 microns has 95 % of particles of the additive entered into polymer. Compositions can contain and other additives complicating burning. Such compositions melt down, homogenized at $230^0$ C also granulate. Granulous (bulk density 650 g/l) pressing at $150^0$ C process in plates thickness of 3,2 mm, fire resistance on method UL-94 V-0, having (in brackets for the compositions containing PE and 25 % melamine cyanurate) bulk density 650 g/l and fire resistance (UL-94) V-0 (V-2).

Fire-resistant hardening frothy the composition contains [26] as the basic component hardening the polymeric system (HPS) including nonsaturated polymer (nonsaturated polyester) and the catalyst (peroxide compounds), or epoxy olygomer or polymer and the sewing agent, reacting with epoxy group. Mark, that the composition can contain 10-95 % HPS, 1-60 % of a phosphate component, 1-50 % polyhydroxide compound and 1-50 % nitrogen containing compounds.

The fire-resistant textile material representing by [27] self a textile cloth with put on him fire resistance by structure, differs that fire resistance the structure is put as a discrete polymeric layer in the form of hemispherical elements with in diameter 0,8-2 mm at a degree of overlapping of a surface of a cloth structure of 15-45 %, and in quality fire resistance structure it contains polyvinylchloride plasticate with the target additive in quantity 10-45 w.p. on 100 w.p. polymer, and as the target additive it contains, at least, one component from the group of substances containing trioxide antimonies, carbamide, tris-(2,3-dubromopropyl)phosphate, phosphate of ammonium.

In work [28] fire-resistant additives to polymers are submitted as a mix which consist from: (w.p.) A) 1-12 $BPO_4$; B) 0,05-1 aromatic sulphonate alkaline metals; C) tetra-fluorinatedethylene (PTFE) and D) 0,06-9 not containing halogens graft a copolymer received by an inoculation a) of 5-90 mixes from 50-95 % of styrene (I), methylstyrene, replaced in nucleus I, methyl metacrylate (II) or their mixes and 50-5 % (met)acrylonitrile, II, maleic anhydride N-replaced maleic imide or their mix on b) 95-10 synthetic rubber with temperature glassy $10^0$ C, and components A, B, C and D it is possible to combine 0,04-1 polymers in various parities. Ascertain, that a component "B" apply in the form of a mix emulsion PTFE with emulsion copolymers, and the parity of a component "B" and "D" makes 5:95-40:60. Mixes apply as the fire-resistant additive in quantity of 0,04-12 % on 100 aromatic polycarbonates which are not containing halogen, sulfur and phosphorus. The samples made of a composition, are characterized by fire resistance (thickness of samples of 1,6 mm) VO (by technique UL 94), impact strength (samples with a cut, temperature $260^0$ C) 23,1 kJ/$m^2$, thermostability on Vica $143^0$ C.

Fire-resistant thermoplastic compositions consist [29] (w.p.) 5-94 polyphenylene an ether; 5-94 vinyl aromatic polymer; 0-60 various additives, for example fibrous and - or disperse, and 1-30 FRA, for example containing (%) 50-93,67 phosphororganic compounds, 0,2-48 derivative triazine (melamine and/or melamine cyanurate) and 0,03-2 PTFE. For reception of KM water dispersion polytetra-fluorinatedethylene put on granular polyphenylenoxide at 0-$130^0$ C or enter her in melt other components. From received in the compositions form block products, a film or fibers.

For giving to polymeric compositions of the raised fire resistance in them add [30] products of interaction (w.p.) 1-25 acid compounds of a pine forest (a boric acid, halogenides a pine forest, borates metals, preferably alkaline and alkaline-earth metals) and 1-100 acid connections of phosphorus (a phosphoric acid, oxychloride, trichloride phosphorus, phosphates). Reaction carry out at presence (w.p.) 100 basic salts (connections of alkaline and alkaline earth metals, amines, ammonium compounds) and 1-500 synthetic polymers (phenoplasts, aminoplasts, polyesters, polyamides, polysiloxanes, epoxide resins, copolymers olefin's and it is possible dienes) and/or natural polymers (wood, cellulose, rubbers, lignocellulose, a paper, starch). A composition use for manufacturing building, insulator, integumentary, decorative and moulded materials.

The fire-resistant composition which is not containing halogens, prepare [31] of nonsaturated polyester, a polyvinyl ether, polyurethaneacrylates etc. with addition as the modifier phosphor organic for compounds. Also it is possible to enter into a composition hydroxide metal in quantity of 1-400 %, novolac phenol-formaldehyde pitch in quantity of 1-50 % and phosphate ethers in quantity of 1-30 % from weight of polymer.

The composition with the improved adaptability to manufacture and high mechanical durability on the basis of regenerated PETF contains [32] 0,5-100 % lactone polymer and it is possible polyolefins and inorganic or organic additives. The fire-resistant composition contains 0,5-100 % FRA, glycerin representing mix polyester or polyglycerin and maximum fat acid polyester and 100 % fire resistance additions. The composition in addition contains one inorganic either organic connection or other additives. The attitude combined with FRA aliphatic or aromatic polyester to glyceric polyester 85-99,5:0,5-15 aliphatic. In quality fire resistance addition use chlorine-containing compound, a phosphoric or polyphosphoric acid, nitrogen containing or inorganic compound.

Studying of compatibility of lines epoxy-dianes oligomers and glycidyl ethers of acids of phosphorus (GE) and properties oligomeric systems has shown [33], that glycidyl ethers of acids of phosphorus are effective thinners epoxy oligomers. It is shown, that epoxy-amine the polymers modified GE, have high heat, thermo- and fire resistance. The analysis of the given thermomechanical and electron-microscopic researches has allowed to draw a conclusion on a determining role of chemical structure GE. It is established, that for all used systems in the field of 10-15 weights % GE the not additive increase measurable parameters is observed.

Thermoplactic a composition having high adaptability to manufacture, shock durability and thermostability receive [34] mixture (w.p.) 100 mixes (%) 30-95, 5-70 polymers or copolymers styrene, methyl or halogen-styrene (only not ABS), modified by rubber (the contents of rubber of 15-25 %, the size of particles of rubber 0,3-5 microns), and gel of 15-70 % having the contents, and 0,5-20 special bloc-copolymers. In quality bloc-copolymer use epoxidating on double compounds a block-copolymer epoxide equivalent 140-2700 structures A-B-A where A- the block of polymer of aromatic vinylic compound (styrene), B - the block of polymer conjugated, is probably in part hydrogenated, diene compounds (butadiene). A fire-resistant composition receive mixture above-stated thermoplactic compositions from 1-40 w.p. organo-phosphorus compounds and if it is necessary with 0,05-5 w.p. fluoro-ethylene polymers.

The fire-resistant polymeric composition contains [35] mix of (%) 20-100 styrene containing graft copolymer from 18-50 % acrilonitrile (I) (ABS) and 0-80 styrene containing copolymer from 18-50 % I and 5-60 polyphenylene an ether, 2-40 styrene containing copolymer and styrene containing graft copolymer from 5-18 % acrilonitrile; 5-30 triphenyl-phosphate and 0-50 other traditional additives (softeners, stabilizers, dyes, fillers and others).

Fire-resistant polyfoam on basis polystyrene or copolymers styrene, the phenolic pitch containing up to 20 % FRA (melamine-polyphosphate ammonium with additive polyvinyl acetate and tripentaerithritol) and frothing graphite, receive [36] mixture of beads foamy-polystyrene with a mix of phenolic pitch, a hardener which is making foam the agent (pentane) and fire resistance additions, then with graphite or all over again with graphite, then with phenolic pitch and make foam the ferry in the form. It is revealed, that processing by a flame with temperature $1000^0$ C results only in dimness of a surface of a product.

The fire-resistant film (thickness 30-1000 microns) is formed [37] basically crystallizable thermoplactic polymer (polyester, it is preferable PETF), distinguished by presence of one structured external surface, contains one fireproof additive (organic compounds of phosphorus, in particular dimethylmethylphosphonate), soluble in polymer.

As fire-resistant additives to nonsaturated polyester pitches are offered [38] phosphororganic compounds $ROPO(OH)_2$, $(RO)_2P(O)OH$, $(RO)_3PO$, where R - 2,4,6-tribromphenyl.

The fire-resistant composition consists [39] of PS or copolymers styrene and 0,01-10 % neutralizer type oxides, hydroxides or carbonates of metals II or III groups $Mg(OH)_2$, MgO, $Al_2O_3$, $Al(OH)_3$, $MgCO_3$, $CaCO_3$, ZnO, and also 1-25 % FRA of group of phosphates of formula $R^1OP(O)(OR^2)OR^3OP(O)(OR^4)O_nR^5$, where $R^{1-5}$ - aryl or hydroxyl containing aryl $\tilde{N}_{6-12}$; n = 0-1, both 0-5 % $BPO_4$ and phosphates.

Fire-resistant thermoplastic a forming composition are going [40] of not containing halogens graft polymer (not containing halogens thermoplastic a copolymer of styrene) with the contents of 0,5-20 % FRA formulas $(C_6H_5O)_2P(O)OC_6H_2(R_2)SO_2C_6H_2(R)_2OP(O)(OC_6H_5)_mOC_6H_5$ (I) where R - it is independent from each other H or alkyl, m=1-5, or 0-15 FRA formulas $R^1(O)_aP(O)(O)_bR^2$, $OXOP(O)(O)_cR^3{}_p(O)_dR^4$ (II) where $R^1$, $R^2$, $R^3$ and $R^4$ - it is independent from each other alkyl, cycloalkyl, aryl or arylalkyl, X - one or a multinuclear aromatic radical, p=1-5, a, b, c, and d it is independent from each other 0 or 1,0-10 means, interfering formation of drops and 0-50 % of other additional means.

The composition proof to ignition on a basis graft a copolymer vinyl aromatic a monomer and vinyl cyanide can include [41] 1-200 % of phenolic pitch and 1-200 % organic phosphorus containing substances (trixylenil- and triphenylphosphits) which does not react with pitch in a fire-resistant composition.

Forming weight from polyphenylen oxide and vinyl aromatic polymer (PS, copolymers styrene), 1-20 % of the fireproof additive [42] (resorcinol - and - or hydroquinonediphenyl phosphate) and 0,5-10 % froth graphite, deposited is offered at normal pressure and temperature $240^0$ C, a component of 10 % from weight of graphite. Weights apply to reception of fire-resistant fibers, films and moulding products.

Thermoplastic compositions contain [43] (%): A) 40-85,98 not containing halides of a polycarbonate; B) 5-25 not containing halides graft polymer (GP), constructed from 40-80 % of a basis from rubber-elastic polymer with glass temperature < $0^0$ C, for example, copolymer from 70-99,9 % $C_{1-8}$-alkylacrylate, 0-30 % copolymerizating monoethylene nonsaturated a monomer and 0,1-5 % copolymerizating multifunctional sewing a monomer, and 60-20 % graft environments from 50-95 % of the styrene (I), replaced styrene (RS) of general formula $C_6H_{5-n}R_n{}^1CR{=}CH_2$, where R - H or $C_{1-8}$-alkyl and $R^1$- $C_{1-8}$-alkyl, n=0, 1, 2 or 3, or methylmetacrylate (II) or their mixes and 50-5 % acrylonitrile (III), metacrylonitrile (IV), II, maleic anhydride (V) or their mixes; C) 5-40 not containing halides, thermoplastic copolymer from 50-95 % I or RS or II or their mixes and 50-5 % III, IV, II, V or their mixes; D) 4-20 not containing halides of connections of general formula $R_2OP(O)(OR^4)OR^3$ where $R^2$, $R^3$ and $R^4$ - independent from each other $C_{1-8}$-alkyl or $C_{6-20}$-aryls which can be I or 2 times are replaced $C_{1-4}$-alkyls; E) 0,01-5 perfluoride- or polyfluorinealkanesulfoacids or derivative carbonic acids of general formula $F(CF_2CF_2)_x(CH_2CH_2)_yX_nM$, where X - $SO_3$ or C(O)O, M - atom H or alkaline or alkali-earth metal, x=1-6, y=0-1 and n - valence of M; K) 0,01-5 PTFE with the average size of particles (ASP) 0,05-20 microns; L) 0-40 fibrous either corpuscular filler or their mixes. The component can consist of 30-70 % RS with ASP 200-700 нм and 70-30 % RS with ASP 50-180 нм. Such compositions, being fire-resistant, differ high temperature constancy and impact strength with a cut.

Offered [44] fire-resistant polymeric composition on the basis of a polycarbonate or his mixes with other polymer (for example PA, ABS, polysulfon etc.) with high mechanical parameters contains 0,1-40 % phosphor organic compounds (for example, ethers phosphoric or phosphorous acids); 0,001-40 % boron containing compounds (for example a boric acid, oxide a pine forest, tetraborates zinc); 0,01-5% polyorganosiloxane (for example, polydimethyl-, polymethylphenylsiloxanes) and - or fluorine containing polymer (for example PTFE, polytriftorochlorethylen). In other work [45] composition make also on a basis of a polycarbonate not containing haloids and includes 40-80 % elastoplastic with glass temperature < $0^0$ C and 20-60 % of a mix of styrene (I), methylstyrene (II), the styrene replaced in a nucleus and (or) alkyl(met)acrylate (alkyl $C_{1-8}$), (met)acrylonitrile (III) and (or) maleic anhydride (IV) and 1-50 % phosphor organic compounds of general formula $CH{=}CR^1\text{-}CO\text{-}X^1\text{-}(CH_2\text{-}CHR^1\text{-}X^2)_n\text{-}P(OR^2)(OR^3)X^3$, where $R^1$- H or $CH_3$; $R^2$ and $R^3$ - independent from each other alkyls ($C_{1-4}$) or not containing haloids aryl ($C_{6-20}$); $R^{1-3}$- O, S, NR, where R- H or not containing haloids alkyl ($C_{1-8}$) and n=0-10, 5-40 % not containing haloids thermoplastic copolymer 50-95, I, II, the styrene replaced in a nucleus and (or) acrylomethylacrylate, 4-50 % III and (or) IV, 1-46 not containing haloids phosphororganic compounds, 20 % not containing haloids of compound of formula $R^1OP(O)(OR^3)OR^2$, where $R^1$, $R^2$ and $R^3$ - not containing haloids alkyl ($C_{1-8}$) either not replaced or replaced aryl ($C_{6-20}$) and 5 PTFE (the average size of particles 0,05-20 microns). The composition has the raised fire resistance at high temperature constancy and impact strength with a cut. Process a composition extruded, it is possible with fanning, moulding under pressure, pressing or sintering. The composition is applied to reception moulded details for motor industry, with constructions, office machines, electric and household devices.

Other fire-resistant composition includes [46] mixes of an aromatic polycarbonate on a basis diphenylolpropane and styrene polymer, 5-50 % FRA, an ether of an aromatic phosphoric acid of formula $ROR(O)(OR^1)OAOR(O)(OR^2)_nOR^3$, where R and $R^{1-3}$ - (independent) phenolic groups replaced in position 2 and 6 on alkyls ($C_{1-4}$) (methyl, ethyl), and 1,3-phenylen and n=1-2 and 0,05-5 % PTFE. Compositions extruded, granulate and cast at $240^0$ C about samples with durability at a stretching 68 MPa, lengthening of 38 %, fluidity melt 26 g/10 mines, deformation temperature constancy $77^0$ C, a class of combustibility on UL - 94 (samples thickness of 1,6 mm) V-O.

For reception of a fire-resistant polymeric composition on the basis of a polycarbonate (or mixes of the personal computer and ABS, PFO) in a polymeric sample enter [47] 5-30 w.p. FRA such as a phosphoric ether of formula $R^3OP(O)(OR^4)OR^5OP(O)(OR^6)OR^7$, where $R^3$, $R^4$, $R^6$, $R^7$ - (identical or various)-alkyl ($C_{1-10}$) or aryl, probably, replaced; $R^5$ - aliphatic or aromatic biradical; 0,005-5 w.p. salts of metal (Ca, Mg, Zn) and an organic phosphoric ether, 0,05-5 w.p. The agent interfering formation flows, and 0,001-5 w.p. oxide Zr, Sb and - or Ge.

Shock-resistant at low temperatures, fire-resistant, containing graft polymers, compositions with combustibility (on UL-94) v=0, impact strength (on samples with a cut) $A_d$ 20 kJ/$m^2$ at -20° contain [48] (w.p.) 40-99 aromatic polyethercarbonate or mixes from personal computers, 1-40 graft the polymer received from 5-95 % vinyl monomer, graft on 95-5 % elastic plastic with glass temperature < 10°, mainly <20°, 0-30 thermoplastic vinyl copolymer or polyalkylenterephthalate; 0-30 phoshororganic comounds [oligo(amino)phosphate, phoshazen]. The composition contains fluorinated polyolefines, filler: talc, mica, vollastonit, quartz, dioxide the titan. From a composition form cases of devices, details of planes, the ships, automobiles.

The structure of a fire-resistant composition includes [49] polycarbonate and aromatic vinylic polymer (PS, copolymers styrene and acrylonitrile, ABS), containing as the assistant in an aromatic ring the acid basis, for example salt metals, alkaline-earth metals or ammonium and sulfonic, a boric or phosphoric acid and 0,01-5 % inhibitor (phenol-phormaldegide resine, fluorine or siliconorganic the polymer, in particular forming fibrils PTFE). The composition not containing halogens has high shock durability and temperature constancy in the usual both damp environment and stability formation of a static electricity. Use such compositions for manufacturing cases or details of electric or electronic devices.

Fire-resistant thermoplastic the forming weights processed in fibers, a film, moulded products, contain [50] thermoplastic PA, 1-30 % of phosphorus, 0,5-30 % rubber-like polymer, 0,05-15 % PE with density 0,94-0,98 g/$sm^3$ or software with molecular weight $1{\cdot}10^5$-$2{\cdot}10^6$ or their mixes, 0-60 % наполнителя fibrous and - or disperse type. Mixes apply in the electronic and electro technical industry. Moulding under pressure receive the standard samples characterized by fire resistance on method UL-94 V-0 (V-0) by impact strength of 43 kJ/$m^2$, concentration of soluble phosphorus after endurance of 10 day in water at $60^0$ C 10 w.p. on 100 w.p. compositions.

Fire-resistant, not soluble in water and not containing halogens of a composition are prepared [51] of thermoplastic PA, 0,2-40 % of compounds of iron with the size of particles < 50 microns: oxides, not soluble salts in water (silicates, borates, phosphates, the clay, containing 3 % Fe) and 0,2-50 % forming at burning coal thermoplastic polymer with aromatic groups in the lateral circuit, characterized by stability and absence of flying substances at temperature of processing of polyamide (>$200^0$ C) (polyether imides, polyphenylensulphide, polyimides, polyarylates, polyphenylensulphones, polyphenylene ethers). Mark, that it is possible to enter into a composition promoting stabilization of coal the additive on basis Zn, Fe (if in a composition of them there were no), water-insoluble borates manganese, low-melting glasses, inorganic fibres (vollastonite), and also 0,2-20 % of compounds of phosphorus (di- or triarylphospates, diarylhydrocarbilphosphonates, aryldihydrocarbylphosphinates, having not replaced or replaced with halogen, methoxy- or cyan groups hydrocarbyl group). Received samples, are characterized by a limiting oxygen index of 35,4 (24,8) % and time of burning on method UL-94 7 (36) with.

Fire-resistant thermoplastic forming compositions on the basis of the polyamide, processed in a film, a fiber and the products used in the electronic and electrotechnical industry, contain [52] 1-30 % of phosphorus, 1-15 % of compound mono- and - or diphenolic such as formula $C_6H_3(R^1)(R^2)OH$, where $R^1$, $R^2$ - (identical or various, are in about ortho-position to OH-group) H, $C_{1-16}$ - alkyl, $C_{6-18}$ - aryl, (saturated $C_{1-12}$- alkyl, or halogen) $C_{7-20}$ - aralkyl in which alkyl can be connected through O, $C_{1-3}$-alkylene or $SO_2$; or $R^1$ together with $R^2$ - aromatic or cyclic aliphatic a ring; or formulas $HOC_6H_{4-m}(R_m)XC_6H_{4-n}(R_n)OH$, where R-

Me, MeO or Et, m and n=0-4, X - simple chemical bonds, $C_{1-7}$- alkylene, $C_{5-6}$- cycloalkylene, O, S, $SO_2$, CO or $O(CH_2)_pO$, where p=2-10, and number of carbon atoms in the replaced phenolic ring > 6; 0-60 % fibrous and - or disperse filler, 0-30 % rubber-like polymer. In quality PA apply in part crystal, in part aromatic PA (with the contents triamines <0,5 %) on a basis of (%) of 20-90 parts hexamethylenediaminetherephtalate, 0-50 caprolactam, 0-80 hexamethylenediamineadipinate, 0-40 parts of others monomers.

Also in a composition on the basis of polyamide enter [53] 5-35 % calcium or aluminium salt (di) phosphinic acids of the formula $(R)(R^1)P(O)\text{-}O_mM$ and $RP(O)OR^2P(O)(R^1)O_mM_x$ where R and $R^1$-alkyl $C_{1-16}$, mainly linear or ramified alkyl $C_{1-8}$, for example, H, $CH_3$, $C_2H_5$, (iso)propyl, butyl, tertiary butyl, penthyl, octhyl, phenyl; $R^2$ - linear or ramified alkylene $C_{1-10}$, for example, methylene, ethylene, isopropylene etc., it is especial phenylene, naphthylene, their replaced methyl(ethyl)(butyl)phenylene, methyl(ethyl)(tert.butyl)naphthylene; M - ions Ca, Al; m=2-3, n=1-3, x=1-2. The composition can be strengthened by fiber glass. The named phosphoric salts apply in quality FRA for PA and polyamide containing plastics. From compositions form electric details, plugs, cases of electric devices, apply for lamination, coverings. Offered composition and control with the additive of the same quantities of Zn-salt ethyl-methyl phosphinic acids have a class of combustibility (on UL-94, on samples thickness of 1,6 mm and 1,2 mm) accordingly V=0, V=0 and V=2, V=2.

Free from halogens and phosphorus fire-resistant compositions on the basis of one thermoplastic PA effective quantity of additives giving [54] fire resistance, namely 0,1-40 % of connections from group: $NH_2SO_3H$, imidodisulphon an acid of formula $NH(SO_3H)_2$, mono- or diammonic, mono-, di- or polyvalent metal salt $NH_2SO_3H$ or imidodisulphonic acids, sulphamide of formula $H_2NSO_2NH_2$, including replaced aliphatic or aromatic groups, sulfate either imidosulphonate multifunctional spirit or sultone, 0-50 % carbon-forming the agent polyol, for example, pentaerythritol, including polyololygomeric the type, replaced or etherified polyol, sugar or polysaccharide, or aromatic polymer, for example polyphenylene oxide, polyphenylene sulphide, polysulphone, polyarylate or polyetherimide, 0-50 % of compounds of metal (oxide, borate, silicate, sulphamate, phosphate, zirconate or titanate alkaline, alkaline-earth other di- or polyvalent metal; for example, oxide or borate Zn or Mg, $Mg(OH)_2$, $MgHPO_4$, $Mg(H_2PO_4)_2$, $ZrO_2$, $TiO_2$, fluor-titanate or zirconate Na, MgO, oxide or borate vanadium), 0-50 % filler or the amplifier (fiber glass, carbon-fibre, aramide, thermo tropic polyester, mineral fibres, vollastonit, mica, plaster, talc, low-temperature glass). Compositions have a limiting oxygen index (OI) of 35 %, a class of fire resistance on method UL-94 V-1 and average (from 10 samples) time of burning after consecutive, through 10 seconds, a double present of a flame to a sample 16 seconds. For initial polyamide material OI makes 26,4 %, at a present of a flame in the same conditions the sample burns.

Though polymers on a basis chlorendic acids and its derivatives in themselves without introduction FRA non-combustible and self-extinguishing [55], their fragility and colouring under action of light stimulated searches alternative monomers to which belong for example, 1,2,8,4,9,9-hexachlor-1,4,4a,5,6,7,8,8a-octahydro-1,4-methanonaphthaline-6,7-dicaronic acid. The acid, its anhydride and their derivatives are used as additives or comonomers for reception nonflammable nonsaturated polyesters, polyamides, polyimides, polyurethanes. Effective FRA are received by interaction of polymer on the basis of the named acid both N-alkyldialkanol amines with alkyl- and thiophosphates, phosphinates or phosphonates.

Fire-resistant pitches represent [56] themselves a mix of the pitch component consisting from thermally hardened of compound, containing 1-oxa-3-aza-tetralinic groups (CAG), or containing CAG, it is possible in quantity >5 % (from weight CAG) epoxy compounds, and the second component containing not mixing up with pitch component FRA or consisting of him, for example, hydroxide of aluminium in quantity of 40 % (from weight of a pitch component), hydrateding Ca-Mg-carbonate in quantity of 50 %, $Mg(OH)_2$ in quantity of 5 %, elementary red phosphorus, oxygen containing acids of phosphorus, their inorganic or organic salts, for example, $(NH_4PO_3)_n$ or $(NH_4)_{n+2}P_nO_{3n+1}$ in quantity 20 w.p., monomelamine phosphate formulas $C_3H_6N_6H_3PO_4$ or dimelamine phosphate formulas $(C_3H_6N_6)_2H_3PO_4$ in quantity of 20 %, a boric acid or borate of zinc. Assert, that CAG can be received from phenol and amine from which one component has functionality >1. Pitch components at hardening with heat treatment form middle-combustible and higher-temperature firm polymers; for example, the pitch component containing $Al(OH)_3$, hardened at 180-220$^0$ C. For achievement of fire-resistant properties at application CAG it is required much less FRA, than at use epoxy pitches.

In the conclusion it is necessary to note, that the given brief review of the scientific literature in the field of creation of fire-resistant polymeric materials allows to look after last achievements of science and technology on the basic directions developing in modern polymeric chemistry.

## REFERENCE


[1] Flammhemmende Kunststoffe. *Galvanotechnik*.-1993.-Vol.84.-No 7.-P. 2381.

[2] R.M. Aseeva., G.E. Zaikov. *Burning of polymeric materials*.- Moscow.: Nauka. -1981.- 280 p.

[3] R. Rangaprasad. Fire and polymers.-*Popul. Plast. and Pack*.-1993.-Vol. 38.-No 6.- P.55-58.

[4] Ch. A. Wilkie. Polymer degradation and stabilization.- *Polym. News*. - 1993. - Vol.18. - No 4. -P.107-108.

[5] 4th meeting of fire retardant polymers, Freiburg i. Br., Sept. 9-11.-1992.- *Makromol. Chem. Macromol. Symp*.-1993.-Vol. 74.- P.1-357.

[6] D. Boolley. Fire retardant polymers in the future: safety by tests of engineering?: Pap. 4th Meet. "Fire Retardant Polym.", Freiburg i. Br., Sept.9-11,1992. - *Makromol. Chem. Macromol. Symp*. - 1993. - Vol.74.- P.1-10.

[7] M. Bernic. Nylon RIM makes strong showing at conference.- *Plast. World*.- 1994.-Vol. 52.- No 7.- P.38-40.

[8] Ch.A. Wilkie. Polymer Degradation and Stabilization. - *Polym. News*.-1995.- Vol. 20.- No1.- P.19-20.

[9] E. Fenyvesi. Ujszeru muszaki es textilipari szalasanyagok 1.- *Gac. med. Caracas*.- 1992.- Vol. 100.- No 3.- P. 278-283.

[10] G. Tesoro. Recent conferences on fire resistant Aircraft interior materials. - *Polym. News*.-1995.-20.-No 6.- P.192-194.

[11] M. Bernie. New technology unveiled at FRCA fall meeting.- *Plast. World*.- 1995.-Vol. 53.- No1.-P. 38-40.

[12] Malaysia under the microscope.-*Polym. and Rubber Asia*.-1994.- Vol.9. - No 51.- P. 56-58.

[13] Improving materials formulations. - *Mater.+ Manuf*.-1993.-Vol.10-No1.-P.14.

[14] Challenging times for performance polymers.-*Eur. Plast. News*.-1994.-Vol. 21. - No 5.
[15] G.S. Kirshenbaum, M. Lewin. Flame retardancy of polymers II. - *Polym. News*.-1992. - Vol.17. - No2.- P. 61-63.
[16] T. Hirose, Sh. Miyata. *Flame retardant and flame retardant resin composition*. - Pat. 5034442, USA (1991).
[17] V. Abolins, J. Berts, F.F. Holub, C.F. Lee. *Crosslinkable flame retardant composition of polyolefin and polyphenylene ether*. - Pat. 4945018, USA (1989).
[18] K. Billington, J. Griffiths, G. West. Developments in fire retarded plastics.- Plast. *and Rubber Int*.-1992.-Vol.17.-No 2.- P. 6-9.
[19] I. Sakon, M. Sekiguchi, A. Kanayama. *Method for producing red phosphorus flame retardant and nonflammable resinous composition*. - Pat.5041490, USA (1991).
[20] N.A. Burditt, D. Chundury. *Free-radical modified fire-retardant polypropylenes*.- Pat. 5079283, USA (1992).
[21] Melamine flame retardant. - *Chim. oggi*.-1992.-Vol.10.-No 4.- P. 25.
[22] D.H. Blount. *Flame retardant compounds and compositions*.- Pat. 6054515, USA (2000).
[23] I.V. Dolbin, A.H. Shaov,T.N. Podobueva, A.Y. Bedanokov. Ecological aspect of updating of polyethylene of high density. - Materials 5 science-practical conf. Maikop the state institute of technology, Maikop, 21-24 November, 2000: *Biology, wood economy, ecology*.- Maikop: Publishing house MGTI.-2001.-P. 82-83.
[24] Guo Shao-Hua, R.A. Grey. *Thermoplastic resin based on vinyl aromatic/vinyl phosphonic acid copolymers and amine-terminated polymers*. - Pat. 5328958, USA (1993).
[25] H. Horacek. *Flammfeste Kunststoffe mit einem Gehalt an Trihydrazin, Triguanidinotriazin oder deren Salzen*.- Pat. 401522, Austria (1996).
[26] P.P. Greigger, S.C. Liptak, Th.A. Ward. *Fire retardant composition for composites*. - Pat. 6479574, USA (2002).
[27] A.V. Zhurko, E.V. Shatalov, A.M. Dorohov etc. *The fire-resistant textile material*.- Application of Russia 95119185 (1997).
[28] W. Tischer, W. Brennig. *Mischungen und ihre Verwendung als Flammschutzmittel*. - The Application of Germany 4024667 (1990).
[29] Zeltner. *Flammgeschutzte thermoplastische Formmassen*. - Application of Germany 4105925 (1992).
[30] D.H. Blount. *Flame retardant compositions*. - Pat. 5156775, USA (1992).
[31] K. Komori, K. Kashihara, K. Ogasawara e.t.c. - *Flame retardant resin composition*. - Application EPV 1059329 (2000).
[32] Sh. Endo, G. Ochiai, Y. Mizutani e.t.c. *Reclaimed pet resin composition, molded article thereof, and flameretardant resin composition and molded article thereof*.- Application EPV 1050559 (2000).
[33] V.F. Stroganov, L.M. Amirova, E.V. Sachabieva. Updating epoxy polymers glycidic esters of acids of phosphorus. - Abst. lecture 2-nd Nat. Rus. Kargin a symposium "Chemistry and physics of polymers in the beginning of 21 centuries", *Chernogolovka*, May, 29-31, 2000. - Part 2. - Chernogolovka (Moscow Region).-2000.- P. 4-31.
[34] Katayama Masahiro, Hamano Hiromitsu. *Polycarbonate compositions*. - Pat. 6066686, USA (2000).

[35] Jang Bok-Nam, Kim Se-Jong, Chang Young-Kil, Lee Gyu-Cheol. *Flameproof thermoplastic resin composition.* - Application EPV 1069158 (2001).

[36] T. Marzouki. *Flammhemmend ausgerustete Schumstoffe auf Basis von Styrolpolymerisaten.* - Application of Germany 19910257 (2000).

[37] U. Murschall, G. Crass, W. Dietz, U. Kern. *Amorphe, strukturierte, flammhemmend ausgerustete Folie, Verfahren zu ihrer Herstellung und ihre Verwendung.* - Application of Germany 10002154 (2001).

[38] Liu Zhi-guo, Ding Tao, Jia Xiu-wei e.t.c. - *J. Henan Univ. Natur. Sci.* - 2002.-Vol. 32.-No 1.-P.33-36.

[39] Lo Tung Fai. *Flame-retarded styrene-based polymer and copolymer compositions.* - Application EPV 860470 (1998).

[40] M. Weber, K. Massonne. *Flammwidrige, thermoplastische Formmassen.* - Application of Germany 19626156 (1996).

[41] M. Suzuki, H. Itoh, K. Sami e.t.c. *Flame retarding resin composition.*- Pat. 5605962, USA (1997).

[42] M. Weber, P. Horn, R. Weiss e.t.c. *Flammwidrige thermoplastische Formmassen mit verbesserter Tropffestigkeit.*- Application of Germany 19621732 (1996).

[43] R. Neumann, K. Muhlbach, K. Ruppmich e.t.c. *Chlor-und bromfreie, flammfesve thermoplastische Formmassen auf der Basis von Polycarbonaten und Styrol/Acrylnitril-Copolimerisaten.*- Application of Germany 4100192 (1992).

[44] J. Watanabe, N. Kudo. *Flame-retardant resin composition.* - Pat. 5266618, USA (1993).

[45] N. Niessner, R. Neumann, K. Ruppmich e.t.c. *Flammwidrig ausgerustete Formmasse.*- Application of Germany 4132264 (1993).

[46] *Flame-retardant resin composition.* - Pat. 6022917, USA (2000).

[47] R. Kimura, T. Kamimoto, K. Hayashi e.t.c. *Flame-retardant resin composition.* - Application EPV 1304350 (2003).

[48] Th. Eckel, A. Seidel, B. Keller, D. Wittmann. *Polycarbonat-Zusammensetzungen.* - Application of Germany 10061080 (2002).

[49] M. Kitayama, T. Isozaki, A. Nodera. *Flame-retardant polycarbonate resin composition and molded articles.* - Application EPV 1142956 (2001).

[50] P. Baierweck, B. Gareiss, Ch. Plachetta. *Flammgeschutzte thermoplastiche Formmassen.*-Application of Germany 4100740 (1992).

[51] E.D. Weil, N.G. Patel, R.M. Leeuwendal. *Flame retardant polyamides.* - Pat. 5071894, USA (1991).

[52] B. Gareiss, P. Baierweck. *Flammgeschutzte thermoplastische Polyamidformmassen.* - Application of Germany 4121261 (1993).

[53] H.-J. Kleiner, W. Budzinsky, G. Kirsch. *Schwerentflammbare Polyamidformmassen.* - Application of Germany 19607635 (1997).

[54] M. Lewin. *Flame retardant polyamide compositions.* - Pat. 5424344 , USA (1995).

[55] P. Sivasamy, C.T. Vijayakumar. Flame-retardaut polymers based on hexolic acid/anhydride and its derivatives. - *J. Macromol. Sci.* C-1991.-Vol. 31.-No 2-3. - P.165-213.

[56] W. Saur, H. Schreiber. *Zu schwerentflammbaren und hochtemperaturbestandigen Kunststoffen hartbare Harse und Verfahren zur Herstellung solcher Kunststoffe.*- Pat. 678531, Switzerland (1991).

In: Chemical Physics of Pyrolysis, Combustion and Oxidation ISBN 1-59454-163-9
Editors: A. A. Berlin et al., pp. 33-39 

*Chapter 3*

# INVESTIGATION OF INFLUENCE OF PHOSPHORUS CONTAINING FIRE RETARDANTS ON THE STRUCTURE, PROPERTIES AND PYROLYSIS OF PAN FIBRES

***V. I. Besshaposhnikova***
Saratov State Technical University, Engels Technological Institute,
Engels, Saratov oblast, 17 Svoboda Sq., *admin@techn.rennet.ru.*,
Fax: 8-84511-63533

## ABSTRACT

Polyacrylonitrile fiber has a complex of valuable properties, but it is a highly inflammable material, which is characterized by high speed of combustion and toxicity of pyrolysis products. That's why, in order to reduce fire hazard, the polyacrylonitrile fiber has been modified by the phosphorus containing fire retardants, which intensify the process of structuring during pyrolysis. It leads to the increase in coke output, change in the pyrolysis product composition, and the combustion elimination.

***Key words*** – Polyacrylonitrile fiber, modification, fire hazard, fire retardants, combustibility parameters, physical and mechanical properties, pyrolysis and combustion.

Polyacrylonitrile (PAN) fibers are widely used in the production of textile clothing materials, raw materials for carbon fiber production, and for reinforcement of polymer matrixes in the production of composite materials. However, their combustion power is rather high. Investigation of mechanisms and kinetics of the chemical transformation of polymers into combustion materials makes it possible to foresee their behavior in the fire risk situations, to control the combustion process and to create the flame resistant materials. Accumulated experimental data show that intensification of polymer cyclization and, also, reduction of pyrolysis volatile product output result in decreasing of the PAN fiber combustibility [1]. The decrease in fiber combustibility is achieved by the modification of a newly formed PAN gel fiber by inclusion technique using phosphorus containing materials, namely, phospholane (PHL), sodium polyphosphate (SPPH), and phosdiol (PHD) as fire retardants (FR). The main feature of the newly formed PAN gel fiber is its developed inner porous structure with a specific surface of 4709 $m^2/g$, and therefore its good sorption power. While drying the fiber pores get closed and strongly hold the FR in the inner structure [2].

Fiber structure was investigated by X-ray structure analysis (XRSA) with a Dron-3 diffractometer. Interaction between the FR and the fiber was studied by IR spectroscopy with a Specord-75JR device and its MJR-4 attachment. Fiber behavior at higher temperatures was determined by thermogravimetric analysis (TGA) with a Derivatograph Q-1500 device. Electronic and microscopical analysis was made by means of a Superprob-733 electronic focused beam microscope equipped with a wave spectrometer. The composition of pyrolysis gasiform products was investigated by stage pyrolysis gas chromatography (SPGCH) technique with polysorbent filled columns. Combustibility parameters were estimated by GOST 12.1.044 - 89; strength parameters – by GOST 6611.2 - 73 with PT-3-1 or Fafegraph devices.

Degradation of PAN fibers begins in the temperature range of 521-551 K, which results in the formation of the cyclic structure proved by exothermal peaks on the DTA curves of the PAN fibers (Fig. 1). Introducing the phosphorus containing FR initiates the fiber polymer cyclization, and it begins 20-30° earlier. At the same time, it provides high coke output (Table 1), decrease in energy for the process activation, and reduction of the total volatile product output, including $CO_2$ and HCN (Table 2). All these prove the influence of FR on the PAN fiber degradation, mainly, in the condensed phase. It is also confirmed by the roentgen fluorescent analysis data showing the distribution of phosphorus in the coke of the burned sample (Table 3). It is noteworthy that the quantity of phosphorus in the coke always exceeds 90 %, irrespective of the FR nature. Modified fibers have higher values of oxygen index (OI) (Table 4).

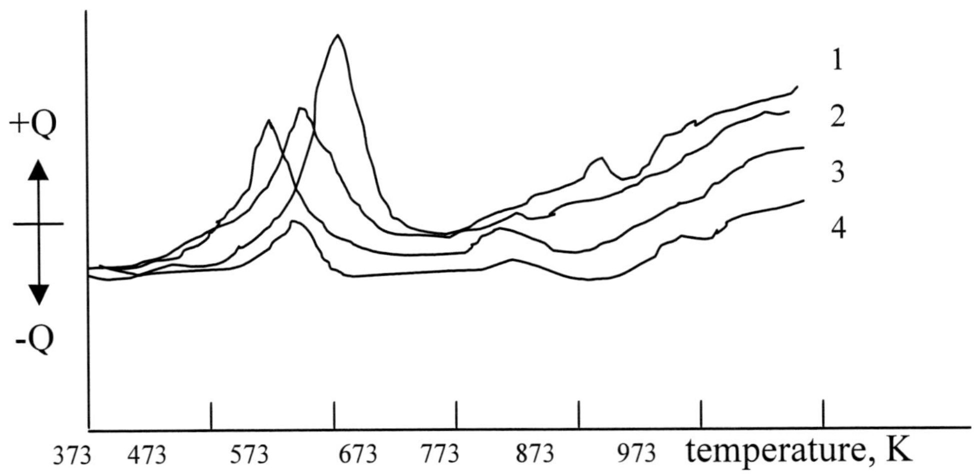


**Fig. 1.** DTA data for PAN fibers: 1 – PAN initial; 2 – PAN + 8 % SPPH; 3 – PAN + 22 PHD; 4 – PAN + 15,8 % PHL.

**Table 1. TGA data for modified fibers**

| № | Sample composition, % mass | $E_a$, kJ/mol | Degradation temperature, °K, Ti – Tf* / Tmax | Mass loss, %, at Tf | Mass loss, %, at temperature, °K | | | | |
|---|---|---|---|---|---|---|---|---|---|
| | | | | | 573 | 673 | 773 | 873 | 973 |
| 1. | 100 PAN | 240 | 533 – 563 / 543 | 15.5 | 16 | 22 | 32 | 41 | 62 |
| 2. | 78PAN+22PHD | 171 | 493 – 568 / 558 | 14 | 18 | 25 | 35 | 49 | 56 |
| 3. | 84.2PAN+15.8PHL | 103 | 495 – 575 / 553 | 10.8 | 25 | 35 | 40 | 52 | 55 |
| 4. | 92PAN+8SPPH | 182 | 523 – 555 / 542 | 9.5 | 10 | 15.5 | 23 | 33 | 46.2 |

* Ti – initial temperature; Tf – final temperature.

**Table 2. Pyrolysis gas composition for modified PAN fibers**

| № | Sample composition, % mass | $H_2O$ | $CO_2$ | HCN | |
|---|---|---|---|---|---|
| | | ΔS, mg | ΔS, mg | ΔS, mg | Ti, K |
| 1. | 100 PAN | 392 | 688 | 507 | 543 |
| 2. | 78PAN+22PHD | 431 | 670 | 323 | 507 |
| 3. | 84.2PAN+15.8PHL | 420 | 620 | 301 | 443 |
| 4. | 92PAN+8SPPH | 480 | 540 | 266 | 507 |

**Table 3. Phosphorus distribution in the sample structure**

| № | Sample composition, % mass | Impulse number | | Phosphorus quantity in coke, % |
|---|---|---|---|---|
| | | In fiber | In coke | |
| 1. | 95 PAN + 5 SPPH | 8960 | 8090 | 90.3 |
| 2. | 92PAN+8SPPH | 15481 | 14097 | 91 |
| 3. | 84.5PAN+15.8PHL | - | - | 93 |
| 4. | 92PAN+8SPPH oxidized | 5879 | - | - |
| 5. | 92PAN+8SPPH carbonized | 2980 | - | - |

Thus, it has been found out that phospholane is a very effective fire retardant for the PAN fibers. Its fire retarding effect is proof against wet treatment. Its OI does not considerably change after 10 washings, that's why the PHL modified PAN fiber can be recommended for the domestic purpose textile materials. At that, the SPPH and PHD modified fibers have high strength values. Relative breaking load increases by 12-19 %, and elongation decreases by 5-12 % as compared with non-modified fiber. Therefore, in this case, modification contributes to the fiber structuring, seemingly at the expense of forming the additional cohesion. The PHL modification leads to a 17% decrease in the fiber strength and 8,5 % decrease in elongation (Table 4).

**Table 4. Properties of modified PAN fibers**

| № | Sample composition, % mass | Relative strength, cent Newton/tex | Elongation at rupture, % | Oxygen index, % | |
|---|---|---|---|---|---|
| | | | | Modified | After washing |
| 1. | 100 PAN | 27.4 | 35 | 19 | 19 |
| 2. | 78 PAN + 22 PHD | 32.08 | 32.8 | 21.5 | 21 |
| 3. | 84,2 PAN + 15,8 PHL | 22.6 | 32 | 30 | 28 |
| 4. | 92 PAN + 8 SPPH | 32.7 | 34 | 23.5 | 23.0 |

Interchange between fibers and FR is confirmed not only by high wet treatment fastness but also by the data of IR spectroscopy. It is known [3] that 1100, 1148, 1172 and 1217 $cm^{-1}$ bands are characteristic for P=O bonds on the IR spectra; a wide band of 900 $cm^{-1}$ and a shoulder of 960 $cm^{-1}$ are characteristic for –P – O- fluctuations. For the fiber modified by, say, SPPH (Fig. 2), the type of the absorption band changes in the range of 900-1000 $cm^{-1}$, and the bands of 968 and 904 $cm^{-1}$ are very intensive as compared to the starting PAN fiber. Absorption bands in the range of 700-900 and 624 $cm^{-1}$ disappear, but the band of 560 $cm^{-1}$ is rather strong; it proves the presence of interaction, which is most likely, takes place due to –P –O –Na groups.

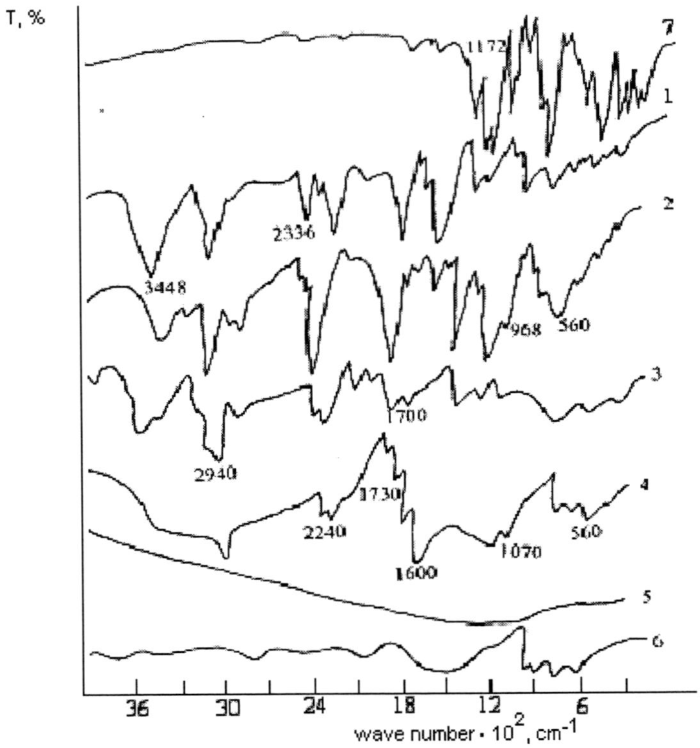


**Fig. 2.** IR spectroscopy data:
1 – 100 PAN; 2 – 92 PAN + 8 SPPH; 3 – 100 PAN oxidized;
4 – 92 PAN+8 SPPH oxidized; 5 – 100 PAN carbonized;
6 – 92 PAN+8 SPPH carbonized; 7 – 100 SPPH.

It seems to be interesting to investigate the influence of the FR on oxidation and carbonization of the modified PAN fiber. Let's take SPPH modified PAN fiber as an example.

On IR spectra of the modified oxidized fiber the band intensity in the ranges of 2940, 3450 and 2200-2240 $cm^{-1}$ typical for $CH_2$, /=N-H/, /C=C-N/ and /C=N/ groups decreases in comparison with oxidized non-modified fiber (Fig. 3, curves 3 and 4). The intensity of absorption bands in the ranges of 1730, 1600, 1680 and 1070 $cm^{-1}$ typical for /C=O/, /-C=C, -C=N, -COO-H, /-C-S-C=O/ and /C-C/ groups increases. It proves catalytic influence of SPPH on cyclization and carbonization of the PAN fibers. Absorption bands in the range of 560-900 $cm^{-1}$ typical for SPPH are the evidence of its presence in the structure of oxidized and carbonized modified fibers (Fig. 2, curves 4 and 6).

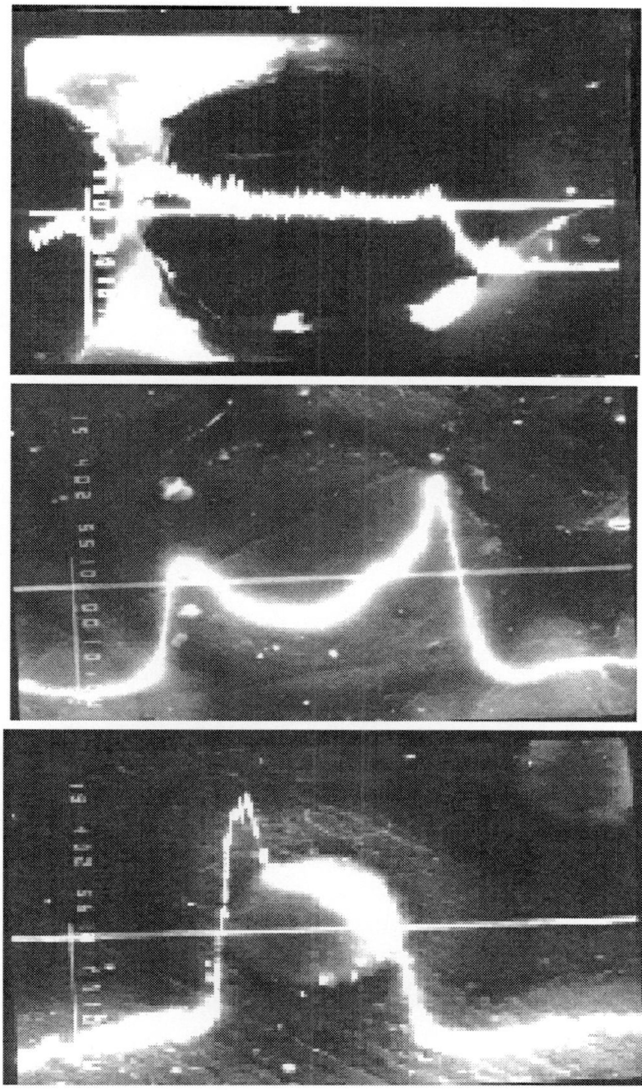


**Fig. 3.** Electronic microscopical and spectral analysis data on phosphorus distribution in the structure of PAN fibers: a – 92 PAN + 8 SPPH; b – 92 PAN + 8 SPPH oxidized; c – 92 PAN + 8 SPPH carbonized.

Also, the presence of phosphorus in the structure of carbonized fiber is confirmed by the roentgen fluorescent analysis data (Table 3).

The data of electronic focused-beam microscopy with microprobe analysis show that phosphorus distributes evenly on the fiber diameter (Fig. 3a). Phosphorus distribution spectra for oxidized and carbonized fibers are more intensive on the surface, and it proves partial FR dewetting at high temperatures.

According to the data of X-ray structure analysis, the modified oxidized fiber has higher intensity, and its peak is taller and larger as compared to non-modified oxidized fiber (Fig. 4, curves 1 and 2). Therefore, modification under these conditions intensifies structuring of the oxidized fiber. As for the carbonized fiber, it's on the contrary – the non-modified fiber has more ordered structure. Apparently, the FR takes the place between the carbon links of the fiber macromolecules; the bias of the diffraction angle maximum into the area of big angles (proving the changes in interplanar spacing in macromolecules of the carbon fiber) testifies to the favor of the fact.

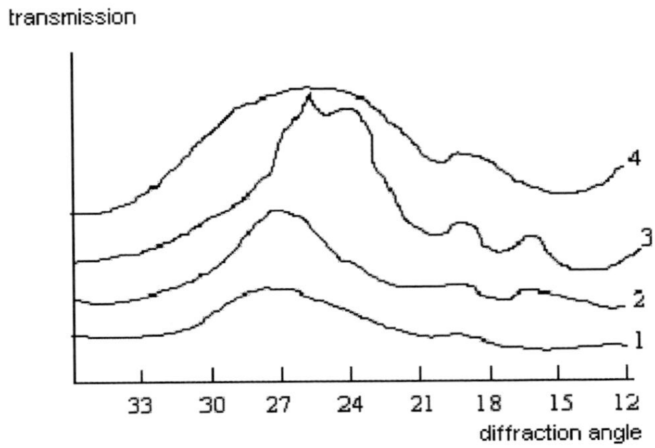


**Fig. 4.** X-ray structure fiber analysis:
1 – 100 PAN oxidized; 2 – 92 PAN + 8 SPPH oxidized;
3 – 100 PAN carbonized; 4 – 92 PAN + 8 SPPH carbonized.

The influence of FR on the carbonized fiber structure results in a 40 % decrease in the elasticity module and a 20 % decrease in the strength of the carbonized fiber.

Thus, the influence of phosphorus containing FR on the structure, properties and pyrolysis has been studied. The regularities of thermal oxidative fiber degradation at the presence of the fire retardants have been determined; they reveal predominance of structuring giving the coke formation over degradation; it leads to the decrease in combustible, volatile and toxic product output.

Obtained modified fibers are proof against wet treatments; they have good fire retarding and high strength properties.

Influence of modification on the PAN fiber structure and properties at high temperature treatment has been investigated.

## REFERENCES

[1] Panova L.G., Artyomenko S.Ye., Besshaposhnikova V.I., in: *Chemical fibers*, 2003, № 6, pp. 56-58.

[2] Geller A.A., Geller A.N. et al., in: *Chemical fibers*, 1990, № 3, pp. 8-11.

[3] Dechant J., Danz R. *Ultrarotspektroskopische Untersuchungen an Polymeren, Akademie*, Verlag, Berlin, 1972, 472p.

In: Chemical Physics of Pyrolysis, Combustion and Oxidation ISBN 1-59454-163-9
Editors: A. A. Berlin et al., pp. 41-44 

***Chapter 4***

# DECREASING COMBUSTIBILITY OF SYNTHETIC MATERIALS

***T. V. Kulikova, V. I. Besshaposhnikova, L. G. Panova, O. A. Grishina, T. G. Nikitina and, O. N. Fridrikh***
Saratov State Technical University, Engels Technological Institute,
Engels, Saratov oblast, 17 Svoboda Sq., *admin@techn.rennet.ru.*,
Fax: 8-84511-63533

## ABSTRACT

The presence of functional groups makes synthetic fiber modification possible; it intensifies the process of structuring during pyrolysis that leads to the increase in coke output as well as combustion elimination. However, there aren't enough effective fire retardants for this purpose. That's why studies of pyrolysis peculiarities for the phosphorus containing fire retardants having different compositions and, also, thermal-oxidative degradation regularities for the modified polymers are of current importance. Finding a solution allows to regulate the pyrolysis and the combustion processes and, also, to lower the fire risk.

***Key words*** – synthetic fiber, modification, fire hazard, fire retardants, combustibility parameters, physical and mechanical properties, pyrolysis and combustion.

Industry produces small and medium tonnage outputs of thermostable and inflammable aromatic fibers and special-purpose threads. They are togylene, tverlane, arselone, arymide and others, which are rather expensive [1].

Large tonnage textile materials based on the thermoplastic fiber-forming polymers of polyethyleneterephthalate (PETPH), polyacrylonitrile (PAN), and polycaproamide (PCA) are comparatively cheap and widely used. But they are from the category of highly inflammable polymers, which are characterized by the high speed of combustion and toxicity of pyrolysis products. These factors put limit on their use in the number of industries. The presence of functional groups makes their modification possible; it intensifies the process of structuring during pyrolysis that leads to the increase in coke output as well as combustion elimination. However, there aren't enough effective fire retardants (FR) for this purpose. That's why studies of pyrolysis peculiarities for the FR having different compositions and, also, thermal-oxidative degradation regularities for the modified polymers are of current importance. Finding a solution allows to regulate the pyrolysis and the combustion processes and, also, to lower the fire risk.

In this connection, to evaluate a possible use of a certain chemical nature FR for different synthetic fibers, this research has been done to determine the influence of phosphorus containing FR of phosdiol (PHD) and phospolyol (PHP) on the pyrolysis procedure, combustion characteristics and physical and mechanical properties of polycaproamide (PA), polyester (PE) and polyacrylonitrile (PAN) fibers and textile materials based on them.

In our previous research [3] aimed to determine the modification parameters we studied cybernetics of FR sorption by fibers under different conditions. The modification was made by PHD and PHP water solutions, at temperatures of 20 and 85°C, with liquor ratio of 10 and FR content from 10 to 50 % mass. The modification of the PA and PE fibers by 10-50 % solution of the PHD and PHP fire-retardant additives at the temperature of the modifying solution of 20°C was found to be made for 120 sec for PE fibers and for 180sec for PA fibers at PHD modification, and for 120 sec for both at PHP modification. It is more effective to modify a newly formed, the so-called "gel", rather than a finished PAN fiber. It allows to include the modification into the technological process. The structure of the gel fiber is not completely formed; it has high porosity and sorptive power. The most effective temperature for the PAN gel fiber modification is 20°C for 120 sec with PHD and for 180 sec with PHP.

With all this going on, the modification results in maximum FR weight increment of 19-22 % mass for PAN, 20-24 % mass for PA and 24-24,5 % mass for PE (Fig. 1).

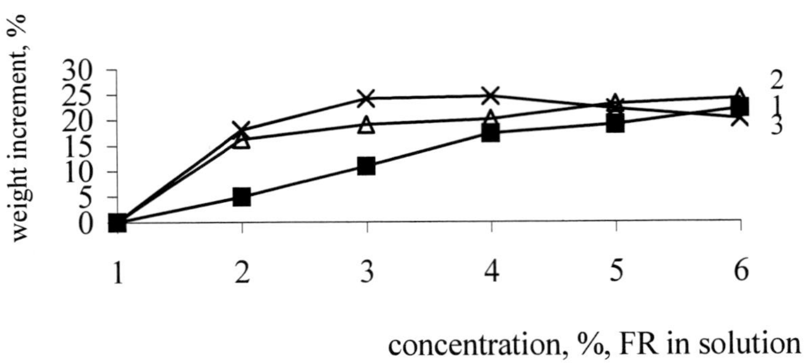


**Fig.1.** FR content in PHD modified fibers:
1 – PAN gel fiber; 2 – PA fiber; 3 – PE fiber.

Studies of the FR influence on the modified fiber pyrolysis reveal coincidence of the degradation initial temperatures (260°C) for both, PHD and PAN fibers (Table 1). These studies were conducted by thermogravimetric analysis (TGA) technique. In the temperature range of 260-350°C the PAN fibers enter the stage of cyclization [2]. The PHD additive initiates the PAN fiber cyclization that is proved by the initial temperature drop of 10-15° and by the presence of exothermal peaks on the DTA curves of the cyclization.

**Table 1. TGA data for modified fibers**

| № | Sample composition, % mass | Degradation temperature, °C, Ti – Tf* / Tmax | Mass loss, %, at Tf | Mass loss, %, at temperature, °C | | | | |
|---|---|---|---|---|---|---|---|---|
| | | | | 300 | 400 | 500 | 600 | 700 |
| 1. | 100 PHD | 260 – 350 / 326 | 74 | 34 | 78 | 84 | 87 | 87 |
| 2. | 100 PAN | 260 – 290 / 270 | 15,5 | 16 | 22 | 32 | 41 | 62 |
| 3. | 78 PAN + 22 PHD | 250 – 280 / 260 | 14 | 18 | 25 | 35 | 49 | 65 |
| 4. | 100 PE | 360 – 455 / 410 | 32 | 0 | 10 | 83 | 95 | 100 |
| 5. | 75,5 PE + 24,5 PHD | 360 – 460 / 420 | 44 | 4,5 | 14 | 82 | 84 | 85 |
| 6. | 100 PA | 345 – 480 / 420 | 95 | 5 | 36 | 98 | 100 | 100 |
| 7. | 76 PA + 24 PHD | 315 – 380 / 360 | 69 | 4,8 | 70 | 82 | 89 | 94 |

* Ti – initial temperature; Tf – final temperature.

The PHD additive influences the PA and PE fiber pyrolysis too, and it increases the coke output (Table 1). Changes taking place during polymer pyrolysis should influence the fiber combustibility.

The combustibility was estimated by oxygen index (OI) (GOST 12.1.044-89). With modification of PHD fibers the OI for PE increases up to 27,5 % vol., and for PA and PAN – up to 20,5 % vol. (Fig. 2). Fibers modified by PHP additive have lower OI values; they are 24,5 % vol. for PE, and 20 % vol. for PA and PAN. It was found that the PE fiber could be modified more effectively from the solution containing 30 % mass FR. Further increase of the FR concentration results in OI reduction by 1-1,5 %. It is most likely caused by the FR molecule aggregation in the solution, where the increase of the FR concentration gives the decrease of the FR weight increment per a fiber; it goes from 24,5 % down to 20 %. Additional thermal treatment does not considerably affect the value of fiber combustibility.

Modification, as a rule, results in structural fiber changes, and it affects the physical and mechanical properties. That's why the influence of PHD and PHP modification on deformation and strength behavior has been studied. Figure 3 shows that strength of the modified PAN gel fibers with a titer of 0,27 tex increases by 8-10 %, and elongation decreases by 12-18 %.

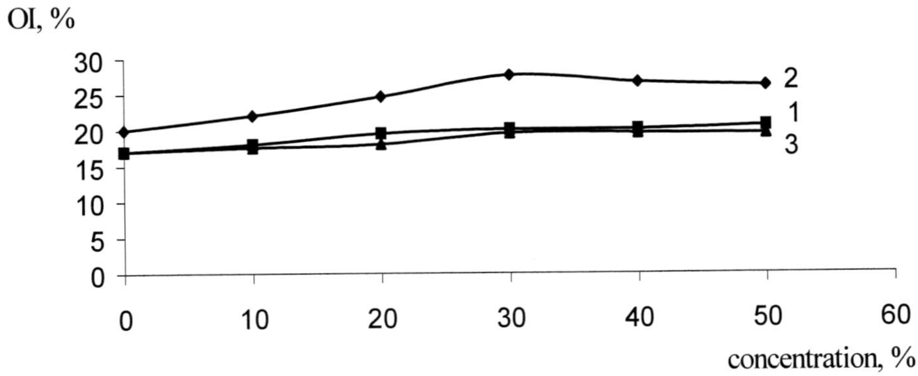


**Fig. 2.** Influence of PHD solution concentration on OI:
1 – PA fiber; 2 – PE fiber; 3 – PAN fiber.

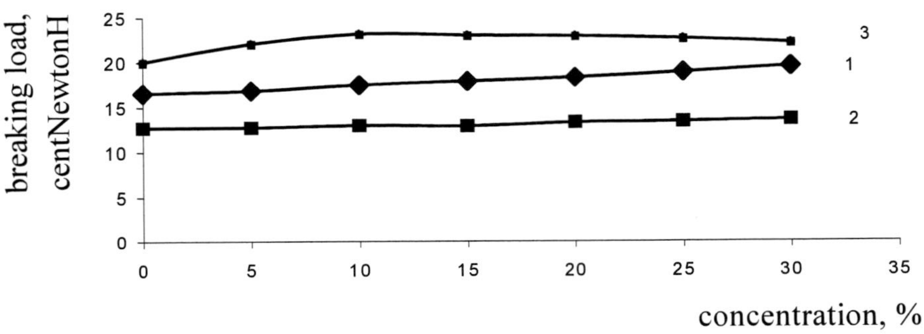


**Fig. 3.** Strength of PHD modified fibers:
1 – PE fiber; 2 – PA fiber; 3 – PAN fiber.

PHD and PHP modified PA fibers with a titer of 0,29 tex have the strength value that is comparable with the value of the starting fiber.

The strength of the PE fiber with a titer of 0,33 tex increases by 12-15 %. The elongation of the modified fibers does not reveal any considerable changes.

Thus, the conducted studies have proven more effectiveness in using phosdiol to decrease the combustibility of the polyester fibers.

## REFERENCES

[1] Perepelkin K., *International symposium on technical textiles, nonwoven fabric and protective clothing*, Moscow, Sept. 24-25, 2003.

[2] Aseeva R.M., Zaikov G.E. *Combustion of polymer materials*, Moscow, Nauka, 1981, 280p.

[3] Nikitina T.G., Besshaposhnikova V.I., *Journal of Chemical fibers*, № 6, pp. 24-26, 2003.

In: Chemical Physics of Pyrolysis, Combustion and Oxidation ISBN 1-59454-163-9
Editors: A. A. Berlin et al., pp. 45-53 

*Chapter 5*

# FACTORS AFFECTING HEAT RELEASE AT THE COMBUSTION OF THE DIFFERENT SPECIES OF WOOD

***R. M. Aseeva, Bui Dinh Thanh* and B. B. Serkov****
Institute of Biochemical Physics, Russian Academy of Sciences,
Bldg.4, Kosygin str., 119334, Moscow, Russia
* State Fire Academy, Ministry of Civil Defence, Emergency Situations, and Disaster Management,
Bldg. 4, B.Galushkin str., 129366, Moscow, Russia

## ABSTRACT

The influence of wood species, including classical hardwoods and softwoods, and also the content of the main chemical constituents on heat release at wood combustion has been studied. The tests cover 11 wood materials from coniferous and foliate trees of the middle Russia region and 5 wood varieties of tropical and subequatorial Viet-Nam region. The linear correlation between the lower values of complete combustion heat and summary content of lignin and extractives in wood samples has been established. The equation is presented that can be used to estimate the lower heat of complete combustion of extractives of different wood varieties. Other heat release characteristics ( peak and average heat release rate, total heat release over fixed time of the combustion) have been determined for the samples of white oak and pine .

**Keywords:** Wood; The chemical composition ; Heat release; The combustion

## INTRODUCTION

Wood is a natural vegetable material used widely in building construction and other technique fields. Wood is a complex composite material in which the fibers are the main structural elements. It is well known that macro- and microstructure, the chemical composition, humidity, density, mechanical and thermal physical properties of wood materials depend on the species of trees, their age, the source of material, processing conditions and other factors [1].

Wood is a combustible material. The heat release is a general cause of the onset and the combustion process development. The regime and the intensity of this process depend on the ratio between the heat release rate and the heat loss rate from the combustion reaction zone of material. During the last two decades, considerable work has been made to determine fire performance of different wood materials. The great attention has been spared to heat release estimation at their combustion .

The heats of combustion and other characteristics of heat release at the combustion of the samples of red oak, Douglas-fir plywood and particleboard were determined with using a heat release calorimeter, developed at an Ohio State University, USA ( OSU-calorimeter), under different external heat flux levels, the sample orientations, pilot and unpilot conditions. To determine heat release characterisics of the wood materials, OSU apparatus was instrumented by both oxygen analyzer and thermocouples, the signals from which were recorded every 3s.[2]. Thus, both standard thermal and oxygen consumption methods were used simultaneously. The standard thermal method gave results 20-35% lower than those by the oxygen consumption method. The definite difference in peak values of heat release rate (HRR) was noted when the same samples were tested in the horizontal or vertical orientations.

Later, H.C. Tran represented the data on using the OSU calorimeter and also the Cone calorimeter , developed at the National Institute of Standard and Technology, USA, to measure heat release characteristics for the set of hardwoods and softwoods, other wood materials [3]. The data represented show that average HRR values of softwoods are generally lower than those of hardwoods. Author identified many factors as key variables that affect the heat release characteritics of wood materials. Density, chemical composition and moisture content are most important ones [3,4].

Macrokinetic parameters of pyrolysis for the main chemical components of Douglas-fir wood and also the combustion heat of their volatile pyrolysis products as function of the residual mass fraction of the samples were determined in the research [5]. In this work the gross heat of combustion of unextracted wood, extracted wood, alpha-cellulose , and lignin at various mass retention fractions were measured in oxygen bomb calorimeter with taking into account the gross heat of combustion of their char residues to calculate the gross complete combustion heat of volatile pyrolysis products.

The present work is focused on the determination of the heat release characteristics at the combustion of different wood varieties from a middle region of Russia and tropical ones of Viet-Nam. The objective of the research was to elucidate the main factors and distinctive signs of wood varieties, affecting heat release at the combustion. First of all, it was important to determine the chemical composition of specimens from hardwoods and softwoods and its influence on lower complete combustion heat of wood. On the first stage of the study, two classic representatiives of hardwoods and softwoods, i.e white oak and pine, were used as object to measure heat release rate and other heat release characteristics at the different conditions of wood combustion.

The published data concerning to above aspects are hardly comparable because of the differences in the experimental methods, testing conditions, lack of minute description of samples and also limited number of studied wood varieties.

## Materials and Test Methods

The tests cover 13 coniferous and foliate wood varieties from the middle region of Russia and 5 tropical wood ones from Viet-Nam. The tropical wood varieties consist of thongkaribe (pinus massoniana), vansham (picea koraiensis nakai), bachdan (eucalyptus camaldulensis), keo thy tion (acacia mangium), keo lay (acacia auriculiformis).

Physical properties and chemical analyses for wood varieties, represented in Table 1, have been determined through the procedures reported in [1].

The values of lower complete combustion heat, $Q_{low}$, for wood samples studied have been evaluated with an oxygen bomb calorimetry in according to Russian standard method of GOST 147-74. Besides, these values have been calculated by D.Mendelejev equation on the basis of elementary composition of wood:

$$Q_{low} = 339.4\ C + 1257\ H - 108.9(O + N - S) - 25.1(9\ H + W), \qquad (1)$$

where C, H, O, N, S – element content into wood composition, %; W- moisture content in the specimens studied, %.

Automatic apparatus of Carlo Erba 1106 (Italy) has been used to determine the elementary composition of oven-dried wood samples. Moisture content in samples has been determined by means of Q-1500 derivatograph (Hungary) at heating rate of $10^oC$/min. Sample weight was 50-100 mg.

The heat release characteristics have been studied in the different combustion conditions of wood with using standard method of GOST 30402-96 and also apparatus of GOST 12.1.044-89, item 4.20, instrumented to measure the change in the chamber of oxygen, CO, $CO_2$ concentrations . The external radiant heat flux has been varied in the range 30-50 $kW/m^2$.

GOST 30402-96 provide for the horizontal orientation of the specimen with exposed size of 100 x 100 $mm^2$ and pilot flame ignitor. The apparatus was additionaly instrumented by weight cell to measure mass loss as a function of testing time.

GOST 12.1.044-89, item 4.20, has been used to determine toxicity of volatile combustion products of different materials. The samples with surface area of 40 x 40 $mm^2$, have been tested under an angle of $45^o$ to the horizontal surface into flame or smoldering regimes of the combustion.. In the present work, the heat release characteristics have been determined at flame wood combustion.

## Results and Discussion

Wood species are usually subdivided on two broad classes: hardwoods and softwoods [6]. However, by the opinion of Tran [3], these terms have little to do with the hardness of the wood. Botanically, the hardwoods are from trees with broad leaves and the softwoods are from conifers with needle-like or scale-like leaves.

The study of the static hardness of tropical wood varieties shown that really coniferous species of thongkaribe and vansham have the lower mean values of static hardness (37.4 and 43.7 MPa accordingly at moisture content of 12%) than foliate ones of bachdan, keo thy tion and keo lay ( 78.8; 74.8 and 65.5 MPa). When moisture content in the samples is increased up to 21%, static hardness of the foliate wood is decreased in 1.35-1.4 time, while that for the coniferous wood is decreased only in 1.1-1.2 time.

The density of classic softwoods [6] is usually in the interval of 400-540 kg/m$^3$, while hardwoods density is over 560 kg/m$^3$ (Table 1). Thus, birch and aspen can be considered as foliate softwoods. Keo thy tion is foliate softwood by density and foliate hardwood by static hardness. Larch amongst conifers must be attributed to hardwood.

Experimental results represented in Table 1 confirm that softwoods are characterized by higher lignin content as compared to hardwoods. The varieties of foliate tropical wood contain the higher proportion of hemicellulose (19.16-21.03%), in the main as lightly hydrolysable pentosanes. Total content of holocellulose in wood samples studied is changed from 61.4% (for cedar) to 74.6% (for aspen).

Extractives are a complex mixture of low molecular weight sugars, inositols, simple fats, carboxylic acids, terpenes, phenolic and other substances [1]. The high content of extractives is found both in coniferous wood varieties and in some foliate ones. For instance, the samples of oak and birch by this index are approached to coniferous wood. The relative proportions of the wood components affect pyrolysis dynamics, the yield of char residue and other products, heat and smoke release at the combustion of different wood varieties.

**Table 1. The Properties of the Different Wood Varieties**

| NN | Specimen | ρ, kgm$^{-3}$ | W, % | $Q_{low}$, KJg$^{-1}$ | Holocellulose/ Hemicellulose, % | Lignin, % | Extractives, % |
|---|---|---|---|---|---|---|---|
| 1 | Spruce | 430 | 9.5 | 18.90 | 62.7 /- | 27.3 | 10.0 |
| 2 | Pine | 450 | 10 | 19.62 | 62.9 / - | 28.0 | 9.1 |
| 3 | Larch | 660 | 9.6 | 18.61 | - | - | - |
| 4 | Cedar | 400 | 10 | 18.84 | 61.4 / - | 30.8 | 7.8 |
| 5 | Beech | 600 | 9.2 | 18.26 | 74.1 / - | 21.0 | 4.9 |
| 6 | Ash-tree | 740 | - | 18.40 | - | - | - |
| 7 | Maple | 610 | 8 | 18.04 | 71.9 / - | 23.2 | 4.9 |
| 8 | Oak | 570 | 7.0 | 18.66 | 68.7 / - | 23.6 | 7.7 |
| 9 | Birch | 540 | 6.5 | 18.08 | 73.3 / - | 20.6 | 6.1 |
| 10 | Hornbeam | 595 | - | 18.42 | - | - | - |
| 11 | Aspen | 480 | 6.8 | 18.14 | 74.6 / - | 21.5 | 3.9 |
| 12 | Thongkaribe | 430 | 4 | 18.62 | 65.5 / 10.5 | 27.0 | 7.93 |
| 13 | Vansham | 400 | 4 | 18.84 | 65.91 / 11.24 | 27.5 | 5.06 |
| 14 | Bachdan | 595 | 5 | 18.55 | 67.26 / 19.16 | 25.4 | 6.40 |
| 15 | Keo thy tion | 420 | 4 | 18.11 | 69.98 / 21.03 | 24.75 | 5.09 |
| 16 | Keo lay | 560 | 4 | 18.53 | 69.12 / 20.10 | 25.16 | 5.56 |

Table 1 shows that lower values of complete combustion heat with account of the latent heat of water vaporisation for all wood varieties studied are in the range of 18.04–19.62 kJg$^{-1}$. The near content of elements into the composition of different wood species (Table 2) is reflected in near values of the complete combustion heat of the samples.

**Table 2. Elementary composition of tropical wood samples**

| NN | Specimen | C , % | H , % | O , % |
|---|---|---|---|---|
| 1 | Thongkaribe | 50.43 | 6.14 | 43.43 |
| 2 | Vansham | 51.01 | 6.10 | 42.80 |
| 3 | Bachdan | 51.19 | 5.81 | 44.10 |
| 4 | Keo thy tion | 50.06 | 5.84 | 45.06 |
| 5 | Keo lay | 50.34 | 6.11 | 43.55 |

Chemical composition of wood affects lower heat of complete combustion. Fig.1 shows the graphic dependence of lower combustion heat on summary content of lignin and extractives for different wood varieties. There is a linear correlation between these parameters. The equation of the linear regression for data presented in Fig.1 is estimated as follows:

$$Q_{low} = 15.45 + 0.1\ (X_l + X_{ext}),\ \mathrm{kJ\ g^{-1}} \qquad (2)$$

where $X_l$ and $X_{ext}$ - the content of lignin and extractives in oven-dry wood, %

Coefficient of the correlation for the regression equation obtained is R=0.86. The biggest scattering of the values of lower complete combustion heat is found for coniferous wood varieties with a high content of extractives.

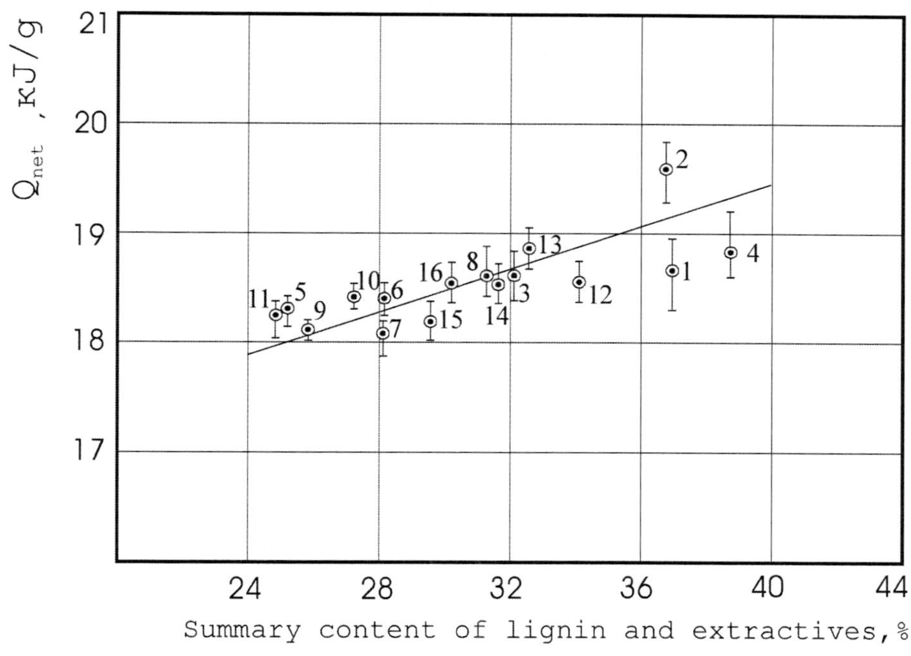


**Figure 1.** The dependence of lower complete combustion heat of wood samples on summary content of lignin and extractives. Numbers correspond to that mentioned in Table1.

The equation (2) can be used to assess the relative importance of the contribution of different wood components into net (lower) complete combustion heat of wood materials.

If summary content of lignin and extractives is equal to zero: $X_l + X_{ext} = 0\%$, in this case, a lower heat of complete combustion for rest part of wood, that is, holocellulose, has to be equal to 15.45 kJ/g. In other extreme case, when $X_l + X_{ext} = 100\%$, the value of a lower heat of complete combustion for a mixture of lignin and extractives has to be equal to 25.45 kJ/g. Such conclusions are just, if to adopt the following assumption:

1. The values of lower complete combustion heat for cellulose and hemicellulose are the same or very near ones.
2. The relative energy contribution of lignin and extractives into total heat release at wood combustion is proportional to their content in wood.

The functional dependence established in the present work on summary content lignin and extractives shows that chemical composition of wood rather than wood category is the main factor which affects on heat release characteristics.

Elementary composition of lignin from different wood species is practically identical and in average it includes 63.54% C; 5.54% H and 30.92% O (by the difference) [1]. Then, lower heat of complete combustion of lignin, calculated by Mendelejev equation (1), must be equal to 23.91 kJ/g. Because of information lack about chemical composition of extractives, their role in heat release is fairly uncertain. It has been supposed that the effect of terpenes and resins in extractive composition on heat release is especially considerable [1].

Let make use of the results obtained to evaluate the values of lower complete combustion heat for wood extractives.

For instance, summary content of lignin and extractives into cedar wood is 38.6 % (Table 1). The fraction of lignin is 0.798 and the fraction of extractives is 0.202. Since a lower complete combustion heat for summary content of lignin and extractives is 25.45 kJ/g (see above), the following equation can be presented:

$$25.45 = 0.798\, Q_{low}^{l} + 0.202\, Q_{low}^{ext}, \qquad (3)$$

where $Q_{low}^{l}$ and $Q_{low}^{ext}$ are the complete combustion heat values for lignin and extractives.

From the equation (3) it is easy to calculate a lower complete combustion heat for extractives of cedar wood:

$$Q_{low}^{ext} = (25.45 - 0.798\, Q_{low}^{l}) / 0.202 = (25.45\text{-}0.798\text{x}23.91) / 0.202 = 31.85 \text{ kJ/g}$$

It can be evaluated by like way the values of lower complete combustion heat for the compounds extracted from other wood species.

So, for pine wood extractives we have $Q_{low}^{ext} = 30.2$ kJ/g. The quantities obtained in the work are in qualitative agreement with the published values for gross complete combustion heat of pine resins (35-37 kJ/g) [7].

The examples given for two wood samples show that the combustion heats of extractives from different wood varieties are not identical. This fact is probably the result of the distinctions in chemical composition and the main constituent's proportions of extractives.

In real fires, the dynamics of the heat release at the combustion is complicated by charring wood. The char layer forming on wood surface executes the role of barrier delaying heat transfer into material and thus protecting wood from excessive increase of temperature.

The calculation of heat release characteristics at wood combustion has to be based on an effective heat of combustion of wood material with taking account of the combustion efficiency coefficient.

There is fundamental relationship between HRR and mass loss rate of a material, m'' , its effective heat of combustion, $\eta Q_{low}$ :

$$HRR = \eta\, m''\, Q_{low} \quad (4)$$

where $\eta$ is the combustion efficiency coefficient or simply the combustion efficiency.

All factors affecting mass loss rate of solid at the combustion (i.e. burning rate) have to affect HRR.

To determine effective combustion heat and other heat release characteristics of wood, it can be used the different approaches. Apart from standard methods of OSU and Cone Calorimeters utilizing the principle of the response on the temperature rise or oxygen consumption during the combustion process accordingly, it is possible to use principle of the response on the production of $CO_2$ and CO at the combustion of organic materials.

This method was developed by A. Tewarson [8]. He showed that HRR at the combustion of fuel can be calculated from the following equation:

$$HRR = (Q_{low} / k_{CO2})\, m''_{CO2} + [\, (Q_{low} / k_{CO} - Q_{CO})]\, m''_{CO} \quad (5)$$

where $Q_{low}$ is lower heat of complete combustion of material; $k_{CO2}$ is the theoretical stoichiometric mass of $CO_2$ produced per unit mass of the fuel converted completely to $CO_2$ ; $m''_{CO2}$ and $m''_{CO}$ are the mass generation rate of $CO_2$ and CO accordingly.

$Q_{CO}$ is the heat of combustion of CO ( It is equal to 10,24 kJ /g CO); $k_{CO}$ is the theoretical stoichiometric mass of CO produced per unit mass of the fuel converted completed to CO.

It was shown that for most materials the value of ( $Q_{low}/ k_{CO2}$ ) = 12.3 kJ/g $CO_2$ and ( $Q_{low} / k_{CO}$ ) = 18,5 kJ/ g CO .

Thus, the equation (5) may be transformed to the following equation:

$$HRR = 12.3\, m''_{CO2} + 8.26\, m''_{CO} \quad (6)$$

Since the concentration of carbon oxides in apparatus of GOST 12.1.044-89 is measured by the analyzers in percentages, appropriate coefficients to translate % in mass units are used:

$$HRR = [\, 12.3 \times 1.89 \times (\%CO_2/ s) + 8.26 \times 1.2 \times (\, \%CO/ s)\, ]/ A \quad (7)$$

where A is area of studied sample, $m^2$ .

It should be noted that the concentration of carbon oxides can be measured much more accurately than oxygen concentration.

Table 3 lists the heat release characteristics at wood combustion obtained by such method.

Data calculated on the basis of mass loss measurement with the correction on the combustion efficiency [9], are given for comparison.

**Table 3. The Effect of External Radiant Heat Flux and Testing Conditions on Heat Release Characteristics at Wood Combustion**

| Sample, Testing conditions | $q_e$, kW/m² | $\tau_{ign}$ s | $\tau_{max}$ s | $HRR_{max}$ KW/m² | $HRR_{av}$ kW/m² | $THR_{fl}$ MJ/m² | DHRR/dτ kW/m²s | $Q_{eff}$ KJ/g | η |
|---|---|---|---|---|---|---|---|---|---|
| White oak, Coil ignitor 45° orient. | 38<br>52.5 | 35<br>10 | 144<br>96 | 315<br>480 | 179<br>315 | 32.27<br>56.7 | 2.89<br>5.58 | 7.95<br>12.26 | 0.43<br>0.65 |
| Pine, Unpiloted, 45° orient. | 38<br>52.5 | 30<br>15 | 77.6<br>73.8 | 420<br>580 | 125<br>275 | 11.24<br>24.72 | 8.8<br>9.8 | 6.0<br>13.04 | 0.3<br>0.68 |
| Pine,piloted horizontal | 40<br>50 | 8.5<br>4 | 140<br>85 | 275<br>466 | 104<br>284 | 30.21<br>59.76 | 2.09<br>5.75 | 9.81<br>12.95 | 0.5<br>0.66 |

Analysis of published and own data permit to distinguish two groups of factors affecting heat release characteristics.at wood combustion.

1 group – factors associated with material nature:
density, moisture content, chemical composition, charring ability, morphological structure of wood, thermooxidative stability of char, mechanical and thermal-physical properties of char and initial wood .

2 group – factors associated with the conditions of the combustion process:
heat flux level, the ignition mode, specimen orientation, the direction and the rate of oxidizer flux, sizes of specimen and other.

Identical conditions of testing permit to display the effect of material nature.

Upon pilot ignition or self-ignition, wood undergoes flaming combustion and releases heat. HRR curves have common features. Following ignition, a peak HRR occurs, followed by HRR decrease and then an almost steady-state period to test end. Time to peak value of HRR depends on thickness of sample. In the case of oak sample with thickness of 10 mm this time is larger than that at thickness of 5mm.

Peak values of HRR for pine as it is seen from Table 3, are more high as compared with oak. However, average HRR and THR for period of flame combustion of pine wood are lower than that for oak. At the calculation of these characteristics for all test time (15min) at heat flux of 52.5 kW/m$^2$ , average HRR equal to 75 and 98 kW/m$^2$ for pine and oak accordingly.

Thus, when average HRR are calculated for longer period of the combustion, these quantities are decreased.

The results obtained in the present work permit to conclude that pine wood burns out and is charred faster than oak The quality of char layer on wood surface has to affect heat release characteristics.

## CONCLUSIONS

The effects of variety, chemical composition of wood and also the intensity of external radiant heat flux on heat release characteristics at wood combustion have been investigated. It has been established that chemical composition and relative proportions of main components of wood (holocellulose, lignin and extractives) rather than wood category (softwood or hardwood) affect lower heat of complete combustion of wood samples.

The values of lower complete combustion heat for 16 coniferous and foliate wood varieties, determined by oxygen bomb calorimetry and also calculated on the basis of elementary composition, are within the range from 18.04 to 19.62 kJ/g. The linear correlation between the combustion heat and summary content of lignin and extractives in wood samples has been found.

The estimation of the values of lower complete combustion heat for lignin and extractives ]\=from different coniferous species of wood has been presented. For lignin $Q_{low}$ is equal to 23.91 kJ/g. For extractives from cedar and pine wood $Q_{low}$ are equal to 31.85 and 30.2 kJ/g accordingly. The difference in chemical composition and main constituent's proportion of extractives from different wood species is probably reflected on $Q_{low}$ values of extractives.

The heat release characteristics at flame combustion of white oak and pine are determined under two heat flux levels, pilot and unpilot conditions , different orientations of the samples. At identical conditions of test, oak samples show higher values of THR and average HRR for fixed time of flame combustion than that for pine wood. The achievement rate of peak HRR is smaller for oak.

## REFERENCES

[1] Nikitin, V.M., *Chemistry of Wood and Cellulose*, 1951, Goslegbumizdat, Leningrad

[2] Tran H.C., Rates of Heat and Smoke Release of Wood in an Ohio State University Calorimeters , *Fire and Materials* 1988, vol. 12, pp. 143-151

[3] Tran H.C., Experimental Data on Wood Materials, In: *Heat Release in Fires* by Eds. V Babrauskas, S.J. Grayson, Elsevier Applied Science, New York, 1992, pp.357-372

[4] Tran H.C., White R.H. , Burning Rate of Solid Wood Measured in a Heat Release Rate Calorimeter, *Fire and Materials*, 1992, vol. 16, pp.197-206

[5] Parker, W.J., LeVan, S.L., Kinetic Properties of the Components of Douglas-Fir and the Heat of Combustion of Their Volatile Pyrolysis Products, *Wood and Fiber Sci.* 1989, 21, No.3, pp.289-305

[6] Kalninsh, L.Ya., Wood, In: *Encyclopaedia of Polymers*, Soviet Encyclopaedia, Moscow, 1972, v.1, p.764

[7] Howard, E.T., Heat of Combustion of Various Southern Pine Materials, *Wood Sci.* 1973 , vol.5 , No.3, pp. 194-197

[8] Tewarson A., Experimental Evaluation of Flammability Parameters of Polymeric Materials, In: *Flame Retardant Polymeric Materials*, by Eds. M. Lewin, S.M. Atlas, E.M. Pearce, Plenum Press, New York, 1982, vol.3, pp.97-153

[9] Sivenkov A.B. *The Retardancy of Fire Hazard of Materials Based on Cellulose*, Ph.D. Dissertation, State Fire Academy, Moscow, 2002.

In: Chemical Physics of Pyrolysis, Combustion and Oxidation ISBN 1-59454-163-9
Editors: A. A. Berlin et al., pp. 55-80 

*Chapter 6*

# CHLORINATED POLYMERS AS THE BASE FOR MATERIALS WITH REDUCED COMBUSTIBILITY

***G. E. Zaikov, A. A. Donskoi*[*] *and M. A. Shashkina*[*]**
N.M.Emanuel Institute of Biochemical Physics, 4 Kosygin st., Moscow 119991, Russia
[*]Institute of Aviation Materials, 17 Radio st., Moscow 119105, Russia

One of the tasks of the polymeric materials industry is to provide the national economy supply of synthetic polymers and materials based on them. A series of demands are imposed on these materials, among which the determining ones are inertness to the effect of aggressive media and ozone combined with resistance to heat and fire. The rapid development of organic polymer production in the second half of the twentieth century and wide application of material made from them have raised the question about creation of new polymers with reduced combustibility. As no new large-tonnage productions of polymers have been planned recently, modification of existing ones has been a priority. Though synthesis of new polymers was preferable, modification of already existing ones was undertaken in order to impart the required properties to them [1 - 7]. Chemical modification of polymers changes the properties of initial products and broadens the field of their application [8]. From the practical point of view, chlorine-containing polymers possessing a complex of valuable properties increased in importance. These polymers possess increased fire and heat resistance, resistance to effect of oils, gasoline, ozone, etc. Materials derived from them are capable of operating in aggressive media for long time and under extreme conditions. Russian industry produces various chlorine-containing polymers: homopolymers and copolymers of vinyl chloride and vinylidene chloride, chlorinated butyl rubbers and polyisoprene, poly(vinyl chloride), chlorinated and sulfochlorinated polyethylenes, etc. Chlorinated polymers can be obtained by polymerization of chlorinated monomers or by chlorination of polymers, the properties of obtained chlorinated products being different in these cases. Products with higher heat resistance are obtained by polymer chlorination, but not by modification of a monomer with subsequent polymerization [9]. Polymers can be chlorinated in a medium of organic solvents, in aqueous suspension, and in the condensed phase by reaction with gaseous or liquid chlorine. Sulfonation may be performed simultaneously with chlorination. The process is initiated by visible light, ultraviolet or ionizing radiation, with the help of substances giving free radicals at degradation (organic peroxides and hydroperoxides, diazo-compounds, metal-organic compounds [8]). Properties of chlorinated polymers depend on the origin of the initial polymer, chlorination degree, and method and conditions of its performance [9]. Chlorinated and sulfochlorinated polyolefins, the properties of which are significantly different from these of initial polyolefins, are widely applied in technology [8].

Modification by halogenation possesses some features stipulated by the chemical inertness of initial polymers, and by the presence of amorphous and crystalline areas in the structure. Halogenation reactions can be accompanied by degradation and cross-linking of backbones of macromolecules [10, 11].

Large-tonnage production of polyolefins now exists. In relation to this, great attention is paid to modification of polymers of the present class.

# 1. Current State of Chlorinated Polyolefin Production and Application

Structure and reactivity of polyolefins are members of the homological sequence of paraffin hydrocarbons in chemical structure, and are quite similar to themin their chemical properties. Chlorination of paraffins is studied quite well. At the present time, chain radical mechanism of chlorination of paraffin hydrocarbons and polyolefins has been investigated most [2, 4]. Reaction initiators can be temperature, ultraviolet radiation, gamma-radiation, as well as compounds capable of releasing free radicals.

Contrary to paraffin hydrocarbons, modification during chlorination of polyolefins is affected by macromolecular chain branching, but such side reactions as degradation and cross-linking, steric factors, and problems with solvent selection may occur. Because of reduced solubility of polyolefins and structure crystallinity, special chlorination methods were developed [5 – 8, 12 - 14].

## 1.1. Chlorination Methods

Polyolefins can be chlorinated in solution in a solvent inert to chlorine, in both aqueous and solvent suspensions, in melt, and in the solid phase. Combined methods can also be applied to chlorination, for example, a combination of dissolved and suspension methods (block-chlorination). Modification in the solid phase includes chlorination of a powder-like or granulated polymer, as well as prepared articles, fibers or films [5, 13, 15]. The modifying agent (chlorine) can be used both as gas and solution. Gaseous chlorine is the most widely applied to modification in solution or suspension in the presence of initiators. Carbon tetrachloride, chlorobenzene, and ethane tetrachloride are most often used as solvents in chlorination of polyolefins.

The most uniform chlorine distribution and, consequently, optimal properties of polymer are obtained by chlorination of polyolefins in diluted solutions at concentration below or equal to 3-8% [13]. The disadvantage is high consumption of solvent, which cannot be regenerated in the majority of cases. At room temperature, polyethylene (PE) does not dissolve in usual solvents. With regard to crystallinity degree, polyethylene begins dissolving in chlorinated aliphatic and aromatic hydrocarbons at temperature above 53°C. Low-boiling chlorinated hydrocarbons, the best among which is carbon tetrachloride, are used in chlorination of polyolefins. Low density polyethylene (LDPE) dissolves in carbon tetrachloride at temperature close to its boiling point. High density polyethylene (HDPE) dissolves at higher temperatures (80 – 100°C) [16]. The method of polyethylene chlorination in carbon tetrachloride solution in the presence of radical initiators is described in the USA patent [17]. High-crystalline polyolefins swell insignificantly in carbon tetrachloride. That is why higher boiling solvents should be used for their solution, for example, chlorobenzene or ethane tetrachloride, which increases the modification rate [18]. Chlorination begins at 100 – 120°C, gradually reducing the reaction temperature as chlorine concentration in the polymer increases, because solubility of chlorinated polyethylenes (CPE) in organic solvents depends on chlorine content. As chlorine concentration is below 30 wt.%, solubility increased, thereafter decreases, and then the polymer becomes insoluble at 50 – 60 wt.% [8]. To increase solubility of the polymer, mixtures of solvents are applied.

Application of initiators allows modification of the initial polymer at comparatively low temperatures. Controlling conditions of polyolefin chlorination, such as the solvent type, the presence of reaction initiator, the state of halogenating agent (gaseous or dissolved), or the state of polymer to be modified (in solvent, suspension or solid), one can perform preselective chlorination of polyolefins. In turn, this enables control of composition, structure and properties of obtained polymers [19].

Thermally initiated chlorination in darkness is of the lowest efficiency, but increases when the process is initiated by azoisobutyronitrile or ultraviolet radiation. Not only individual initiators of the radical type [20, 21], but mixtures of initiators are also used for chlorination [89]: dinitrilisobutyric acid with benzoyl or lauryl peroxide, chloral dioxyperoxide together with acetylcyclohexylsulfonyl peroxide. Combination of two or several types of initiators increases their efficiency [7].

LDPE chlorination in suspension in carbon tetrachloride or acetic acid in the presence of metal chlorides was performed for the first time by JCJ Company (Great Britain) [8]. Liquid chlorine was suggested as the chlorinating agent [23], but gaseous reagent is more frequently used. The product with chlorine content up to 40 – 44 wt.% was obtained by polyethylene chlorination in carbon tetrachloride suspension by gaseous chlorine [24]. Chlorination of high density polyethylene in aqueous suspension requires application of surfactants [25].

Chlorinated polyethylenes produced by many companies in different countries, with regard to chlorine concentration, are amorphous (at 25 – 50 wt.%) or low-crystalline materials. Tensile strength varies from 8 to 17 MPa and relative elongation at rupture ranges from 350 to 800%. The multiplicity of polymers from thermoplasts to friable glassy-like ones are obtained in wider range of chlorination (from 14 to 70%) [26]. The structure of the final polymer is changed by modification.

Investigations of chlorinated low and high density polyethylenes indicate that polyethylene molecules lose rigidity and ability to form regulated supermolecular conglomerates with increase of the chlorination degree in solution. Modification by chlorination or sulfochlorination in solution proceeds identically in initial high-crystalline and low-crystalline polyethylenes. Therewith, crystallinity degree decreases linearly and disappears completely at total concentration of methyl groups and chlorine atoms of 185 – 190 per 1,000 carbon atoms in the polymer chain. Regulated structure and crystallinity of low density polyethylene disappear completely at chlorine concentration of 27% and, in the case of high density polyethylene, at 32%. Polymers become completely amorphous, elastic products with rubber-like properties. Figure 1 shows data on changing the crystallinity degree of polymer during chlorination of low and high density polyethylenes, performed in solution and suspension. Crystallinity of polymers was estimated by X-ray structural analysis [19].

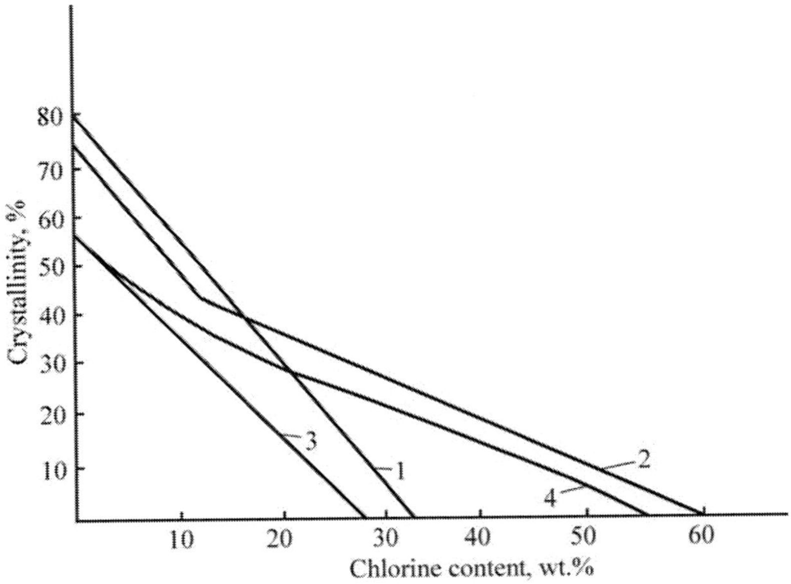


**Figure 1.** Polyethylene crystallinity degree change with regard to chlorine content in the polymer

Crystallinity of low and high density polyethylenes changes equally during chlorination. The apparent difference in the final result is associated with branching degree of low density polyethylene chains, i.e. the effect of total chlorine and side methyl group contents in the polymer shows up in the crystallinity degree change.

Block low density polyethylene possesses a fibrillar-spherulitic structure, which begins changing at injection of 13% of chlorine only. Crystalline and amorphous phases are distributed uniformly throughout the volume. Destruction of spherulites with preservation of fibrillar fragments is observed above 13% chlorine content. When chlorine concentration reaches 23.6%, the polymer obtains the structure of nonvulcanized crystalline rubber. Its surface becomes somewhat smooth and molecules become unable to form crystalline structures. Therewith, the polymer possesses the optimum of rubber-like properties. Homogeneous, completely amorphous structure is formed at chlorine concentration of 31.5%.

Infrared spectroscopy of polymer containing 33.6% of chlorine shows that 93.7% of chlorine are contained in secondary radicals, 3.9% only in primary radicals, and 2.4% in tertiary ones.

The elementary cell is deformed at low chlorine concentration [19]. If chlorine content is 10 – 15%, the structure transforms from rhombic to hexagonal. All supermolecular formations size 10 nm are smooth and disappear at chlorine concentration over 20 – 30% [27].

Additional injection of chlorocarboxylic groups ($CCl_2$, COOH) into chlorinated polyethylene increases adhesive properties. The analogous effect was observed for polyethylene sulfochlorination.

As the reaction proceeds in suspension and in the solid phase, crystallinity of polymer is partly preserved up to chlorine concentration of 55 – 60 wt.% [28].

The reaction rate, maximum concentration of injected chlorine and its distribution along the macrochain depend not only on the chlorination method, but also on the initial polymer structure. More branched amorphous low density polyethylene is chlorinated more rapidly than linear high density polyethylene [8].

Investigations of supermolecular structure of chlorinated polyethylene by the electron-microscopic method show that the structure of chlorinated low density polyethylene becomes similar to that of nonvulcanized crystallizing rubber at chlorine concentration of 24%. Ability of polyethylene to form crystalline formations is abruptly reduced with chlorine content increase above 18% [19].

X-ray diffraction patterns taken at large angles show that nonoriented chlorinated polyethylene with 10.6% of chlorine preserves the rhombic cell but with increased volume ($a = 0.787$ nm, $b = 0.502$ nm, $c = 0.254$ nm, $V = 10.3$ nm$^3$). The initial polyethylene possesses the following values: $a = 0.754$ nm, $b = 0.504$ nm, $c = 0.254$ nm, $V = 9.65$ nm$^3$. For oriented chlorinated polyethylene with the same chlorine concentration, these values are the following: $a = 0.815$ nm, $b = 0.521$ nm, $c = 0.254$ nm, $V = 10.07$ nm$^3$ [30].

Under tensile stress, amorphous chlorinated polyethylene with chlorine concentration of 27% crystallizes, and a regular structure with well-oriented hexagonal lattice is formed. However, the crystalline structure disappears as the sample contracts [19].

At the initial stages of polyethylene chlorination, chlorine atoms are distributed along the chain substituting hydrogen atoms at tertiary carbon atoms. At higher levels of chlorination, hydrogen atoms at secondary atoms of carbon are substituted.

Chemical behavior of polyethylene in the homogeneous reaction of chlorination in solution is practically fully described in terms of the neighboring chain effect and is not complicated by conformational effects [4, 30]. Polyethylene chlorination is decelerated after injection of more than 60% of chlorine and is practically terminated at content rise up to 73%. This is associated with the deactivating influence of substituents, chlorine in the present case, on the carbon-hydrogen bonding [4].

At thermal initiation of chlorination, occurrence of $CCl_2$-groups is observed from the very beginning of the process, and CHCl-groups occur at chlorine concentration over 20%. In the case of photochemical initiation, occurrence of CHCl-$CH_2$-CHCl- and $CH_2$-CHCl-$CH_2$-groups foregoes $CCl_2$-groups.

Three main mechanisms of polyethylene chlorination are suggested in literature [31]:

1. Hydrogen substituting by chlorine, which reaches 64 wt.% at thermal initiation and 68 wt.% at photochemical one.
2. Dehydrochlorination of poly(vinyl chloride) sequences with recurrent chlorination of double bonds.
3. Dehydrochlorination of structures of the –$CCl_2$-$CH_2$–CHCl–$CH_2$-$CCl_2$- type with consecutive hydrochlorination and formation of –$CCl_2$–$CH_2$–CHCl–CHCl–CHCl-structure.

The type of substitution in the polyethylene chain during chlorination in solution is independent of the molecular mass, crystallinity degree and macrotact [4, 32].

At chlorination in suspension, the reaction rate and substitution degree depend on the grain-size composition, crystallinity and temperature of modification execution [33]. Therewith, chlorination proceeds first in amorphous areas. Halogenation proceeds selectively on the surface in amorphous areas, more accessible for chlorine, and on defects of monocrystallite surface. Chlorination of crystalline areas is initiated at chlorine concentration over 20%. In the solid state, polyethylene is chlorinated by gaseous chlorine in a wide range of temperatures [6].

At heterogeneous chlorination, the reaction kinetics depends upon the initial supermolecular structure of polyethylene, and reduced permeability of chlorinated polyethylene shows up in this case, which causes irregular chlorination.

The method of reaction should be taken into account in estimating the structure and properties of the polymer obtained. The variety of chlorination methods created many structure modifications at one and the same chlorine content. Moreover, one and the same polymer can exist in various conformations: linear, spiral-shaped coils (in solution), amorphous, crystalline (in suspension), swollen polymer, dispersed powder, and article, which also shows up on the polymer structure.

At polyethylene chlorination in solution, no noticeable amount of $-CCl_2$-groups is formed, and up to high degree of chlorination the polymer consists of $-CH_2$-CHCl- units. At higher chlorination degree, -CHCL-CHCl-groups occur.

Both crystalline and supermolecular structures of polyethylene are changed at chlorination. Crystallinity degradation is associated with substitution of hydrogen by chlorine possessing greater size, which decreases the ability of macrochains to form regulated supermolecular associates [26, 29].

Structure and thermal stability of saturated polyethylene chlorination products are close in properties to chlorinated poly(vinyl chloride).

Study of chlorinated polyethylene structure shows that $-CH_2$-$CHCl_2$ groupings are absent, and chlorination proceeds mainly up to $-CH_2$-CHCl-groups [28]. Amorphous areas of the polymer are first attached during chlorination, and then more ordered areas are involved in the reaction [8]. Reactions of hydrogen atom substitution by chlorine proceed statistically leading to an arbitrary distribution of chlorine in the polymer chain. Supermolecular structure of the polymer changes from spiral-shaped for polyethylene to spherulitic (monocrystalline) at chlorine content of 8%. As its concentration rises up to 14 – 18%, short fibrils are observed, and at 22 – 30 wt.% of chlorine, packs and globules occur [28]. Complete amorphization of the polymer appears at chlorine content over 55%.

Chlorination of polyolefins under various conditions gives polymers with predetermined properties over a wide range of variation [6, 14, 15]. At modification in solution, chlorine is uniformly distributed in the chain. In this case, polymers are elastic and frost-resistant, and possess low temperature of friability. The absence of unsatiration and high chlorine content provides for high resistance to oils and petroleum. Materials derived possess reduces combustibility. The polymer does not melt when a flame is applied, but is covered with ash preventing flame spreading, which increases fire and heat shield properties. In their resistance to fire, chlorinated polyethylenes exceed elastomers, and are second only to fluorinated ones [6].

Table 1 [34] illustrates the change of physical state of polymer with increasing chlorine concentration.

However, this reference shows the averaged data with no regard to properties of the initial polymer. If this is considered, the physical state will change in the manner described in Table 2, presented in refs. [15, 35, 36].

**Table 1. Dependence of the physical state of polyethylene on chlorination degree**

| Polymer state | Chlorine concentration, wt.% |
|---|---|
| Semi-elastic thermoplast | 10 |
| Thermoplastic elastomer | 20 |
| Thermoplastic, flexible, viscous elastomer | 30 |
| Thermoplastic elastomer | 40 |
| Semi-elastic thermoplast | 50 |
| Rigid thermoplast | 60 |

**Table 2. Types of polymers obtained by polyolefins' chlorination in solution**

| Polyolefin | Change in type of polymer with chlorine concentration, % | | | | |
|---|---|---|---|---|---|
| | Plastic | Thermo-elastoplast | Elastomer | Rigid polymer | Friable resin |
| LDPE | 0 – 12 | 13 – 24 | 25 – 45 | 46 – 61 | 62 – 73 |
| HDPE | 0 – 14 | 15 – 30 | 31 – 46 | 47 – 61 | 62 – 73 |

## 1.2. Structure and Properties

Chlorination of polyethylene induces considerable changes in crystalline and supermolecular structure and, consequently, in physicomechanical properties.

Density of polymers increases monotonically and characteristic viscosity decreases with increase of the chlorination degree, which indicates degradation of polymer chains during modification. The degradation temperature changes, but no direct dependence on chlorine concentration is detected. Study of mass loss kinetics [30, 37] at temperatures of 120 – 200°C (393 – 473 K) shows that it is negligible below the degradation onset temperature, and above it an intensive detachment of hydrogen chloride is detected. Intermolecular structuring proceeds with dehydrochlorination, which is more noticeable at increase of chlorine content in the polymer.

Temperature of phase transitions in the polymer also depends on the chlorine concentration and the method of chlorination (Figure 2). The glass transition temperature of the polymer increases from –20 to 30°C (253 – 303 K) with chlorine content rise from 25 to 40 wt.%. As chlorine concentration becomes 68 – 73 wt.%, the temperature reaches 100 – 180°C (373 – 453 K) [8].

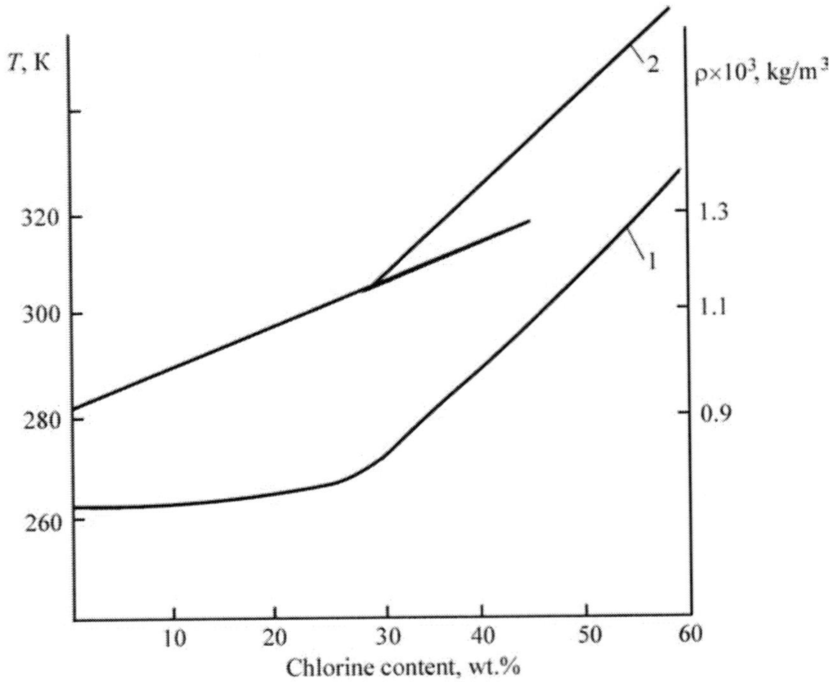


**Figure 2.** Dependencies of density (1) and glass transition temperature (2) on chlorine concentration in the polymer

The method of modification does also show up on the glass transition temperature. At the same chlorine concentration, a polymer modified in suspension possesses higher glass transition temperature than one modified in solution. At chlorine content of 63 wt.%, the chlorination method causes no effect on the glass transition temperature [8].

The softening temperature is of the parabolic type with the minimum at chlorine concentration of 35 – 40 wt.%, which is associated with an effect on the softening temperature of crystallinity reduction, and increase of intermolecular interaction of chains containing polar groups [8]. According to data of thermomechanics, optimal rubbery properties and the lowest glass transition temperature are displayed by chlorinated low density polyethylene (240 – 248 K) in the range of chlorine concentration of 27.5 – 39.5%, and for high density polyethylene of 29 – 42% (240 – 241 K). Then the glass transition temperature increases. The flow temperature and conditional strength decrease at chlorine concentration of 48.6%, and then increase. The polymers become more rigid, the plateaus of viscoelasticity become narrower and the flow temperature increases.

The extreme change of properties typical of all crystalline polyolefins proceeds with increasing [38] intermolecular interaction of polar atoms in the polymer chain with an excess compensating the absence of crystallinity.

At the same chlorine concentration, the crystallinity degree of chlorinated polyethylenes modified in the solid phase and in suspension is reduced more than the one modified in solution. These polymers possess lower elasticity, higher conditional strength and elasticity modules at smaller relative elongation due to the presence of residual crystallinity [27].

Chlorinated polyethylenes are resistant to acids, weak alkali, salt solutions, and possess reduced combustibility. Contrary to the initial polyethylene, the chlorinated polymer does not melt in flame but carbonizes [1].

A considerable number of works have been devoted to the study of thermal degradation of chlorinated polyethylenes [15, 34, 39 - 41]. Compared with the initial polymer, the degradation mechanism of chlorinated polyethylene is more complicated. The process becomes multistage including degradation of chains, detachment of substituents with formation of polyene fragments with their future transformation into cross-linked aromatic structures. Kinetics of chlorinated polyethylene thermal degradation displays a varying rate: 10 – 20 min after heating begins, the degradation rate decreases abruptly and thermal degradation terminates, which is associated with formation of spatially cross-linked thermoresistant structures. Chlorinated HDPE is the most stable [1]. Studies of chlorinated polyethylene by differential thermal analysis [37, 42] showed the presence of expressed peaks: the endothermal peak corresponded to crystalline phase melting and the exothermal one describing the recurrent crystallization at 180 – 200°C (453 – 473 K). Temperature regions of the melting range of crystalline formations in high density polyethylene and chlorinated high density polyethylene is narrower than for polymers based on low density polyethylene. This is associated with the branching degree of polymers. The peaks are flattened with increase of the chlorine concentration. The first peak disappears at chlorine content of 24%, and the second one at 31%. Thermal degradation proceeding in two stages: dehydrochlorination and thermal degradation of the polymer chain. Polymer chain degradation proceeds at 678 – 778 K and is independent of the chlorine concentration. Temperature of 10% mass loss increases with the chlorination degree and the coke residue yield increases, too.

Of interest are investigations of changes in optical properties with the increase of chlorination degree of polyethylene [48]. These data are shown in Table 3.

**Table 3. Optical properties of chlorinated polyethylenes**

| Chlorine content, % | Coefficient (index) | | |
|---|---|---|---|
| | Reflection | Permeation | Absorption |
| 3.4 | 0.121 | 0.842 | 0.037 |
| 6.5 | 0.120 | 0.833 | 0.047 |
| 23.8 | 0.118 | 0.780 | 0.102 |
| 31.3 | 0.121 | 0.764 | 0.115 |

Chlorinated polyethylenes possessing a series of quite valuable properties are thermoplasts, the spatial cross-linking of which is technologically complicated and is associated with changes in chlorination degree. This disadvantage is not observed for sulfochlorinated polyethylenes (SCPE), the general chemical formula of which can be presented as follows [49]:

$$[\text{-CH}_2\text{-CH}_2\text{-}]_m\text{-}[\text{-CH}_2\text{-CHCl-}]_n\text{-}[\text{-CH}_2\text{-CH(SO}_2\text{)Cl-}]_p.$$

To study the influence of chlorination degree and method on fire resistance of materials based on SCPE, a series of modified polymers based on HDPE with density of $0.966 \cdot 10^3$ g/cm$^3$ was synthesized. Chlorine concentration varied from 0 to 59 wt.%. Sulfochlorination was performed by gaseous chlorine and polyethylene sulfurous anhydride dissolved in a chlororganic solvent. Designation and chlorine content in the products obtained are shown in Table 4.

The traditional modification method involves application of an inert solvent and the presence of an initiator. High density polyethylene was modified in the medium of individual ethane pentachloride in the presence of initiator, diisobutyronitrile (polymers from this synthesis are named SC HDPE). The reaction was carried out in the mixture of ethane pentachloride with ethylene trichloride in the presence of the same initiator (the polymers were named SC HDPE R), as well as in the mixture of the same solvents in the absence of initiator (polymers from this group were named SC HDPE IL). Sulfochlorination does not proceed in ethane pentachloride medium in the absence of initiator. In the mixture of solvents, the reaction rates are equal for both cases of the presence and the absence of initiator. This indicated active participation of ethane trichloride in the reaction of PE modification. Study of infrared (IR) spectra of the polymers obtained displayed the presence of absorption bands responsible for oscillations of ethyl chloride substituents, which indicates the active role of this solvent in the reaction of PE modification.

**Table 4. Composition of sulfochlorinated polyethylenes**

| Reference designation | Chlorine content, % | Density, g/cm$^3$ |
|---|---|---|
| HDPE | 0 | 0.966 |
| SC HDPE R-20 | 10 | 1.018 |
| SC HDPE R-25 | 24.8 | 1.086 |
| SC HDPE R-30 | 39.2 | 1.195 |
| SC HDPE IL-18 | 18 | 1.057 |
| SC HDPE IL-20 | 23.1 | 1.054 |
| SC HDPE IL-25 | 24.8 | 1.099 |
| SC HDPE IL-40 | 31.5 | 1.184 |
| SC HDPE-30 | 33.2 | 1.146 |
| SC HDPE-45 | 35.0 | 1.195 |
| SC HDPE-40 | 36.6 | 1.195 |
| SC HDPE-50 | 52.0 | 1.359 |
| SC HDPE-60 | 59.0 | 1.418 |
| SC LDPE-20 | 29.2 | 1.127 |
| SC LDPE-40 | 31.2 | 1.161 |
| PSC HDPE-60* | 53.3 | 1.456 |

* contains 0.1 wt.% of phosphorus.

Infrared spectra of sulfochlorinated polyethylene obtained by the traditional method are thoroughly described in several refs. [35, 45, 46]. Chlorination and sulfochlorination of polyethylene is accompanied by typical changes in IR-spectra of the initial polymer: the range of 400 – 700 cm$^{-1}$ corresponds to valence oscillations of carbon and chlorine atoms, the ranges of 610 – 615 and 660 – 665 cm$^{-1}$ are typical of CCl-groups.

Intensity of 600 – 665 cm$^{-1}$ band increases and broadens simultaneously with the chlorination degree. Moreover, absorption intensity at 670 – 680 cm$^{-1}$ associated with oscillations of closely located groups, substituted by chlorine, or units of –CHCl-CHCl- type in the *trans*-configuration increases. Absorption bands at 688 $cm^{-1}$ associated with –C-S-valence oscillations and at 588, 565 and 538 $cm^{-1}$ associated with deformation oscillations of –C-S-groups are also observed. The band at 720 $cm^{-1}$ present in the spectrum of initial polyethylene splits into two bands at 720 and 730 $cm^{-1}$ [47]. The band at 720 $cm^{-1}$ is responsible for pendulum oscillations of $CH_2$-sequences. On further chlorination, the component at 730 $cm^{-1}$ disappears first, which indicates transition of crystalline areas in polyethylene into amorphous ones, and then intensity of the band at 720 $cm^{-1}$ decreases, which is associated with random attachment of chlorine atoms to PE chain and reduction of $CH_2$-group part participating in long blocks. Bands at 1160 and 1368 $cm^{-1}$ relate to valence oscillations of $SO_2$-groups.

All changes of the absorption spectrum of PE associated with introduction of Cl- and $SO_2$-groups into the chain were observed during the absorption study of SC HDPE samples. IR-spectrum of SC HDPE R possesses a series of maximums in the range of 700 – 800 $cm^{-1}$, which can be caused by oscillations of $CCl_3$-$CCl_2$- units, which can be grafted in small amounts to the PE chain as substituents.

The influence of the degree and method of modification on low- and high-temperature properties was studied on these polymers and, for comparison, on SCPE produced by the Russian industry. This provides high-quality information about the influence of chlorination degree on the phase and physical properties of SCPE [43] and allows kinetics of high-temperature transformations to be studied.

Ref. [43] describes studies of influence on the physical state and properties of polymers by methods of dynamic mechanical spectrometry, linear dilatometry and differential scanning calorimetry over a broad range of temperatures. Kinetics of thermal degradation was studied by methods of integral and differential analysis in air and inert atmosphere.

Dynamic shear modulus and mechanical dissipation factor were measured in accordance with the methodology described in ref. [5] with the help of inverse torsion pendulum operating under the mode of damped free vibrations in the temperature range from –200 to +130°C at 1 Hz frequency.

Linear thermal expansion coefficient was determined on dilatometer DL-1500L of Ilvac-Rico Company (Japan) under the mode of linear change of sample temperature at the cooling–heating at the rate of 3 deg/min. Thermal effects in the temperature range of 27 – 180°C were studied on automatic differential scanning calorimeter DSC-1500 of Ulvac Rico Company.

Density was measured by the method of hydrostatic weighing at 20°C.

Thermal degradation kinetics was studied on the device TGS-1 at the heating rate of 160 deg/min in the inert medium and on derivatograph at the heating rate of 5 and 20 deg/min in the air medium.

Studies of mechanical dissipation factor show that in HDPE, a crystalline polymer, devitrification of the amorphous phase is displayed as a low maximum at –10°C (Figure 3, curve 1). Intense $\alpha'$-peak indicated by mobility defrost in folds of crystallites is superimposed on this maximum. Loss maximums are observed that describe the spectrum of relaxation and phase transitions associated with display of various types of molecular motion correlating well with literary data [5, 44]. Sulfochlorination changes significantly the spectrum of relaxation and phase transitions. Polymer amorphization proceeds with increase of the chlorination degree. Therewith, intensity of the glass transition peak increases by an order of magnitude at chlorine content of 10 wt.%, and by two orders at 27 wt.%. At low chlorination degrees, the crystalline phase exists still in the chlorinated polymer (Figure 3, curve 2), which is indicated by the presence of tg$\delta$ maximum at temperatures exceeding the glass transition temperature of the amorphous interlayer. Further chlorination induces a noticeable increase of the $\alpha_1$-peak and a jump increase of $\Delta\beta_1$ is observed. Heights of $\alpha_2$- and $\alpha_3$-maximums of mechanical losses decrease, which indicates polymer amorphization.

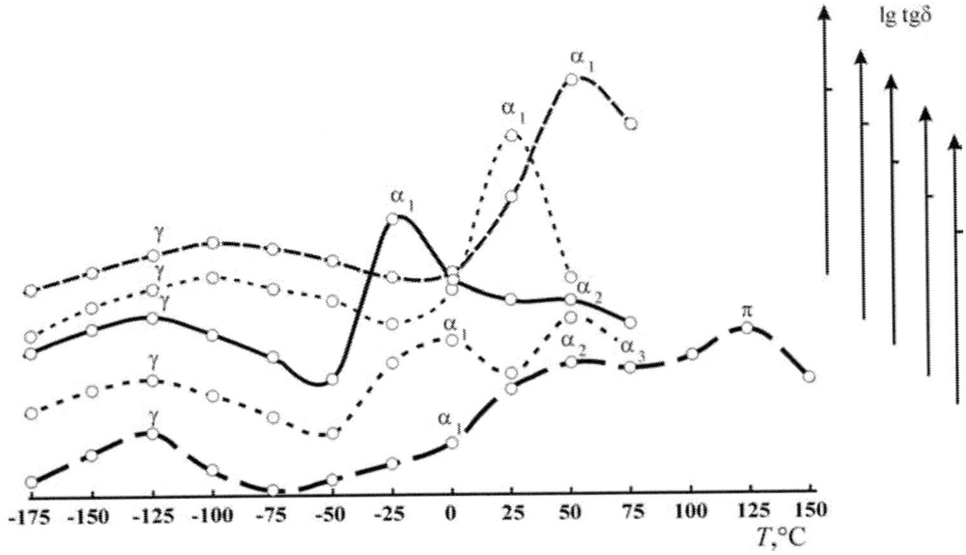


**Figure 3.** Temperature dependence of tg$\delta$ of HDPE (1) and SCPE with chlorine concentration of 10 (2), 18 (3), 52 (4), and 59 wt.% (5)

Dependence of linear expansion thermal coefficient on temperature is more sensitive to molecular mobility display and confirms conclusions of investigations performed by the method of dynamic pendulum. The decrease of the crystallinity degree with chlorine concentration increase is also confirmed by the DSC (differential scanning calorimetry) method (Figure 4). For HDPE, endopeak of the crystalline phase melting is observed at 133°C. It changes shape with chlorine concentration increase, shifts to the side of lower temperatures, reduces the height and degenerates completely at about 30 wt.% chlorine concentration.

Polymer structure variation is indicated by studies of the shear modulus at 20°C, which is the highest for PE (Figure 5). As the chlorine concentration increases dependence of the shear modulus on chlorine concentration in semilogarithmic coordinates (Figure 5) displays a linear lowering of the shear modulus up to 30 ± 0.3 wt.% chlorine content. Further on, the shear modulus becomes independent of the chlorine content in polymer. The chlorine concentration of 30 ± 0.3 wt.% is probably the critical one determining the limit of the crystalline phase existence and at which increase the polymer becomes completely amorphous. This is in complete agreement with the data shown above with regard to sulfur concentration in the polymer.

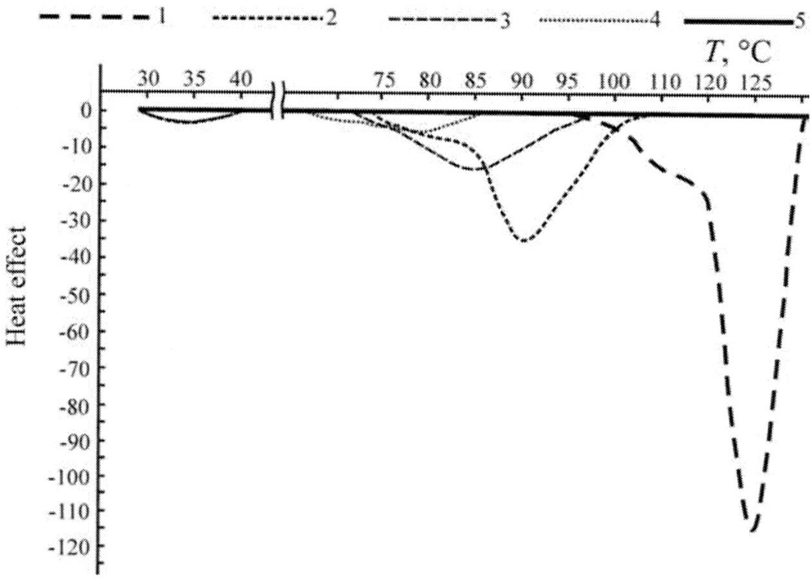


**Figure 4.** DSC curves in the area of the crystalline phase melting of HDPE (1) and SCPE with chlorine content of 10 (2), 18 (3), 12 (4), 29 (5), 31 (6), and 59 wt.% (7)

Dependence of the shear modulus measured at room temperature on the chlorination degree up to 30 wt.% concentration can be expressed by the following relation:

$$G' = G_0 \cdot e^{-RW},$$

where $G_0$ is the dynamic shear modulus of HDPE; $W$ is the chlorine mass concentration; $R$ is a constant equal to 9.33±0.03.

Experimental and theoretical values of the shear modulus for polymers with different chlorine concentrations coincide well, which gives a possibility to calculate it for the given chlorine concentration.

As the chlorination degree exceeds the critical level, the shear modulus increases insignificantly (Figure 5).

The critical chlorination degree is also clearly expressed by the dependence of polymer density on chlorine concentration, which changes with the chlorine content as follows:

$$\rho = \rho_0 + qw,$$

where $\rho_0 = 0.966 \cdot 10^3$ kg/m$^3$ (HDPE density); $q$ is the coefficient determining the straight line slope in $w$–$\rho$ coordinates (on the initial part corresponding to crystalline polymer, $q = 0.485 \cdot 10^3$; after complete amorphization of the polymer $q = 1.0 \cdot 10^3$).

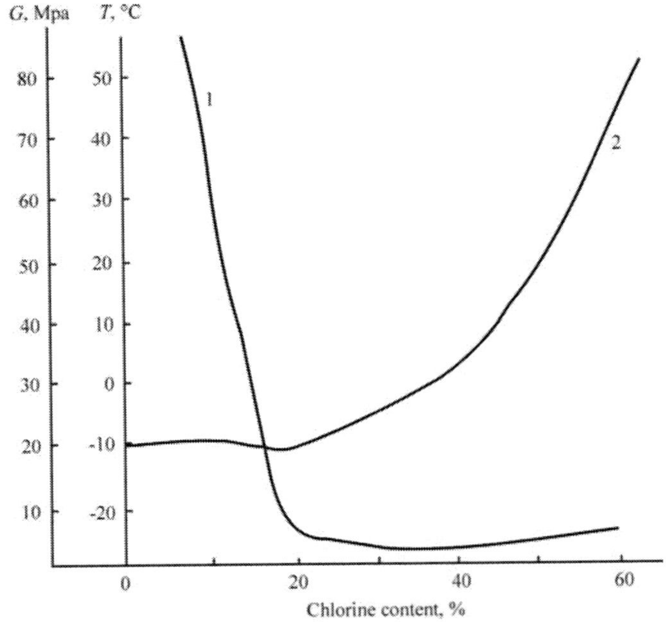


**Figure 5.** Dependence of the elasticity modulus, temperature of α-peak and loss tangent on chlorine concentration (see Figure 6)

Figure 5 also displays temperature dependence of α-peak on chlorination degree. Temperature does not change up to 30 wt.% concentration and at further chlorination increases sharply.

Hence, as the chlorination degree increases, the polymer phase changes. At chlorine concentration over 30 wt.% partly crystalline HDPE transforms into a completely amorphous state. Types of dependencies of some macroscopic characteristics change qualitatively at transition over the critical concentration.

It is known from several works [8, 50, 51] that in the range of high chlorination degree, the glass transition temperature of PE increases monotonically with it. Such dependence is observed in experiments by both the results of dynamic mechanical measurements and dilatometrically, which has been described already for partially crystalline modified PE [52, 53]. Hence, being optimal, the critical level of chlorination determined by a series of investigations provides for low density and a rubbery state at room temperature. These criteria represent convincing evidence for choosing the chlorination degree near 30 wt.% for future investigation and creation of materials.

The study of thermal degradation kinetics of modified polymers in the inert medium at the heating rate of 160 deg/min shows that chlorine concentration has the determining effect on the type of SCPE degradation. The initial HDPE degradation proceeds in a comparatively narrow temperature range (430 – 450°C) and represents a classical example of the one-stage process described by a single selection of kinetic parameters (the activation energy $E = 169$ kJ/mol, the reaction order $n = 0.85$, the pre-exponent multiplicand $Z = 1.23 \cdot 10^{21}$ $s^{-1}$).

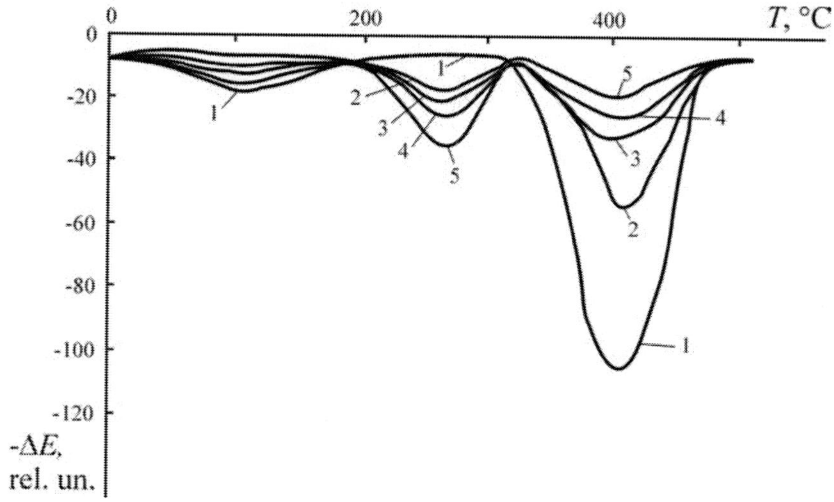


**Figure 6.** Differential thermal studies of polyethylene (1) and sulfochlorinated polyethylene with chlorine concentration of 10 (2), 20 (3), 30 (4), and 50 wt.% (5)

SCPE begins losing mass at much lower temperature (100 – 200°C with regard to chlorine concentration). This process is characterized by three stages, the last of which corresponds in temperature to degradation of the initial HDPE. Leveling of integral curves in Figure 6 indicates a strong chlorine effect on both the temperature range of the process and the relative height of separate stages, and change of the peak heights on differential curves reflects the appropriate variation of the gas release rate on each stage. Degradation termination temperature is the same for all polymers studied and equals to 550 – 553°C. Similar to the initial PE, the majority of them releases up to the moment of reaching this temperature, excluding samples with high chlorine concentration. The latter, even heated up to 1,000°C, remained on the surface of the platinum crucible furnace as a dense black-colored film with metallic luster. The mass of this residue, to remove which a second crucible furnace annealing in air at 1,000°C was required, gave from 2 to 6 wt.% with chlorine concentration of 38 – 59 wt.%, respectively. Moreover, a dull brown residue occurred in the area neighboring the cooled down surface of quartz retort of thermal scales, which may be an indirect indication of the presence of relatively large fragments of macromolecules in the composition of volatile products.

The features of SCPE thermal degradation were analyzed more thoroughly by groups of materials differing by production methods. More clear separation of the first and the second degradation stages was observed for polymers synthesized without initiator in a mixture of solvents [54], although the general character of thermogravimetric curves and their change with chlorine concentration increase is preserved. Materials of the SC HDPE R group are characterized by a shear of the second stage to the side of lower temperatures, which leads to clearer separation of the second and the third stages. Polymers synthesized from LDPE display a tendency to fusion of the second and the third stages and relatively high contribution of the second one. Figure 7 shows temperature dependencies of degradation initiation and maximum degradation rates at different stages on the chlorine concentration in the polymer.

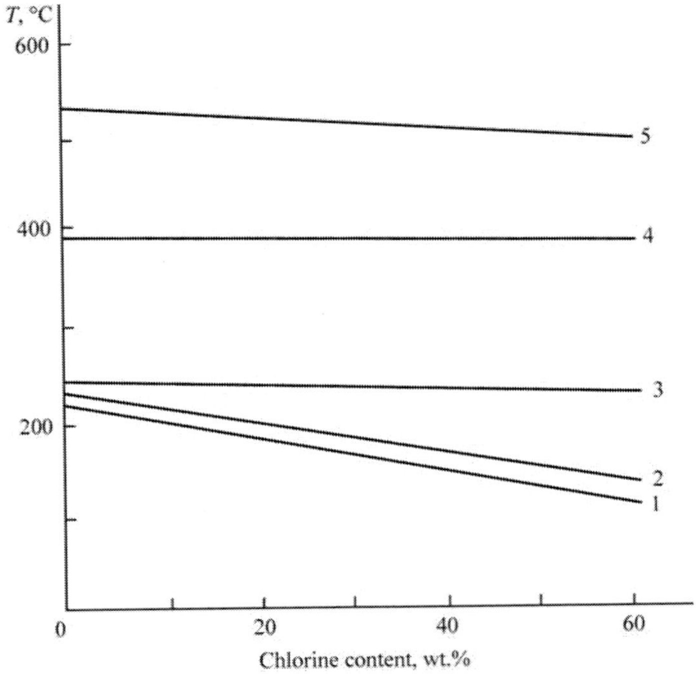


**Figure 7.** Temperature dependence of degradation initiation and maximum degradation rates on chlorine concentration

Figure 8 shows changes in mass parts of every stage.

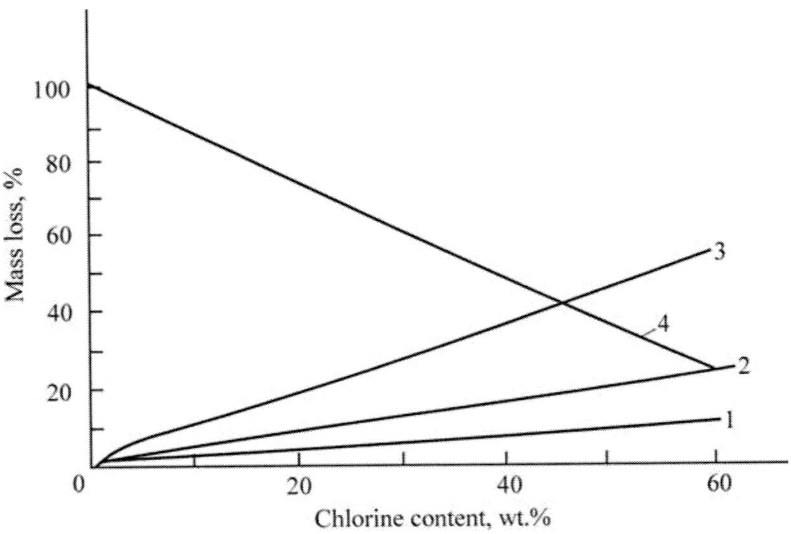


**Figure 8.** Influence of chlorine concentration in polymer on mass loss during thermal degradation

Figures 7 and 8 show that the degradation of chlorinated polymer is initiated at lower temperatures than in the initial one and, contrary to the latter, is of the multistage type. Temperatures of maximum degradation rates on the first and the third stages change insignificantly with the chlorine concentration and can be described by the following equations:

$t_{m1} = (243 - 0.27[Cl])°C$;

$t_{m3} = (526 - 0.63[Cl])°C$.

The second maximum temperature is practically independent of the chlorine concentration in the polymer and gives:

$t_{m2} = (382 \pm 9)°C$.

Temperature of conditional degradation onset is under stricter dependence on the chlorine concentration and can be represented for all polymers as follows:

$t_{onset} = (239 - 1.70[Cl])°C$.

For a group of polymers synthesized without initiator, this temperature is the following:

$t_{onset} = (229 - 1.95[Cl])°C$.

Dependencies of the mass part of the substance degrading on every stage on the chlorine concentration, shown in Figure 9, are close to linear and approximately described by the equation for polymers obtained with initiator:

$m_{01}/m_0 = 0.199 \cdot 10^{-2}$ [Cl].

For polymers obtained in the absence of initiator, the equation is reduced to the following form:

$m_{01}/m_0 = 0.430 \cdot 10^{-2}$ [Cl].

For the second and the third stages, dependencies are general for both types of polymer and are of the following form:

$m_{02}/m_0 = 0.965 \cdot 10^{-2}$ [Cl];

$m_{03}/m_0 = 1 - 1.264 \cdot 10^{-2}$ [Cl].

The chlorination method shows up on the degradation kinetics, which is possibly associated with the structure of obtained polymer. For example, synthesis in the mixture of solvents in the absence of initiator gives a polymer with branched structure possessing chlorinated side substituents. At degradation, the first stage is shifted to the side of lower temperatures, and the degradation rate increases, therewith reducing on the second stage. Temperature of the thermal degradation termination is independent of the modification method. The dependencies obtained can be recommended for predicting thermal degradation characteristics of SCPEs of the types considered. Thermal degradation kinetics was studied on a derivatograph at the heating rate of 5 and 20 deg/min in air medium. It is shown that degradation regularities are the same as in the inert medium: degradation of chlorinated polymers is initiated much earlier than that of the initial PE and is characterized by the presence of two low-temperature stages, a part of which increases with chlorine concentration in the polymer. Contrary to thermal degradation in the inert medium, degradation in air displays one more high-temperature stage with loss of from 5 to 12% of mass. The first two stages of degradation correspond to low endothermal peaks, and the third and the fourth ones to exothermal peaks. Analysis of these data and their comparison for air and inert medium shows that the medium composition causes no effect on proceeding of the first two stages of degradation. Mass parts do practically coincide with dependencies obtained for the inert medium. Temperatures of maximum degradation rates are practically independent of the chlorine concentration. The only factor affecting these temperatures is the heating rate. Mass parts of the third and the fourth exothermal stages of thermooxidative degradation depend on the heating rate: at slow heating, the relative contribution of the latter stage increases while the third stage contribution decreases, which is much clearer for materials with lower chlorine concentration.

Besides determination of qualitative characteristics of the processes and comparison of material behavior, combined analysis of the results of SCPE thermal degradation in inert medium and in air indicated the influence of chlorine concentration on temperatures and mass parts of separate stages, and on degradation rates. Studies of these dependencies allow prognosis of thermal properties for polymers of this class at different heating rates. The conclusion is very important that parameters of the initial two stages are independent of the medium composition.

According to the data of gas chromatography of SCPE degradation products at temperatures above 165°C, hydrogen chloride was detected among pyrolysis products. Bands corresponding to unsaturated bonds were detected for them by infrared spectroscopy.

Hence, the investigations held show that PE sulfochlorination affects significantly low- and high-temperature properties of obtained polymers depending on both chlorination degree and the modification method. Crystalline high density polyethylene loses its crystallinity with increase in the degree of chlorination and transforms into completely amorphous polymer at chlorine concentration over 30 wt.%. This is the critical concentration, because types of dependencies of such properties as density, elasticity modulus, and mechanical dissipation factor on chlorine concentration change after reaching it.

Chlorination degree appreciably shows up also on kinetics of thermal degradation, which can be described mathematically for each type of polymer by stages of thermal degradation.

Chlorinated polyolefins, including sulfochlorinated polyethylenes, are characterized by reduced combustibility, the indices for which depend on the chlorination degree of the polymer. Figures 9 and 10 show changes of combustion heat, time of free combustion after igniter remove, oxygen index and temperature of self-ignition [55, 56].

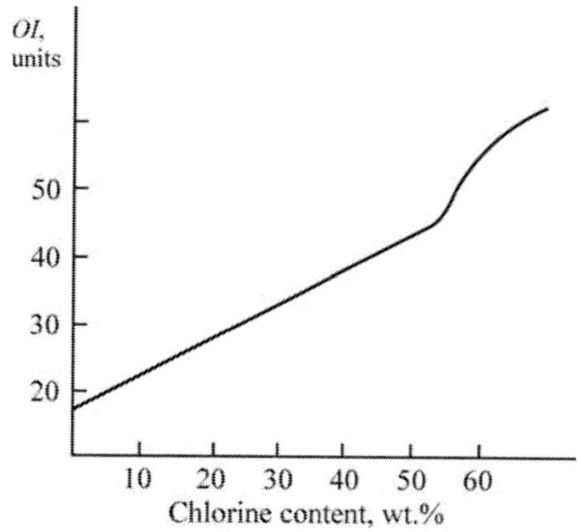


**Figure 9.** Dependence of sulfochlorinated polyethylene oxygen index on chlorine concentration, wt.%

Polymers become incombustible at chlorine concentration exceeding 50%. Spontaneous ignition temperature and the oxygen index increase with the chlorination degree.

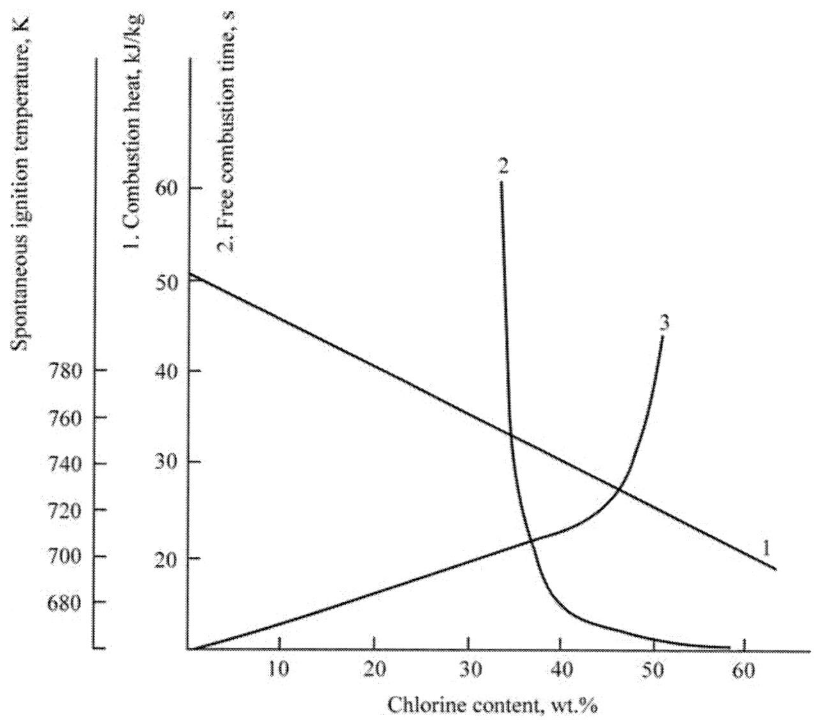


**Figure 10.** Characteristics of fire-resistance of sulfochlorinated polyethylene with different chlorine concentrations

Similar to CPE, SCPE is a fire-resistant polymer possessing all the properties of a chlorinated polymer but with a major advantage – the ability to be vulcanized by sulfochloride groups. Optimal sulfur concentration does not exceed 1.5 wt.%. Rise of sulfur content increases rigidity, i.e. reduces elasticity, although increasing strength and resistance of vulcanizates to aging. Lower sulfur concentration reduces ability for vulcanization and strength characteristics of the materials. As it is a saturated polymer, SCPE is quite stable during storage.

During sulfochlorination, crystallinity varies similar to chlorination and adheres to the same laws [8].

Reactivity of chlorinated polymers is associated with high mobility of chlorine atoms, which enter into substitution reactions. Reactive centers of SCPE are chlorosulfone groups and chlorine atoms present in *n*- and *o*-position in relation to one another and displaying capability of catalytic dehydrochlorination.

National industry produces 6 trademarks of SCPE based on LDPE [41]. There are many foreign varieties of polymer based on HDPE and LDPE: for example, Du Pont Company (USA) produces 8 trademarks of SCPE based on both HDPE and LDPE.

National SCPE trademarks with chlorine concentration of 26 – 36 wt.% and sulfur content of 0.85 – 2.2 wt.% possess a degradation initiation temperature in the range of 115 – 150°C, a tensile strength limit of 15.0 MPa, and relative elongation at rupture of 300 – 350% [15].

Thermal degradation of SCPE proceeds in three stages:

1. At low temperatures, sulfochloride groups are detached and decay forming sulfur dioxide and hydrogen chloride. No molecular chlorine is released.
2. Dehydrochlorination at medium temperatures proceeds similar to CPE. First, chlorine atoms in *p*- and *o*-positions to $SO_2Cl$-groups are detached. Mobile chlorine gives about 22% of the total chlorine amount, which is about 6% of the polymer mass. Already at 150°C up to 70 – 80% of sulfur is detached in the form of sulfur dioxide.
3. Further temperature increase initiates chain breaks.

SCPE degradation is initiated by peroxides and metal chlorides of varying valence. Oxygen causes no effect on desulfonation but initiates dehydrochlorination. Cyclic structures are also formed during thermal degradation of SCPE.

Chlorination of polyolefins substantially reduces combustibility of polymers and changes all indices associated with it [57]. The main factor providing for fire-resistance of chlorinated polyethylenes is chlorine concentration. In the case of pure polymer, it should be 52%, i.e. 0.85 chlorine atom per $CH_2$-group. The oxygen index of chloropolyethylene increases with the chlorine concentration reaching a maximum at 39 – 42% chlorine content. In this case, combustion heat is reduced to 20 – 22 kJ/kg from 48 kJ/mol for the initial polyethylene. Fire-resistance increases with the chlorine concentration. When the latter reaches 50.35%, chloropolyethylene becomes hardly combustible. However, sulfochlorinated polyethylene belongs to the class of combustible polymers. Figure 9 shows change of the oxygen index on polyethylene chlorination. Figure 10 shows some characteristics of sulfochloropolyethylene characterizing fire resistance of chlorinated polyethylenes with regard to chlorine concentration in the polymer. Fire-resistance of the polymeric material is determined by the following factors: combustion heat, spontaneous ignition temperature, and time of free combustion after heat source removing [58 - 60].

Analysis of Figures 9 and 10 indicates that the chlorination and sulfochlorination increase fire-resistance of polymers. However, pure sulfochlorinated polyethylene is a combustible polymer.

Table 5 shows data on determination of the combustibility group of chlorinated and sulfochlorinated polyethylenes with respect to time of free combustion after the heat source has been removed.

**Table 5. Combustibility of polymers with regard to chlorine concentration**

| Polymer | Chlorine concentration, wt.% | Combustion duration outside the furnace, s | Combustibility group |
|---|---|---|---|
| Polyethylene | 0 | 65 | Combustible |
| Chlorinated polyethylene | 27.1 | 25 | Combustible |
| Chlorinated polyethylene | 33.0 | 35 | Combustible |
| Chlorinated polyethylene | 35.2 | 15 | Combustible |
| Chlorinated polyethylene | 35.6 | 15 | Combustible |
| Chlorinated polyethylene | 43.3 | 8 | Combustible |
| Chlorinated polyethylene | 50.4 | 3 | Hardly combustible |
| Chlorinated polyethylene | 58.9 | Does not combust | Hardly combustible |
| Chlorinated polyethylene | 62.1 | Does not combust | Hardly combustible |
| Chlorinated polyethylene | 71.3 | Does not combust | Hardly combustible |
| Chlorinated polyethylene | 72.5 | Does not combust | Hardly combustible |
| Chlorinated polyethylene | 73.5 | Does not combust | Hardly combustible |
| Sulfochlorinated polyethylene* | 28.6 | 65 | Combustible |

* sulfur concentration is 1.65%.

Solid material can combust after gasification only, i.e. combustion proceeds in the gas phase. Gaseous products of polymeric material gasification, formed by physical and chemical transformations in the polymer at high-temperature, form diffusional flame contacting the air. Combustibility decrease of chlorinated polyethylenes in the air flow with increase of the chlorination degree can be explained by comparison with behavior of monomolecular chlorohydrocarbons, which undergo endothermal pyrolysis (dehydrochlorination) before the stage of gasification product oxidation. Alternation of these processes and reduced concentration of combustible gases in gasification products make continuous combustion impossible.

However, it is difficult to use highly chlorinated polyethylenes as the polymeric base for fire and heat shield materials due to technological and operational properties of these polymers, because they are thermoplastic resins melting at increased temperature and are able to flow, i.e. to lose the preformed shape. Reduced elasticity in a wide temperature range and ability to flow at increased temperature lead to limitations in application of chlorinated polyethylenes as fire and heat shield materials themselves. That is why combined polymers containing thermoplastic and thermoreactive components simultaneously are used as the base for fire and heat shield materials. Thermoplastic chlorine-containing polymers provide for reduced combustibility of a composite material, and the thermoreactive component for preserving the article shape at increased temperature. However, application of the polymeric base consisting of two or several polymers makes production technology of protective articles more complicated. This disadvantage is not displayed by sulfochlorinated polyethylene.

In accordance with all the above-said, fire and heat shield properties of chlorinated and sulfochlorinated polyethylenes were estimated on the one-side heating (thermal impact) unit. However, protective properties of samples prepared from pure polymers cannot be objectively estimated because of short-term operation of the materials under these conditions. That is why fire and heat shield properties of various polymers were estimated on samples of molded and vulcanized material. Estimation of the protected surface temperature change rate and time of reaching 400°C on it under the effect of the heat flux with 1,100°C on the sample surface show that fire and heat shield properties depend directly on chlorine concentration in the polymer and increase with it (Figure 11, Table 6). This dependence is clearly observed by both the change of protected surface temperature and time of reaching 400°C on it. Type of the initial polymer and the modification method show up insignificantly on these indices.

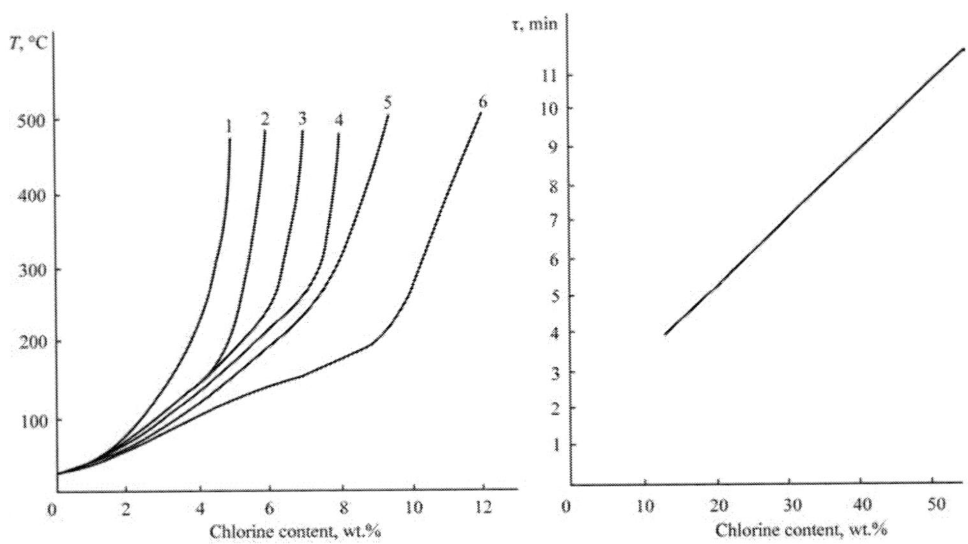


**Figure 11.** Dependence of fire and heat shield properties of sulfochlorinated polyethylene vulcanizates on chlorine concentration: a) temperature of protected surface; b) time of obtaining 400°C on the protected surface

**Table 6. Time taken to reach 400°C on the protected surface under samples of sulfochlorinated polyethylene with various chlorination degrees**

| Polymer | SC LDPE | SC HDPE | | | | | | SC HDPE IL | | |
|---|---|---|---|---|---|---|---|---|---|---|
| Chlorine content, % | 28 | 24 | 33 | 35 | 35 | 36.6 | 63 | 21.3 | 24.8 | 31.2 |
| Time taken to reach 400°C, min | 7 | 7 | 8 | 8 | 9 | 10 | 11 | 5 | 6 | 9 |

In the case of applying materials possessing no spatial cross-linking to fire and heat shielding, the operation time of such protection is abruptly reduced, and the materials become ineffective even if they absorb a great amount of heat. Spatial three-dimensional cross-linking of polymer molecules stretches heat expenditure in time up to required duration. That is why problems of vulcanizing the polymeric base of fire and heat shield material are of great importance.

## REFERENCES


[1] Ronkin G.M., Chlorinated Polyethylene, *Plasticheskie Massy*, 1980, No. 8, p. 16. (Rus)

[2] Sirota A.G., *Modification of Structure and Properties of Polyolefins*, Leningrad, Khimia, 1984. (Rus)

[3] Krentsel B.A., *Chlorinated Paraffinic Hydrocarbons*, Moscow, Nauka, 1964, 158 p. (Rus)

[4] Plate N.A., Litmanovich A.D., and Poa O.V., *Supermolecular Reactions*, Moscow, Khimia, 1977, 255 p. (Rus)

[5] Ronkin G.M., *Sulfochlorinated Polyethylene*, Moscow, TsNIITEKhim, 1977, 101 p. (Rus)

[6] Ronkin G.M., Modern State of Production and Application of Chlorinated Polyolefins. *Reviews on the Most Valuable Scientific and Technical Problems of Chemical Industry*, Moscow, NIITEKhim, 1979, 82 p. (Rus)

[7] Ronkin G.M., Halogenated Polymers, the Main Tendencies in Production and Application. *Reviews on the Most Valuable Scientific and Technical Problems of Chemical Industry*, Moscow, NIITEKhim, 1980, 103 p. (Rus)

[8] Dontsov A.A., Lozovik G.A., and Novitskaya S.P., *Chlorinated Polymers*, Moscow, Khimia, 1979, 232 p. (Rus)

[9] V.I. Abramov and Krasheninnikova A.A., *Production of Chlorinated Polymers Abroad*, Mosocw, NIITEKhim, 1978, 78 p. (Rus)

[10] Lecont J., *Infrared Radiation*, Izdat. Fiz.-Mat. Literatury, Moscow, 1958, 584 p. (Rus)

[11] Carslaw H.S. and Jaeger J.C., *Conduction of Heat in Solids*, Oxford University Press, London, 1959.

[12] Dontsov A.A., Lozovik G.A., and Novitskaya S.P., *Chlorinated Polymers*, Moscow, Khimia, 1979, 232 p. (Rus)

[13] Ronkin G.M., Polymer Modification by Halogen-containing Compounds, *Ser. Obshcheotraslevye Voprosy*, Iss. **11**(205), Moscow, NIITEKhim, 1982, 79 p. (Rus)

[14] Ronkin G.M., Chlorinated Polyolefins. *Polymer Encyclopedia*, Moscow, Sovetskaya Entsiklopediya, 1977, vol. 3, pp. 18 – 24. (Rus)

[15] *Industrial Chlororganic Products*, Ed. L.A. Oshin, Moscow, Khimia, 1978, pp. 560 – 592. (Rus)

[16] *USA Patent* No. 3,819,577.

[17] *USA Patent* No.2,640,048.

[18] *USSR Patent* No. 583,139. (Rus)

[19] Khromenkov L.G., Ronkin G.M., Tsvankin D.Ya., et al., On Structure of Chlorinated Polyethylene, *Vysokomol. Soedin.*, 1977, vol. **19B**(4), pp. 308 – 310. (Rus)

[20] *USA Patent* No. 3,347,835.

[21] *USSR Patent* No. 364,634. (Rus)

[22] *USSR Patent* No. 547,456. (Rus)

[23] *Canada Patent* No. 471,037.

[24] Kaprawan H.S., Mechanism of chemical inhibitions, Fire Eng., 1974, vol. 1(1), pp. 103 – 108.

[25] *Japan Patent* No. 27,605.

[26] Shibaev V.P., Plate N.A., Grushina R.G., and Kargin V.A., Structuring in Chlorinated Polyethylene and Its Solutions, *Vysokomol. Soedin.*, 1964, vol. **6**(2), pp. 231 – 236. (Rus)

[27] Ronkin G.M. and Khromenkov L.G., Change of Supermolecular Surface Structure of Crystalline Polyolefins During Chlorination, *Plasticheskie Massy*, 1987, No. 5, pp. 59 – 60. (Rus)

[28] Levin A.S., *Chlorinated Polyethylene and Its Application*, Moscow, TsNIITEIssl. Legkoi Promyshlennosti, Ed. A.M. Savranski, 1973, 120 p. (Rus)

[29] Patrikeev G.A., The Effect of Structural Cleavage of Chain Molecules in Liquid Isoalkanes, *Zh. Fiz. Khim.*, 1978, vol. **52**(1), pp. 247 – 248. (Rus)

[30] Ronkin G.M., Chlorination and Structure of Chlorinated Polyolefins. *Kinetics and Mechanism of Macromolecular Reactions, Thesis All-Union Meeting*, January 10 – 12, 1984, AS USSR, Chernogolovka, 1985, pp. 18 – 19. (Rus)

[31] Quenum B.M., Bertricat P., and Vallet C., Chlorinated Polyethylene, *Polym. J.*, 1975, vol. **7**(3), pp. 277 – 311.

[32] Krentsel L.B., Litmanovich A.D., Pastukhova I.V., et al., Determination of Individual Rate Constants of Polyethylene Chlorination, *Vysokomol. Soedin.*, 1971, vol. **13A**(11), pp. 2489 – 2495. (Rus)

[33] Quenum B.M. and Bertricat P., Etude comparatve de la chlorination des polyethylene liniair et remitie, *Eur. Polym. J.*, 1971, vol. **7**(11), pp. 1527 – 1535.

[34] Osipchuk V.S., Rumyantseva N.D., Sokolova N.I., Yablokov A.G., and Yatsenko E.B., Physicochemical properties of materials based on modified sulfochlorinated polyethylenes, *Plastmassy*, 1992, No. 4, p. 8. (Rus)

[35] Ronkin G.M., Structure and Properties of High and Low Density Chlorinated Polyethylene, *Plasticheskie Massy*, 1981, No. 6, pp. 17 – 20. (*Intern. Polym. Sci. and Technol.*, 1981, vol. **8**(10), pp. 31 – 35.)

[36] Ronkin G.M., Chlorinated Polyethylene Properties, *Plasticheskie Massy*, 1984, No. 11, pp. 20 – 21. (Rus)

[37] Maevskaya B.M., Simulin Yu.N. et al., Thermal Analysis of Chlorinated Carbochain Polymers. *Thes. Proc. 8$^{th}$ All-Union Conference AS USSR on Thermal Analysis*, Moscow, Kuibyshev, 1982, pp. 174 – 175. (Rus)

[38] Gul' V.E., *Polymer Structure and Strength*, Moscow, Khimia, 1978. (Rus)

[39] Vasilieva T.A., Golovan E.N., and Korotneva L.A., Prospects in production development of fire-resistant polyolefins and polystyrene plastics, *Plastmassy*, 1986, No. 3, pp. 51-53. (Rus)

[40] Halido Carlos J., A method for estimating limits of flammability, *J. Fire and Flammab.*, 1975, vol. **6**(2), pp. 130 – 139.

[41] *Sulfochlorinated Polyethylene*, TU 6-01-715-80, MKhP, GNIIKhP. (Rus)

[42] Evlanov S.F., Frolov Yu.A., and Ronkin G.M., Thermodynamic Composition of Combustion Products of Chlorine-containing Polymers, *Thes. Proc. 6$^{th}$ All-Union Conference on Combustion of Polymers and Creation of Incombustible Materials*, Suzdal, Nov. 29 – Dec. 1, 1988, Moscow, Nauka, 1988, p. 77. (Rus)

[43] Startsev O.V., Vapirov Yu.M., Ovanesov A.S., Donskoi A.A., and Shashlina M.A., *Vysokomol. Soedin.*, 1987, vol. **29A**(12), p. 2473. (Rus)

[44] Dyunina V.G., Studies of Obtaining Porous Footing Material Based on Sulfochlorinated Polyethylene, *Dissertation*, Moscow, MTILP, 1972, pp. 35 – 60. (Rus)

[45] Mansfield J.A. and Riccitillo S.R., An effective and practical fire production flammability, *J. Fire and Flammab.*, 1975, vol. 6(4), pp. 492 – 498.

[46] Ronkin G.M., Korotyanski M.A., Gershenovich A.I., and Jagatspanyan R.V., Low density Polyethylene Optimal for Sulfochlorinated Elastomers, *Kauchuk i Resina*, 1980, No. 1, pp. 5 – 8. (Rus)

[47] Ronkin G.M., Korotyanski M.A., Gershenovich A.I., and Jagatspanyan R.V., Properties of Sulfochlorinated LDPE Derivatives, *Plastmassy*, 1980, No. 3, pp. 26 – 29. (Rus)

[48] Ronkin G.M., Khromenkov L.G., and Iliin B.A., Thermal Analysis of Carbochain Polymers, *Thes. Proc. 18$^{th}$ All-Union Conference on High-molecular Compounds*, Kazan, 1973, p. 125. (Rus)

[49] Makhlis F.A. and Fedyukin D.L., *Terminological Reference Book on Rubber*, Moscow, Khimia, 1989, pp. 356 – 358. (Rus)

[50] Dyunina V.G., Pavlov S.A., and Dinzburg B.A., Sulfur and Organic Accelerator Effects on Porous Material Properties Based on Sulfochlorinated Polyethylene, *Kozhevenno-obuvnaya promyshlennost*, 1970, No. 5, p. 115. (Rus)

[51] Blokh G.A., *Organic Accelerators of Rubber Vulcanization*, Moscow, Khimia, 1972, 559 p. (Rus)

[52] Blokh A.G., *Organic Accelerators of Vulcanization and Vulcanizing Systems for Elastomers*, Leningrad, Khimia, 1978, 240 p. (Rus)

[53] Tugov I.I. and Kostyrina G.I., *Polymer Chemistry and Physics*, Moscow, Khimia, 1989, 431 p. (Rus)

[54] Dontsov A.A., Novitskaya S.P., Kochanov I.M., and Stepanova L.N., Chlorinated Polyethylene Cross-linking by Hexammonium Sebacate, *Kauchuk i Rezina*, 1976, No. 3, p. 30. (Rus)

[55] Maltseva A.S., Ronkin G.M., Froliv Yu.V. et al., Regularities of Chlorinated Polyolefins Oxidative Gasification, *Doklady AN SSSR*, 1989, vol. **307**(2), pp. 397 – 400. (Rus)

[56] Maltseva A.S., Rozlovski A.I. et al., Initiation of Chlorinated Polyethylene Combustion, *Fizika Goreniya i Vzryva*, 1987, No. 5, pp. 38 – 41. (Rus)

[57] Rohde E., Chlorietes Polyathylen für technike Gummiartikel, *Kaut. und Gummi Kunst.*, 1982, Bd. **35**(6), S. 478 – 487.

[58] Ronkin G.M., Masanko A.F., Romashin O.P. et al., New Chlorinated Polyolefin Thermoplastics, *Proc. Rub. Conf. JRC-95*, Robe, Japan, 1995, pp. 629 – 632.

[59] Ronkin G.M., Romashin O.P., Belousov P.K., and Zheltukhin I.A., Combustibility of Chlorinated Elastomers, *Proc. 3rd Scientific and Applied Conference of Rubber Specialists. Raw and Materials for Rubber Industry: Present and Future*, May 13 – 17, 1996, Moscow, 1996, pp. 49 – 51. (Rus)

[60] Ronkin G.M., Mazanko A.F., and Romashin O.P., New Chlorinated Polyolefines and Thermo- and Fire Resistant Materials Based on Them, *Proc. Scientific and Research Institute "Sintez"*, Moscow, 1996, pp. 449 – 508. (Rus)

In: Chemical Physics of Pyrolysis, Combustion and Oxidation ISBN 1-59454-163-9
Editors: A. A. Berlin et al., pp. 81-85 

*Chapter 7*

# THERMODINAMICS OF PMMA-PHOSPHATE PLASTICIZER SYSTEMS

***L. V. Adamova, E. V. Rusinova and S. A. Vshivkov***
Ural State University.
Lenin avenue 51, Ekaterinburg, 620083 Russia.
E-mail: *lydiya.adamova@usu.ru*
FAX: (343)3507401

## ABSTRACT

The boundary curves of the poly(methyl methacrylate) – phosphate plasticizer systems and thermodynamic parameters of interaction between the components are determined. The relationship between the UCST and the chemical structure of plasticizers is shown.

**Key words** Poly((methyl methacrylate), plasticizer, phase diagram, Gibbs energy of mixing, second virial coefficient.

## INTRODUCTION

To produce the fire – resistant polymer materials the special plasticizers are used [1]. Such plasticizers are halialkyl- and phenyl phosphates. A necessary condition for the prolonged use of these systems is the thermodynamic compatibility of the components. Investigation of phase diagrams provides information about this problem [2, 3]. The purpose of this work was to study the phase transitions in the PMMA – plasticizer systems and to determine the thermodynamic parameters of interaction between the components: the second virial coefficients $A_2$ and the Gibbs energy of mixing $\Delta g_x$.

## EXPERIMENTAL

Poly(methyl methacrylate) with $M_\eta$=4,8x10$^4$ and plasticizers enumerated in the table were used as objects for this study.

**Table. Characteristic of plasticizers**

| Plasticizer | Structure formula | $T_{boil}$, K [2] | Density, g/cm$^3$ [2] |
|---|---|---|---|
| 1 | $Cl(CH_2)_3O$<br>$CH_3(C_2H_5)CH(CH_2)_3CH_2O$– P→O<br>$CH_3(C_2H_5)CH(CH_2)_4CH_2O$ | 194 (270) | 0,9997 |
| 2 | $CH_3(C_2H_5)CH(CH_2)_4CH_2O$<br>$Cl(CH_2)_3O$–P→O<br>$Cl(CH_2)_3O$ | 170 (270) | 1,1040 |
| 3 | $Cl(CH_2)_2O$<br>$Cl(CH_2)_2O$– P→0<br>$Cl(CH_2)_2O$ | 175 (270) | 1,4225 |
| 4 | $Br(CH_2)_2O$<br>$Br(CH_2)_2O$– P→O<br>$Cl(CH_2)_2O$ | 205 (270) | 1,8015 |
| 5 | $ClCH_2(ClCH_2)CHO$<br>$ClCH_2(ClCH_2)CHO$– P→O<br>$C_6H_5O$ | 223 (270) | 1,4143 |
| 6 | $CH_3(C_2H_5)CH(CH_2)_4CH_2O$<br>$C_6H_5O$– P→O<br>$C_6H_5O$ | 196 (600) | 0,9610 |

Quantities in brackets – residual pressure, Pa.

The temperatures of phase separation $T_{ph}$ of blends were determined by the cloud point method [4] and rephractometric method, proposed for the polymer – solvent systems by Rechage [5] and used later for the polymer – plasticizers [6].

The Gibbs energy of mixing for the systems was determined by a procedure described elsewhere [2]. The sorption isotherms of common solvent (chloroform) vapors were determined for individual components and their blends using a gravimetric variant of the method using a coil balance with 4x10$^2$ – 6x10$^2$ m/kg sensitivity. The experiments were performed at 298 K and under a residual pressure of 10$^{-3}$ Pa. Then, changes in the chemical potential $\Delta\mu_1$ of chloroform during its sorption by individual polymers and their blends were calculated using conventional procedure.

Using the approximate integration method and the Gibbs – Duhem equation, we determined the change in the chemical potential $\Delta\mu_2$ of the individual polymer and its blends with the plasticizers. The average Gibbs energy of mixing of sorbate with polymer and blends was calculated by the equation

$$\Delta g_m = \omega_1\Delta\mu_1 + \omega_2\Delta\mu_2,$$

Where $\omega_1$ and $\omega_2$ are the weight fractions of the solvent and the polymer (or polymer blend), respectively, in the solution formed during sorption. The portions of the $\Delta g_m - \omega_2$ curves corresponding to dilute solutions were approximated by straight lines, which were drawn up to their intersection with the ordinate at $\omega_2=1$, and the values of the Gibbs energy of mixing per g of polymer ($\Delta G_I$ and $\Delta G_{II}$ ) or per g of a blend ($\Delta G_{III}$) were obtained. The values obtained were introduced into the relationship

$$-\Delta g_x = \Delta G_{III} - (\omega_1 \Delta G_I + \omega_2 \Delta G_{II}),$$

where $\omega_1$ and $\omega_2$ are the weight fractions of polymer in a blend, and $\Delta g_x$ is the Gibbs energy of mixing of polymer with the plasticizer.

The values of $A_2$ were determined using the light scattering method [5].

## Results and Discussion

Fig. 1 represents the phase diagrams of the studied systems. These systems are separated upon cooling; that is, they exhibit the upper critical solution temperature UCST. The average values of the maximum on the coexistence curves obtained on heating and cooling were assumed to be the values of critical solution temperatures, which are presented below:

| | | | | | |
|---|---|---|---|---|---|
| Plasticizer | 4 | 5 | 3 | 6 | 1 |
| UCST,K | 295 | 296 | 318 | 323 | >393 |

For the system PMMA – plasticizer 1 (curve 5) the value of UCST was not determined because of plasticizer destruction at high temperatures. The least values of UCST are obtained for the systems with the plasticizers that contain two Br atoms and Cl atom or phenyl radicals per molecule. The substitution of Br atoms for the Cl atoms leads to the increase of the UCST values.

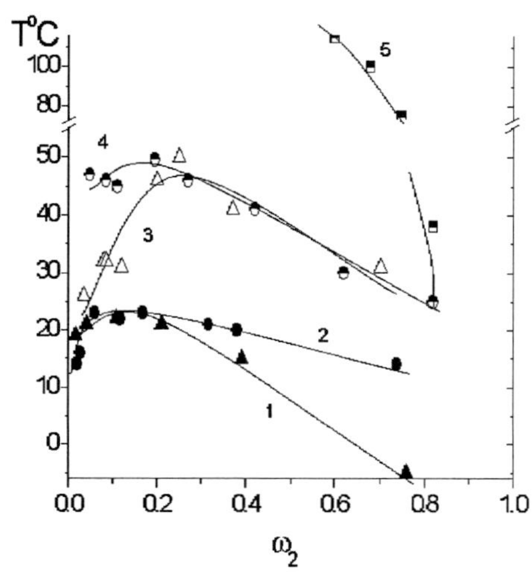


**Fig. 1.** Phase diagrams of the PMMA – plasticizer systems: 1 –(4), 2 –(5), 3 – (3), 4 – (6), 5 –(1). Number in bracket is the plasticizer number in the table of the text.

These data are in good agreement with the results about the concentration dependence of $\Delta g_x$ shown in Fig. 2. As is seen, for the PMMA – plasticizer 2 and PMMA – plasticizer 5 systems the $\Delta g_x - \omega_2$ curves are concave; that is, $(\partial^2 \Delta g_x / \partial \omega_2^2)_{P,T} > 0$, thus demonstrating the thermodynamic compatibility of components within studied composition range. Indeed, the UCST of the PMMA – plasticizer 5 is below 298 K. For the system PMMA – plasticizer 1 at the concentration of PMMA $\omega_2$ = 0,8 the value of $\Delta g_x$ = 0 and for the smaller PMMA concentrations $\Delta g_x > 0$, that indicating that these components are incompatible. The results obtained are in good agreement with the phase diagram of the PMMA – plasticizer 1 system.

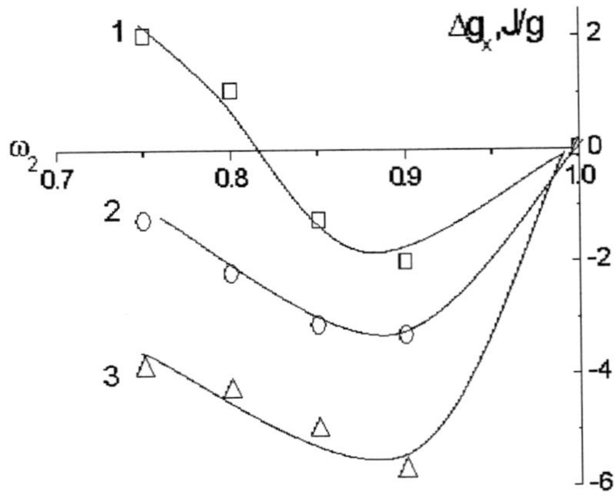


**Fig. 2.** The concentration dependence of $\Delta g_x$ for the systems PMMA – plasticizers: plasticizer 1 (1), 2 (2), 5 (3). T = 298 K.

Upon cooling of the PMMA – plasticizer systems (systems with UCST) the solvents become poor. The temperature dependence of $A_2$ for the PMMA – plasticizer 3 is shown in Fig. 3. As is seen, at the temperature T>318 K (T>UCST) $A_2 > 0$. That is, this plasticizer is a good solvent of PMMA. At the temperature T<318 K (T<UCST) $A_2 < 0$, that is, the plasticizer is a poor solvent of PMMA. At 293 K even the dilute solutions of PMMA separate (Fig. 1).

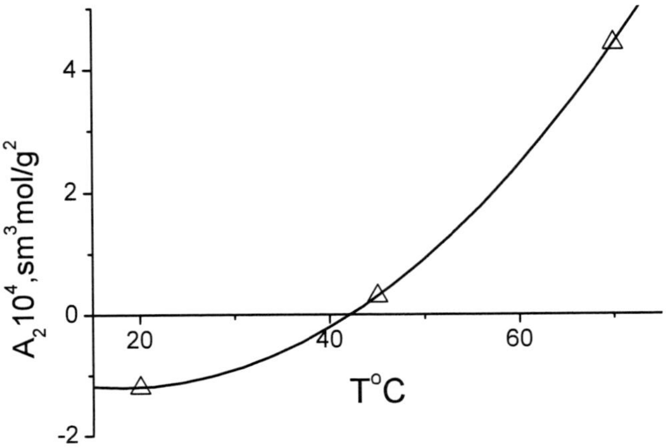


**Fig. 3.** The temperature dependence of $A_2$ for the PMMA – plasticiser 3 system.

Thus, the good correlation between the phase diagrams and the thermodynamic parameters of interaction of components in the PMMA – plasticizer systems is discovered.

The best solvents of PMMA are the plasticizer, that contain in their molecules phenyl radicals and halocarbons. It is caused by: 1 – the dipolar interaction between the C – Cl and C – Br groups of the plasticizers with the C=O groups of PMMA, 2 – formation of the donor-acceptor bindings between of halogen atoms of plasticizers and oxygen atoms of PMMA [7], 3 – interaction of electrons of C=O PMMA groups and plasticizer benzene rings [8].

## REFERENCE

[1] I.I. Razinskaja, V.A, Ageeva, N.I. Ermilina, I.K. Rubtsova, B.P. Shtarkman, *Plast. Massi*. Vol. 1, p. 27, 1997.

[2] A.A. Tager, *Physical Chemistry of Polymers*, Moscow, Khimiya, 1978.

[3] S.A. Vshivkov, E.V. Rusinova, *Phase Transitions in Polymer Systems Induced by Mechanical Field*, Ekaterinburg, Ural Univ., 2001.

[4] S.A. Vshivkov, *Methods for Stadying of Phase Equilibria in Polymer Solutions*, Ekaterinburg, Ural Univ., 1991.

[5] G. Rechage, *Kunststoffe*, Band 1, s. 603, 1963.

[6] A.A. Tager, S.A. Vshivkov, O.E. Polyak, *Vysokomolek. Soed., A*. Vol. 24, .N. 3, p. 661, 1982.

[7] I.N. Rasinskaya, V.A. Isvoschikova, B.P. Shtarkman, L.V. Adamova, B.I. Lirova, A.A. Tager, *Vysokomolek. Soed., A*. Vol. 23, N. 12, p. 2738, 1981.

[8] E.N. Gurianova, , I.P. Golshtein, , I.P. Romm, *Donor-acceptor binding*, Moscow, Khimiya, 1973.

In: Chemical Physics of Pyrolysis, Combustion and Oxidation
Editors: A. A. Berlin et al., pp. 87-91
ISBN 1-59454-163-9


*Chapter 8*

# INTERCONNECTION EVALUATION OF PYROLYSIS AND COMBUSTION INDICATORS

***L. G. Panova*, S. E. Artemenko*, N. A. Halturinskiy** and A.A Berlin *****

* Technological Institute of Saratov State Technical University
Square of Svoboda, 17, Engels 413100, Saratov region, Russia
Tel.: (84511) 63518, (84511)63553; Fax.: (84511) 49584,
E-mail: *xt@techn.renet.ru*
** - Institute of chemical physics RAS

## ABSTRACT

Interconnection of different pyrolysis indicators, obtained by methods of thermal gravimetric analysis with combustibility indicators was evaluated. The main contribution of DC is condensed phase processes change. Data obtained by TGA method together with SPGC (stage pyrolysis gas chromatography) and electronic microscopy let us predict the behavior of polymers during combustion.

**Key words:** polyethylene (PE), viscose (VF), polyacrilnitril (PAN), polyester (PE) fibers, characteristic temperatures of pyrolysis, coke formation, composition of pyrolysis gases, maximum combustibility dimensions (MCD), Thermal gravimetric analysis method.

To evaluate the correlation of different indicators, characterizing decomposition, inflammation and combustibility of polymers, characteristic temperatures of pyrolysis and coke formation, composition of pyrolysis gases, maximum combustibility dimensions (MCD) determining the minimum thickness of polymer layer at which combustibility becomes impossible [1-5], oxygen index (OI) were defined. The experiments were conducted using initial and fireproof polymers of different origin and behavior at pyrolysis and combustion, such as polyethylene (PE), viscose (VF), polyacrilnitril (PAN) and polyester (PE) fibers and composites based on them.

MCD were defined on linens, films and fabrics at 30-100% of oxygen content in oxygen nitrogen mixture with simultaneous these samples' OI determining. (Table 1)

**Table 1. Maximum combustibility dimensions and polymers OI**

| Polymer type | | Maximum combustibility dimension (mm) at oxygen content (%) | | | | | | | OI, volume (%) |
|---|---|---|---|---|---|---|---|---|---|
| | | 30 | 50 | 60 | 70 | 80 | 90 | 100 | |
| PAN | | 3,60 | 2,70 | 2,30 | 2,20 | 2,10 | - | 2,03 | 19,0 |
| PVC with films thickness, (mm) | 0,35 | - | 4,90 | 4,76 | 4,50 | 4,30 | 3,90 | 3,30 | 42,0 |
| | 0,15 | - | 4,70 | 4,53 | 3,64 | 3,49 | 3,36 | 2,60 | 42,0 |
| | 0,1 | - | 4,60 | 4,10 | 3,57 | 3,39 | 3,27 | 2,55 | 42,0 |
| PE with thickness, (mm) | 0,1 | - | - | - | - | - | - | 4,75 | 21,5 |
| | 0,05 | - | 4,69 | 4,61 | 4,46 | 4,30 | 3,71 | 2,53 | 21,5 |
| VF | | 3,74 | 2,76 | 2,42 | 2,10 | 2,02 | 1,89 | 1,60 | 19,5 |
| FPVF – PASPP | | - | 4,70 | 4,40 | 3,72 | 3,03 | 3,10 | 2,01 | 27,5 |
| FPVF – PASPP | | - | 4,80 | 4,70 | 4,14 | 3,81 | 3,50 | 2,71 | 33,0 |
| FPVF – PASPP | | - | 4,81 | 4,80 | 4,70 | 4,60 | 4,10 | 3,20 | 25,5 |
| FPVF – DAMP | | - | 4,72 | 4,20 | 3,81 | 3,20 | 2,61 | 1,92 | 35,6 |

**Remarks:**
**PVC**- polyvinyl chloride; **FPVF** -fireproof viscose fiber; **PASPP** -phosphoric acid salt of polyethylene polyamine; **DAMP** -diamine methyl phosphate.

With the increase of oxygen content MCD of all tested samples decrease. Coking polymers (PAN and VF) have similar values both of OI and MCD. OI of PVC samples is 41,5% and does not depend on film thickness, but MCD ara not changed considerably only for films having thickness of 0,1; 0,5 mm, and with the increase of thickness they increase as well (Table 1). PE films samples have OI – 21,5%, but at 1mm film thickness they support combustion only in pure oxygen, i.e. in this case MCD change in no proportion with OI.

For fireproof VF correlation of MCD with OI having combustibility decelerators of the same origin (PASPP) in a sample was noted. Moreover, OI and MCD increase with those decelerators increasing. At the same time fibers modified by different DC having approximately equal OI values differ in MCD. For VF containing different quantities of the same type DC close values of characteristic destruction temperatures (initial, maximum and final mass loss speed) were noted as well as coke and water outlets.

Consequently, MCD can be more correct, than OI, combustibility indicators for non-coking polymers not containing DC, in all the rest cases OI shows polymers combustibility better.

Interconnection of different pyrolysis indicators, obtained by methods of thermal gravimetric analysis with combustibility indicators was evaluated.

Among non-coking polymers high density polyethylene 277-73 AH02 (SS 16338-78) was used, both initial and multy-functional action filament containing (ammonium polyphosphate (AP) containing P-27% and N-13%), wood ash (WA) and chlorine paraffin CP-1100 and Calcium Carbonate (CC) mixture were also used.

Polyethylene is related to easily inflammable polymers. In the process of combustion it gasifies easily forming combustible volatile products [6]. Thermal oxidation of PE covers a wide range of temperature from 260 to 440°C. This process is accompanied by various thermal effects influencing upon the general reaction of thermal oxidative destruction. In 130-190°C temperature range PE melting with the energy of 178 kJ/mole takes place. Endothermal top point at maximum temperature 450°C corresponds to polyethylene depolymerization(Figure 1). Mass losses at this stage of destruction are 88,6% and activation energy is 170 kJ/mole.

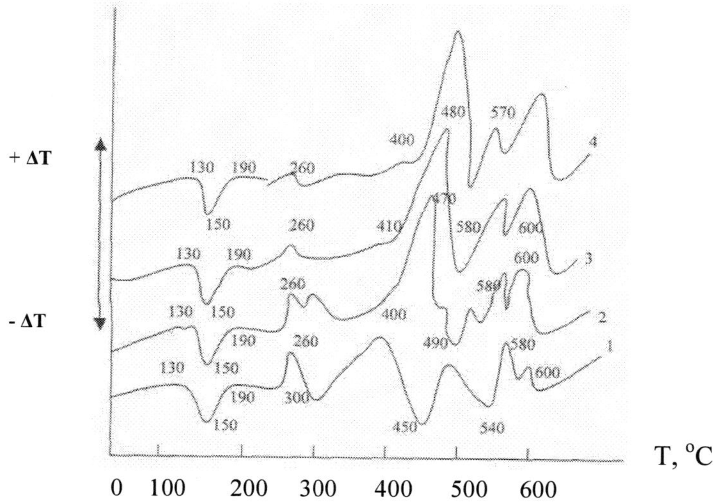


**Figure 1.** Data DTA PE: 1 – PE; 2- PE+CP; 3 – PE+CP+CC; 4 – PE+CP+WA

PE filling with AP increases heat stability (Table 2), thermal stability, but at the same time specific heat decreases, obviously because of PE macro molecules oscillations limit by solid filament particles. Using all investigated filling agent makes notable corrections in pyrolysis process. PE structurization and joining processes are initiated at their presence. Effective energy of destruction activation defined in the interval of main mass losses and carbonized residue formation prove that. (Table 2).

**Table 2. Properties of initial and filled PE.**

| Material composition, $V_{mass.}$ | $T_v$, °C | Specific heat, kJ/kg·K | Thermal stability of the main stages | | Δm, % | $E_a$, kJ/mole | Carbonized residue outlet % at 600 °C | OI vol., % |
|---|---|---|---|---|---|---|---|---|
| | | | I | II | | | | |
| 100 PE | 108 | 3 | 260 | - | 62 | 170 | 0 | 19 |
| 30AP+100PE+2 | 124 | 2,7 | 385 | 430 | 13,0 | 243 | 19,0 | 28 |
| 100PE+20CP | 120 | - | 260 | 430 | 33,0 | 214 | 27,0 | 25,0 |
| 100PE+20CP+40CC | 150 | - | 270 | 460 | 15,8 | 214 | 31,4 | 28,5 |
| 100PE+20CP+40WA | 150 | - | 270 | 430 | 9,2 | 190 | 36,5 | 32,0 |
| 100PAN | - | - | 248 | 320 | - | 240 | 48,0 | 19,0 |
| 100PAN+15.8 Fosfolan | - | - | 248 | 36 | - | 103 | 60,0 | 29,5 |

AP decomposition at increased temperatures occurs with PPA and ammonia formation [7]. PPA creates a protective layer on the surface of destructing material preventing interdiffusion processes of oxidant into polymer and volatile combustible destruction products into combustion zone. Heat exchange between surrounding atmosphere and polymer becomes worse. Heavy ammonia vapours displace an oxidant and combustible gases from the flame zone and lower their content. As the result of it a chain breakage occurs because of the 2 free radicals recombination and the possibility of joint structure formation increases:

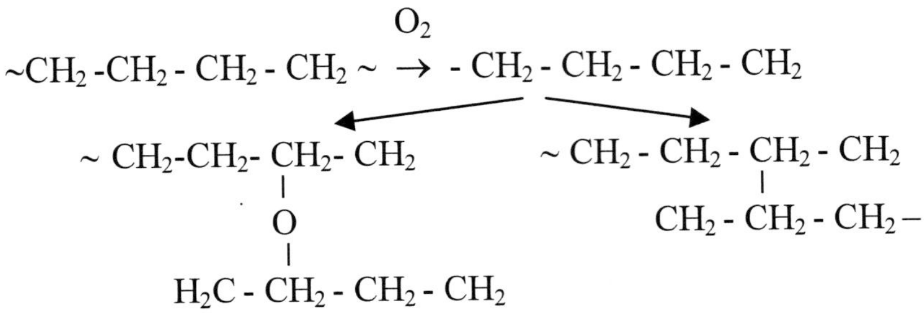

There is no induction transfer while filling PE with CP (Figure 1). Mass losses in 260-300°C temperature range are connected with dehydrochlorization of CP and equal to 7,5 % mass. The second stage of composition destruction begins at 430 °C. Exothermal emergence at 470°C being absent in initial PE and CP (Figure 1) is the result of CP initiation processes of PE sructurization and joining. This is confirmed by effective activation energy increase at this stage being 214 kJ/mole and also by a larger, in comparison with initial PE, CR outlet exceeding its additive outlet. Destruction proceeds with lower speeds.

The influence of CC and WA being CP stabilizers upon CP compositions pyrolisis proceeding is analogous. The initial temperature of PE depolymerization increases (Figure 1), the process proceeds with lower speeds and CR outlet increases.

In WA compositions it is connected with joint action of CP and metal oxides (CaO, $K_2O$, $K_2CO_3$) being a part of ash and leading to metal chlorides formation able to inhibit the destruction chain reactions.

Surface structure of coke is also changed. The formation of foaming carbonized residue with high porosity (Figure 2) that lowers coke heat conductivity and helps to reduce a reverse heat flow on polymer is noted.

Data obtained by TGA method let us talk about PE pyrolysis mechanism change in the presence of filling agents which is proved by such materials behavior during combustion. (Table 2) While ignition of such samples in the air mass losses decrease from 62 (initial PE) to 33-9,2% and OI increases from 19 to 28-32%.

PAN refers to coking polymers, the destruction of which has 2 stages divided by temperature interval. On the first stage (248-278 °C) cycle structures formation starts[8], which is proved by the occurrence of exothermal effect according to DTA data. Fireproof polymer has a 20°C less temperature of cyclyzation beginning than initial polymer and this process energy is 103kJ/mole instead of 240 kJ/mole for non-modified polymer and the process is accompanied by a large CR outlet (Table 2). The formed coke is more heat resistant and has less mass losses at temperature exceeding 500C than initial polymer. During modified polymer pyrolysis a general outlet of products decreases and a HCN content reduces in their composition from 507 mkg/mg to 301 mkg/mg, exothermal top values of cyclyzation process decrease and that reduces heat flow towards polymer.

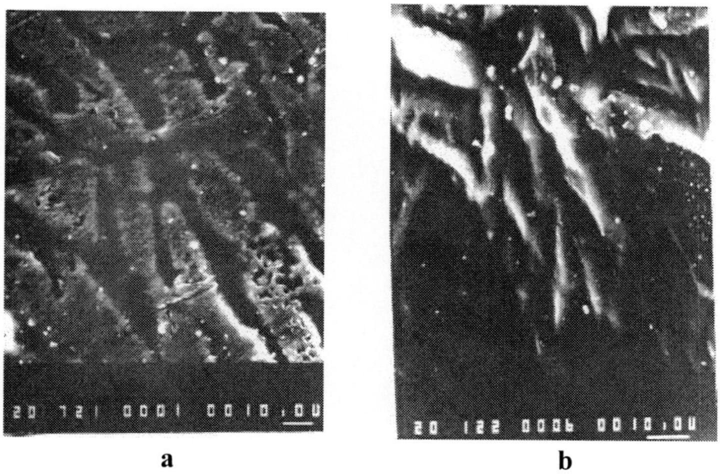


**Figure 2.** SEM data obtained during PE combustion under the following oxygen index conditions: a – polyethylene, b – 100 mass PE+20 mass CP.

Hence, the main contribution of DC is condensed phase processes change. Fireproof polymer is characterized by rather high values of OI (Table 2).

Consequently, data obtained by TGA method together with SPGC (stage pyrolysis gas chromatography) and electronic microscopy let us predict the behavior of polymers during combustion.

## REFERENCES


[1] Bahman N.N., Albaev L.I. Polymer films combustion. Maximum sizes of combustion. / *Physics of combustion and explosion*. 1978. V.14-#3, page 47-52.

[2] Jonson P.R. General correlation of the flammability of natural and scientific polymer./ J.Appl. *Polymer Science and Technology*. 1975. V.10 #1 page 21.

[3] Margolin A.D., Krupkin V.G. Maximum combustion sizes of polymer thin layers. *Physics of combustion and explosion*. 1978. V.14-#2, page 56-63.

[4] Bahman N.N., Albaev L.I. Polymer films combustion. Flame spreading speed. / *Combustion of condensed systems*. Chernogolovka. 1980 page 41-44.

[5] Margolin A.D., Krupkin V.G. The influence of heated layer of condensed phase on combustibility resistance of polymers. / *Chemical physics*. 1982. #7. Page 994-1003.

[6] Aseeva R.M., Zaikov G.E. Polymer materials combustion. Moscow. *Science*. 1981 page 279.

[7] Javonovie Slobodan. Berhalogeni negorivi plastvski kompaund za eneretske Koblove/ *Plastiquma*. 1995 V.15 #3 page 856-857.

[8] Kanovich M.M., Rudenko A.P. Polyacrilnitril conversion in inert and oxidative environment./ *Chemical fibers*. 1983 #3, page 19-21.

In: Chemical Physics of Pyrolysis, Combustion and Oxidation ISBN 1-59454-163-9
Editors: A. A. Berlin et al., pp. 93-96 

*Chapter 9*

# INTERCALATION TECHNOLOGY OF CARBON-, GLASS- AND BASALT-FIBER-REINFORCED PLASTICS

***S. E. Artemenko, J. A. Kadykova and O. G. Vasiljeva***
Saratov State Technical University, Technological Institute, Russia, Saratov region, Engels, sg. Svoboda, 17, fax (845-11), 4-95-84, tel.6-35-18 xt@tehn.renet.ru.

## ABSTRACT

This work contains research results of intercalation technology (IT) of carbon-, glass- and basalt-fiber-reinforced plastics with the usage of monomer mixture of phenol with formaldehyde and NaOH catalyst for polymer matrix formation.
Various methods showed that rather uniform structures of BP,GP and CP are formed as a result of chemical and physical interaction of monomer intercalation into fibre volume. They possess high strength, fire resistance and non-toxicity. Fibre hybridisation in PCM structure provides the possibility of directed regulation of properties and material cost in accordance with functional purposes.

**Key words** Carbon-, glass- and basalt filaments, basalt-, glass- and carbonpolymers, intercalation technology, phenol-formaldehyde oligomer.

Intercalation technology is currently being developed for the production of polymer composite materials (PCM). It is based on the intercalation (introduction) of monomers into the structure of filling agent with the further synthesis of polymer in the pores and on its surface. But for PCM, reinforced with fibers of different chemical nature, many problems, concerning mechanisms of polymer intercalation into the structure of fibers (filaments), their interaction and also the formation processes of such materials are at the stage of solution now.

This work contains research results of intercalation technology (IT) of carbon-, glass- and basalt-fiber-reinforced plastics with the usage of monomer mixture of phenol with formaldehyde and NaOH catalyst for polymer matrix formation.

Carbon-(CF), glass-(GF) and basalt(BF) filaments and phenol-formaldehyde oligomer (PFO) formed during the synthesis have different chemical nature but contain active functional groups. They enter into the condensation reaction with the formation of PCM containing ~50-60% mass. of non-organic component.

For example,

Basalt fiber-Si-OH+R-$CH_2OH$ fiber SiO-$CH_2R$+$H_2O$
PFO

Porous structure of filaments determines their sorptive properties and kinetic and dynamic characteristics of sorption processes as well. According to the size of pores, maximum adsorbed volumes and characteristic energy, the studied filaments form the sequence CF>BF>GF.

The researches establish that the ability to netted structure formation of composites on the BF and CF bases is close. Apparently, cluster (uneven) surface structure of BF increases their specific surface and its sorption capacity, determining ability to form netted structures.

The research of sample sections from obtained materials with the help of scanning electron (SEM) (Hitachi-HU12A) and scanning tunneling (STM) microscopes showed that in carbonpolymer (CP) there is a uniform distribution of polymer film on CF surface and relatively uniform distribution of filaments in the section of material sample. From the study of SEM picture of glasspolymer (GP) sample section it is noted a thicker polymer film on GF surface with characteristic uneven fillings of irregularities in the relief of filament surface compared with CP. Taking into account that the porosity of GF is considerably less than porosity of CF, the greater part of polymer is formed on the filament surface and is a voluminous layer of polymer matrix between filaments. From STM picture of basaltpolymer (BP) cross section it is noted a practical lack of smashed filaments, and a thin polymer film formed on the surface of filaments has a marked orientation by its relief.

Physical-chemical and mechanical properties of PCM obtained with IT are considerably better than the analogous properties of PCM formed by traditional technology of filament impregnation (table 1). Besides, BP is superior GP in all studied characteristics. The important indicator of such materials is a high resistance to burning: oxygen index for BP is 60%, GP – 50%, CP – 70%. The samples don not keep up burning in air. Such materials are attributed to hard to burning ones. These properties are brought into the material structure by the phenol-formaldehyde matrix as well which is a carbon-forming material.

Physical- mechanical properties of CP, GP, and BP formed by IT practically are not changed after two-hour boiling in distilled water. This proves the dense structure of composites, formed by intercalation of monomers into micro defects and pores of filaments with the generation of thin films on its surface and their further solidifying. There is an orientation by relief of pore and filament surfaces.

**Table 1. Comparative characteristics of developed PCM with analogues**

| Polymer composite material | Density, $kg/m^3$ | Shearing breaking stress, MPa | Bending breaking stress, MPa | Modulus of elasticity, GPa | Brinell hardness, MPa | Water absorption, % |
|---|---|---|---|---|---|---|
| Developed by intercalation/ traditional technology | | | | | | |
| basalt-fiber-reinforced plastic | 2030/1710 | 26/22 | 635/520 | 45/37 | 420/376 | 0,21/0,33 |
| glass-fiber-reinforced plastic | 1900/1650 | 28/24 | 400/206 | 28/15 | 400/355 | 0,24/0,38 |
| carbon-fiber-reinforced plastic | 1600/1350 | 23/14 | 840/600 | 56/42 | 632/596 | 0,39/0,55 |
| Russian by traditional (mixture) technology | | | | | | |
| Basalt fabric БТ-1+ЭД-20 | - | - | 430-560 | 33,6 | - | 0,2 |
| Basalt chopped fiber + polyamide binder | 1750 | - | 300 | - | - | - |
| Carbon rope УК-320 (random reinforced)+ЭД-20 | 1400 | - | 190 | - | - | - |
| Carbon fabricЭЛУР-0Б1П+ЭД-20 grade ЭХД | 1350 | - | 240-270 | 18 | - | - |
| Glass fabricЭ1-100+ЭД-20 grade ЭХД | 1950 | - | 300-340 | 27 | - | 0,3 |
| Glass fabric + epoxy-phenol resin | - | - | 100 | - | - | - |

With the help of X-ray diffraction analysis (table 2) it is shown that the crystallinity degree and crystallite size of PCM obtained by IT, are decreased compared with traditionally filled plastics. It is connected with introduction of monomer filaments into the microstructure. The monomer filaments cause some disorder of macromolecules and their orientation especially under the condition of PCM formation at T=155°C and pressure 25 MPa which does not considerably affect the strength characteristics.

**Table 2. X-ray diffraction analysis data**

| PCM | crystallinity degree , % | crystallite size, nm |
|---|---|---|
| basalt-fiber-reinforced plastic | 40/42 | 17,84/25,04 |
| glass-fiber-reinforced plastic | 34/47 | 12,25/29,25 |
| carbon-fiber-reinforced plastic | 77,5/82 | 9,63/9,97 |

**Note:** the nominator - values by IT, the denominator – values by traditional filament impregnation with a resin.

Toxicity study of gaseous substances eliminated at 20°C was carried out by chromatograph "Crystal" (Russia). It testifies that free phenol is absent in samples of BP and CP. The presence of phenol in GP (0,00032 mg/m$^3$) is found, however these values are lower of phenol maximum permissible concentration (0,005 mg/m$^3$).

One of the ways of directed regulation of the PCM properties is the usage of hybrid fibrous fillings. The combination of widely used GF with CF is of great interest. It can increase physical-chemical characteristics of hybrid PCM and give them specific properties. So, the application of hybrid fillings allows to achieve synergism effect and to form PCM with the required set of properties in accordance with their functional purpose by varying of CF:GF ratio (table 3). The cost of PCM is greatly reduced compared with carbon-fiber-reinforced plastics.

**Table3. Comparative characteristics of PCM with hybrid fillings**

| Filling composition | Brinell hardness, MPa | Bending breaking stress, MPa | Water absorption during 2-hour boiling, % |
|---|---|---|---|
| GF | 400 | 400 | 0,24 |
| GF+1 layer CF | 432 | 435 | 0,25 |
| GF+2 layers CF | 447 | 508 | 0,26 |
| GF+3 layers CF | 469 | 542 | 0,29 |
| CF | 632 | 840 | 0,39 |

Comparative analysis of obtained BP,GP and CP with frequently used PCM on the BF, GF and CF bases which are produced in commercial size showed (table1) that developed BP,GP and CP by IT are better in $\sigma_i$ and $E_i$ than well-known analogues.

Various methods showed that rather uniform structures of BP,GP and CP are formed as a result of chemical and physical interaction of monomer intercalation into fibre volume. They possess high strength, fire resistance and non-toxicity. Fibre hybridisation in PCM structure provides the possibility of directed regulation of properties and material cost in accordance with functional purposes.

In: Chemical Physics of Pyrolysis, Combustion and Oxidation ISBN 1-59454-163-9
Editors: A. A. Berlin et al., pp. 97-101 

*Chapter 10*

# MODIFICATION OF WOOD FOR COMBUSTIBILITY REDUCTION

***A. A. Ponomarenko, E. V. Bychkova, L. G. Panova and V. N. Olifirenko***
Technological Institute of Saratov State Technical University
Square of Svoboda, 17, Engels 413100, Saratov region, Russia
Tel.: (84511) 63518, (84511)63553;
Fax.: (84511) 49584, E-mail: *xt@techn.renet.ru*

## ABSTRACT

In the present project an organic chlorine-containing compound (OCC) was used to reduce wood combustibility. Modification was being conducted by means of wood (pine-wood) samples impregnation both of samples being dried for an hour, and air dried ones made from OCC water solution at room temperature in baths containing 20 and 35% OCC of mass. On the basis of sorption kinetics results study the following impregnation parameters are chosen: preliminary thermal treatment; DC concentration in a bath; baths temperature; impregnation. Samples containing modifiers have weight losses 4.2 and 5.7% and OI 29 and 28%, which let us consider these samples as hardly combustible materials.

**Key words:** Wood, Modification, Sorption, Reduction combustibility, Thermal gravimetric analysis method.

Nowadays wood is widely used as a finishing and building material. However, the disadvantage of this regenerative ecologically clean material is its high inflammability and combustibility. In this connection wood materials modification in order to increase its fire resistance becomes a matter of current interest.

In the present project an organic chlorine-containing compound (OCC) was used to reduce wood combustibility.

The samples were made in the form of wooden bars sized 50x50x15 mm and 150x25x4 mm. Modification was being conducted by means of wood (pine-wood) samples impregnation both of samples being dried for an hour, and air dried ones made from OCC water solution at room temperature in baths containing 20 and 35% OCC of mass.

Sorption kinetics was being investigated on samples sized 50x50x15 mm (Figure 1).

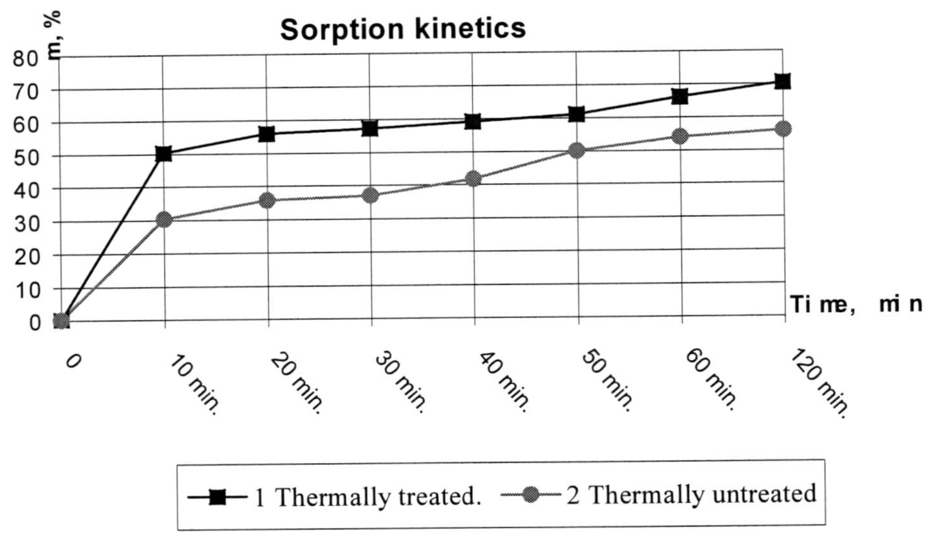


**Figure 1.** Sorption kinetics of OCC by wood.

The maximum content of OCC both on thermally treated and thermally untreated air dry samples is reached in 10 minutes, however, a considerable gain in weight of decelerated combustion (DC) is noted within 60 minutes.

A larger quantity of DC solution is sorbed by thermally treated samples, that is connected with the removal of moisture from samples and additional emergence of micro pores in the wood structure. Equilibrium sorption does not occur within 7 days of experiments. However, taking the process technology into consideration, the impregnation can be conducted within 60 minutes as the further sorption time is increased DC solution content in the samples does not increase considerably.

The content of DC in wood composition after solvent removal has been evaluated (Figure 2). DC mass in samples depends on the thickness of samples, DC content in impregnation bath and on the presence or absence of thermal treatment.

Maximum mass of DC in wood composition is noted in samples having thickness of h=4, preliminary thermal treated during 1 hour, and impregnated by 20% and 35% OCC solutions within 60 minutes. Moreover, decelerated combustion content in a bath does not influence considerably on DC amount in a sample, and DC mass in a wood composition after solvent removal is 60 and 64% accordingly.

Samples having thickness of h=15 mm modified under the same conditions as samples having thickness of h=4, contain considerably less quantity of DC – 23 and 27% after solvent removal. (Figure 2).

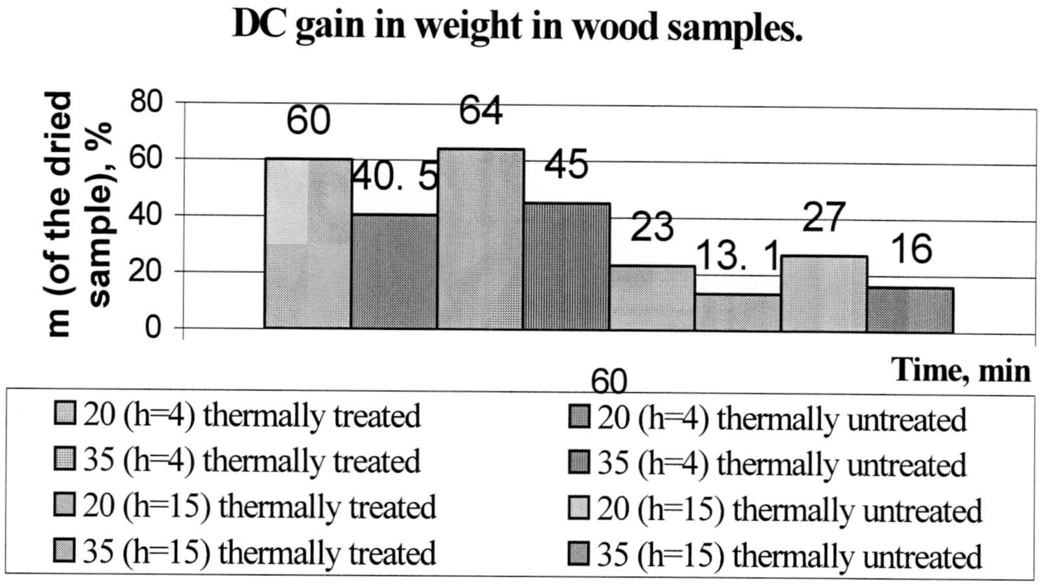


**Figure 2.** DC gain in weight in wood samples.

On the basis of sorption kinetics results study the following impregnation parameters are chosen: preliminary 1 hour thermal treatment; DC concentration in a bath is 35% mass; baths temperature is 20±5C; impregnation duration – 60 minutes.

To define DC preservation in the structure of samples obtained under chosen conditions, the samples were put into distilled water at t=20 ±5C for 60 minutes. The experiments showed that loss in weight of samples was 7.1% mass for thermally treated and 8.3% mass for thermally untreated ones.

The investigation of wood samples behavior at increased temperatures was conducted with the help of thermal gravimetric analysis method (TGA).

On the curves of initial wood 2 top points are noted: 1- in the thermal range of 30-120 $^{o}$C is endothermically conducted with 8.5% loss in weight and corresponds to water content in wood.

Cellulose destruction is conducted in 210-400 $^{o}$C thermal range. Dehydration process proceeding under chain mechanism with cations is actively conducted. As the result of dehydration conjugated non-saturated structures are formed. They form carbonized residue at pyrolysis which increases material hent resistance.

Coke outlet after the destruction of initial wood main stage completion is 18.5%.

OCC used for modification is heat resistant enough. In temperatures range of 40-210 $^{o}$C OCC loses 15% of its mass. General mass loss at 600 $^{o}$C is 26%.

The analysis of modified samples thermal gravimetrics data showed that their decomposition is also conducted on the 2$^{nd}$ stage.

The first stage more likely is connected with the extraction of sorbed water, though we cannot deny OCC decomposition in this temperature range.

The second stage corresponds to the wood destruction. The initiation influence of OCC upon wood decomposition is noted. As the initial temperature of decomposition decreases for 20-30 $^{o}$C in comparison with unmodified wood and the destruction temperature range decreases considerably, the outlet of carbonized residue somehow increases. Coke has a more uniform structure with visible OCC inclusions (Figures 3,4).

**Figure 3.** Surface coke structure of a modified sample

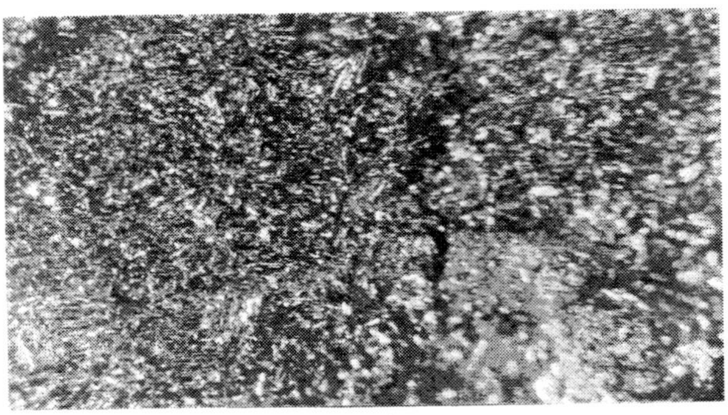

**Figure 4.** Coke structure under removed coke layer.

These data in accordance with cellulose decomposition theory may testify about the change of the decomposition mechanism and about its advantageous proceeding in dehydration mechanism. [1-4].

This is corroborated by combustibility indicators study by methods of "fire tube" and oxygen index.

Weight loss of initial samples at air combustion is 82% mass, and OI is 18%. Samples containing modifiers have weight losses 4.2 and 5.7% and OI 29 and 28%, which let us consider these samples as hardly combustible materials.

As the result of the conducted work the efficiency of OCC usage as a modifying addition agent for wooden materials in order to reduce their combustibility was shown. It was noted that best results were obtained at the impregnation of being preliminary thermally treated for 1 hour wood sample by 35% OCC water solution, and phosphorus content in samples was 6.4% mass. Moreover, the used modifying addition agent shows antiseptic properties.

**Table 1. Pyrolysis indicators of wood samples.**

| Material composition (for 100 mass of wood), mass. | Main stages indicators of destruction | | Loss in weight (%) at T, °C | | | | | |
|---|---|---|---|---|---|---|---|---|
| | $\frac{T_H - T_K}{T_{MAX}}, {}^OC$ | $\frac{m_H - m_K}{m_{MAX}}, \%$ | 100 | 200 | 300 | 400 | 500 | 600 |
| Wood (W) | 210-430<br>350 | 8,5-73,5<br>42,5 | 1,5 | 8 | 15 | 70 | 81,5 | 91,5 |
| OCC | 40-210<br>110 | 0-15<br>6,2 | 4 | 14,5 | 19,2 | 24 | 25 | 26 |
| W+60 OCC (thermally treated) | 190-269<br>250 | 8-39<br>33 | 2,5 | 9 | 45,5 | 61,5 | 75 | 87,5 |
| W+40.5 OCC (thermally untreated) | 200-280<br>240 | 8-42<br>29,5 | 3 | 8 | 47 | 60 | 74 | 87 |
| W+64 OCC (thermally treated) | 185-260<br>240 | 8-41<br>33 | 4 | 11 | 47 | 61 | 73,5 | 87 |
| W+45 OCC (thermally untreated) | 180-260<br>240 | 8-41<br>34 | 5 | 10 | 47 | 60 | 72 | 85 |

**Table 2. Combustibility indicators of wood samples.**

| Material composition (for 100 mass of wood), mass. | Phosphor content in samples, % mass. | Temperature interval of destruction, °C | KO outlet, % at 500 °C | Weight loss at air combustion (Δ m) %(mass.) | OI % |
|---|---|---|---|---|---|
| Wood (W) | - | 220 | 18,5 | 82 | 18 |
| W+60 OCC (thermally treated) | 6,4 | 75 | 26,5 | 4,2 | 29 |
| W+40.5 OCC (thermally untreated) | 4,5 | 80 | 28 | 5,7 | 28 |

## REFERENCES


[1] Aseeva R.M. *Combustion of polymer materials.*/R.M. Aseeva, G.E. Zaikov. Moscow: Science, 1981.-p. 280.

[2] *Decreased combustibility polymer materials.*/ edited by A.N. Pravednikov. Moscow: Chemistry, 1986.- p. 224.

[3] Kodolov V.I. *Combustibility and fire resistance of polymer materials.* Moscow: Chemistry, 1976.-p. 157.

[4] Aseeva R.M., Bui Dinj Thanj, Serkov B.B. Factors influencing the heat emission during wood combustion// *Decreased combustibility polymer materials: Reports Thesis of the 5th International Conference.* Volgograd October, 2003.- Volgograd: VSTU, 2003.-p. 4-5.

In: Chemical Physics of Pyrolysis, Combustion and Oxidation ISBN 1-59454-163-9
Editors: A. A. Berlin et al., pp. 103-109 

*Chapter 11*

# RADICAL (CO)POLYMERIZATION AND PROPERTIES OF METHACRYLIC DERIVATIVE (CO)POLYMERS OF PHOSPHORIC ACIDS

***I. A. Novakov, G. D. Bakhtina and A. B. Kochnov***
Volgograd State Technical University
28 V.I. Lenin Ave., 400131 Volgograd, Russia
Fax: (8442) 344-121
E-mail: *bachtina@vstu.ru*

## ABSTRACT

The data on regularities of radical polymerization and copolymerization of phosphorus-containing methacrylates and the influence of their structure on polymer properties are generalized. New possible pathways of application of polymers derived from phosphorus-containing methacrylates are shown.

***Keywords:*** radical (co)polymerization, phosphorus-containing methacrylates, polymer properties, reduced combustibility

Active phosphorus injection to macromolecules by copolymerization of organophosphorous and industrial monomers is one of the effective methods for combustibility reduction of polymeric materials [1, 2]. Among unsaturated phosphorus-containing monomers, (meth)acrylates are of the highest practical importance [3]. They possess high ability to (co)polymerize under radical polymerization conditions and, in contrast with vinyl and allyl derivatives of phosphoric acids, form high molecular weight products. Their use makes possible production of polymeric materials, possessing high fireproof property and chemical resistance, high thermophysical and mechanical properties.

At present, a series of works devoted to studies of radical (co)polymerization features of phosphorus-containing methacrylates and determination of the influence of their chemical structure on the polymerization rate and properties of derived polymers was published. Some of these results are generalized in reviews [1, 3, 4].

By studies of radical polymerization of phosphorus-containing monomers, it was found that, despite remoteness of phosphoric group from the double bond, its structure affects the polymerization rate. As shown on the example of *β*-methacryl-oyletoxyalkoxy(phenoxy) methylphosphonates and *β*-methacryloyl-*α*-chloromethyletoxyalkoxy(phenoxy) methylphosphonates (alkyl: $CH_3$, $C_2H_5$, *n*-$C_3H_7$), the monomer polymerization rate decreases as the volume of substituting groups at phosphorus atom increases. For phosphorus-containing monomers of the acryl sequence similar regularity is observed. Higher activity in polymerization of three-valence phosphorus is observed [5], all other factors being the same, and monomers with double bonds may polymerize at higher rate rather than monomethacrylates [4]. For studied phosphorus-containing methacrylates, it is found that the reaction order by radical initiator concentration equals 0.5; therefore, the reaction order by monomer may deviate from the theoretical value [6]. The polymerization activity of many studied methacrylic derivatives of phosphoric acids is higher than for methyl methacrylate (MMA). One of the probable reasons for this fact is higher viscosity of phosphorus-containing methacrylates which hinders diffusion of active ends of propagating macroradicals and, consequently, decreases rate constants of the chain termination. Because of this, earlier occurrence of the gel-effect may also be observed for the type of monomers under consideration. The quantum-chemical calculations have proven high importance of steric factors for polymerization of phosphorus-containing methacrylates [7].

Studies of *β*-methacryloyletoxyalkoxy(phenoxy)methylphosphonates and *β*-methacryloyl -*α*-chloromethyletoxyalkoxy(phenoxyl)methylphosphonates (alkyl: $CH_3$, $C_2H_5$, *n*-$C_3H_7$) copolymerization with styrene, MMA copolymerization in bulk and solution with butyl acrylate and in emulsion with vinyl acetate showed the copolymerization rate increase with the part of organophosphorous monomers in the initial mixture. For studied pairs of monomers, copolymerization constants found at low transformations of the reagents are slightly dependent on their structure and fall below 1 [1, 4, 8]. Clearly, this indicates the affinity of monomers to alternation in macromolecular chain formation. Comparison of parameters characterizing ability of unsaturated monomers to polymerize (parameters $Q$ and $e$ in the Alfrey-Price scheme), which were found in the study of phosphorus-containing methacrylates' copolymerization with various vinyl monomers, show that the majority of studied methacrylic derivatives of phosphoric acids are less conjugated monomers than styrene and possess electrically positive double bond. Comparison monomer activities relative to styrene radical (by $1/r_2$ – from data on copolymerization with styrene – $M_2$) indicates affinity of phosphorus-containing methacrylates to methyl methacrylate, somewhat lower or the same level versus diene derivatives of phosphoric acids, and usually exceed vinyl and allyl phosphorus-containing monomers.

The reactivity of phosphorus-containing methacrylates in radical polymerization can be increased by injection of complexing agents, Lewis acids, for example [9]. Besides polymerization rate increase, their application allows obtaining of copolymers with more regular structure.

At the application of organophosphorous monomers in the binder composition for composite material production, of the highest practical importance are phosphorus-containing monomers or oligomers with two (meth)acrylic groups or more. They may be synthesized by processable techniques and work as cross-linking agents at curing. The studies of di(*β*-methacryloyl-α-chloromethyletoxy)alkylphosphates homopolymerization in the presence of azodiisobutyrate dinitrile (DAC) (0.01 – 0.03 mol/l) at 60 - 80°C show the limiting degree of polymerization equal 82.2 – 90.2%, for every monomer slightly dependent on the curing conditions. At the initial stage, the polymerization rate is proportional to square root from the initiator concentration and depends on the monomer structure, the presence of inhibitors and the process degree. From positions of microheterogeneous model of the three-dimensional polymerization of oligoester acrylates, the maximum rate corresponds to initiation of a monolith formation from polymer grains in the system [10]. The values of activation energy obtained increase with the curing degree due to the increase of diffusion factors' influence on polymerization process. Polymerization products possess a complex composition and include a liquid layer identified with unreacted monomers and the polymerization product containing 69.2 – 90.3% gel-fraction with respect to the monomer origin and polymerization conditions. The use of increased amount of a radical initiator or redox system increases the degree of curing and eliminates the liquid phase of unreacted monomers.

Homopolymers and phosphorus-containing monomethacrylate copolymers are usually insoluble but just swell in polar solvents and do not transit to the viscous flow state without degradation. Apparently, this is caused by three-dimensional structure formation by chain transmission reactions to the polymer with participation of alkyl and methylene bridge groups. Moreover, the presence of tetrafunctional cross-linking dimethacrylic compounds in polymerizing monomers must not be ruled out due to insufficient purity of monomers and their possible synthesis in the known disproportionation termination of monomethacrylate derivatives of phosphoric acids. This circumstance imposes some limits on the possibility of application of phosphorus-containing methacrylate (co)polymers. As monomethacrylates are polymerized in solution and in bulk, at low transformation degrees, soluble products may be obtained. Characteristic viscosity obtained for soluble phosphorus-containing methacrylate copolymers with styrene (in dimethylformamide solution) equals 0.20 – 0.45 and confirms their high-molecular structure. The study of physicomechanical properties of $\beta$-methacryloyletoxyalkoxy(phenoxy)methylphosphonate homopolymers obtained by radical polymerization in block indicated their low thermal resistance and mechanical strength.

Physicomechanical properties of phosphorus-containing monomethacrylate (co)polymers essentially depend on the comonomer structure. For $\beta$-meth-acryloyletoxyalkoxy(phenoxy, hydroxy)methylphosphonate and $\beta$-methacryloyl-$\alpha$-chloromethyletoxyalkoxy(phenoxy)-methylphosphonate copolymers with styrene and methyl methacrylate, it is found that the following events happen in the substance as the concentration of phosphorus-containing units in copolymer macromolecules increases: water absorption of samples and coke residue in thermal resistance tests performed by thermogravimetric technique (under dynamic conditions up to $T$ = 600°C) increase; glass transition temperature, thermal resistance and initiation temperature of thermooxidative degradation decrease; mechanical strength of the samples decreases; and relative elongation at break slightly increases (Table 1). As heated in air, with respect to the composition and structure, phosphorus-containing methacrylate copolymers begin degrading at 160 – 240°C.

**Table 1. Properties of $\beta$-methacryloyletoxyalkoxy(phenoxy)methylphosphonate block copolymers (alkyl: $CH_3$ (MPEM), $C_2H_5$ (EPEM), $n$-$C_3H_7$ (PPEM); phenyl (PhPEM)) with styrene and MMA**

| Copolymer | POM amount in polymers | | Water absorption per 1 day, wt.% | $T_g$, °C | Heat resistance | | | Mechanical characteristics | | Fireproof by "fire tube" technique | |
|---|---|---|---|---|---|---|---|---|---|---|---|
| | mol% | wt.% | | | $T_0$, °C | $T_{10\%}$, °C | $T_{50\%}$, °C | $\sigma$, MPa | $\varepsilon$, % | Self combustion, s | Mass loss, % |
| MPEM-styrene | 20.00 | 34.80 | 0.46 | 82 | 210 | 265 | 345 | 32 | 2.0 | < 1 | 1.8 |
| EPEM-styrene | 20.00 | 36.20 | 0.54 | 80 | 205 | 265 | 345 | 30 | 2.2 | < 1 | 1.8 |
| PPEM-styrene | 20.00 | 37.54 | 0.58 | 78 | 205 | 260 | 340 | 30 | 2.5 | ~1 | 2.2 |
| PhPEM-styrene | 20.00 | 40.57 | 0.65 | 75 | 200 | 260 | 340 | 32 | 2.5 | ~1 | 1.9 |
| ditto | 40.00 | 64.54 | 1.01 | 60 | 180 | 220 | 325 | 25 | 6.0 | < 1 | 1.0 |
| ditto | 60.00 | 80.37 | 2.20 | 40 | 175 | 200 | 315 | 15 | 40.0 | < 1 | 0.5 |
| ditto | 80.00 | 91.61 | 3.10 | 30 | 170 | 190 | 280 | 5 | 75.0 | < 1 | 0.4 |
| PhPEM-MMA | 20.00 | 41.52 | 1.15 | 80 | 200 | 255 | 355 | 34 | 2.0 | < 1 | 2.2 |
| ditto | 40.00 | 65.43 | 2.18 | 62 | 180 | 255 | 310 | 26 | 6.0 | < 1 | 1.1 |
| ditto | 60.00 | 80.98 | 2.90 | 40 | 180 | 220 | 305 | 16 | 70.0 | < 1 | 0.6 |
| ditto | 80.00 | 91.91 | 3.30 | 30 | 175 | 215 | 300 | 6 | 85.0 | < 1 | 0.4 |
| Polystyrene | - | - | 0.05 | 90 | 230 | 325 | 375 | 35 | 2.0 | Burn off completely | |
| PMMA | - | - | 0.20 | 100 | 230 | 290 | 335 | 41 | 4.0 | Burn off completely | |

Apparently, a decrease of some physicomechanical indices as the concentration of phosphorus-containing units increases is associated with the plasticizer influence of bulky phosphorus-containing substituting groups at carbon atoms of double bonds in methacrylic monomers. Glass transition temperature, heat and thermal resistance, and mechanical properties of polymers decrease as the volume of substituting groups at phosphorus atoms at equimolar relation of components in copolymers increases.

The effective initiation of phosphorus-containing dimethylacrylate homopolymerization induces synthesis of highly cross-linked polymers possessing high physicomechanical properties. Copolymerization of phosphorus-containing di-$\beta$-methacryloyl derivatives of phosphoric acids with styrene and methyl methacrylate in the presence of DAC (0.5 – 1.0 wt.%) in bulk with 20 – 30 mol% phosphoric components in copolymers produced self-extinguishing polymers with 30 – 50 MPa tensile strength and 110 - 135°C Vicat softening temperature. They possess low water absorption and exceed polystyrene and poly(methyl methacrylate) by grease resistance and resistance to nonpolar organic solvent action. Studied phosphorus-containing mono- and dimethacrylate copolymers with styrene possess higher dielectric constants versus poly(methyl methacrylate) which are then lower versus polystyrene.

One of the problems in synthesis of copolymers is determination of the composition that provides a definite decrease of combustibility and preserves high properties of the modified polymer. It is common knowledge that fireproof of polymers depends on their chemical structure and composition. Analysis of the present data on polymer fireproof tests performed by "fire tube" technique for studied phosphorus-containing methacrylates classifies copolymers with styrene and MMA containing 4% phosphorus or more as incombustible materials. They immediately self-extinguish after flame removal (Table 1). Oxygen index of these copolymers exceeds 23.5%. To provide for this fireproof level, at least 35 – 40 wt.% (~20 mol%) of phosphorus-containing methacrylate units are added to copolymers. Hence, for block copolymers the satisfactory level of physicomechanical characteristics, close to polystyrene and poly(methyl methacrylate) indices, is preserved.

One of perspective applications of phosphorus-containing methacrylates is binders of composite materials, in particular, for structural fiberglass of reduced combustibility, and in compositions of filling and impregnating compounds for electrical engineering purposes. Therefore, they may copolymerize with unsaturated polyesters or other unsaturated monomers or polymerize in other curing compounds, epoxy oligomers, for example. Polyester resins modified by unsaturated derivatives of phosphoric acids are cured by organic peroxides combined with various booster compounds (amines, metal salts of variable valence). Initiator system compositions are developed which allow composite curing at room temperature [11]. This gives an opportunity to prepare large-scale polymeric articles without additional costs.

The studies of phosphorus-containing monomers structure influence on properties of oligoestermaleinatephthalate binders modified by them in the acrylate sequence show that such dimethacrylates serving as active cross-linking agents are the most effective for copolymerization. Their application gave binders with reduced combustibility and a complex of properties similar to cured unmodified oligoestermaleinatephthalate resins (Vicat softening point 150 - 170°C, Brinell hardness 160 – 180 MPa, water absorption during 30 days 1.3 – 2.5%, initiation temperature of thermooxidative degradation 160 - 175°C). Polyesters modified by dimethacrylate derivatives from methyl-phosphonic acid and alkyl(aryl)phosphoric acid possess similar properties [4]. Fiberglass obtained under application of oligoestermaleinatephthalate binder modified by di($\beta$-mathacryloyl-$\alpha$-chloromethyletoxy)methylphosphonate, possessing 2 wt.% linked phosphorus and the filling degree of 49 – 52%, displayed calorimetric index equal 0.55. The self-extinguishing property of the material estimated by shipbuilding branch technique is below 25 mm. Chemical modification of unsaturated polyesters by phosphazene oligomers with methacrylate end groups allows an increase of the initiation temperature of thermooxidative degradation to 180 - 190°C. Copolymerization of oligoestermaleinatephthalates with organophosphorous oligoesteracrylates possessing aromatic fragments produces polymers with somewhat increased hardness (up to 200 – 220 MPa). Phosphorus-containing dimethacrylates are of interest as binders for decorative facing, polymer mortar construction materials (tiles and coatings) and active diluters of rubber mixture modification for obtaining chemically resistant polymeric materials with reduced combustibility. Copolymers and composite materials derived from phosphorus-containing methacrylate monomers are of interest for aircraft engineering, shipbuilding, vehicle building, construction, and electrotechnical engineering.

## REFERENCES

[1] Shulyndin S.V., Levin Ya.A., and Ivanov B.E., 'Copolymerization of unsaturated organophosphorous monomers', *Uspekhi Khimii*, 1981, vol. **L**(9), pp. 1653 – 1677. (Rus)

[2] Vinogradova S.V. and Andreeva M.A., 'Phosphorus-containing polymers', In Coll.: *The Progress In Synthesis Of Elementorganic Polymers*, Moscow, Nauka, 1980, pp. 97 – 118. (Rus)

[3] Khardin A.P., Kargin Yu.N., and Bakhtina G.D., 'Phosphorus-containing (meth)acrylate polymers', In Coll.: *Combustibility Of Polymeric Materials*, VolgTGU, Volgograd, 1987, pp. 197 – 207. (Rus)

[4] Novakov I.A., Bakhtina G.D., Kochnov A.B., and Grekov L.I., 'The features of phosphorus-containing methacrylates (co)polymerization and polymeric materials derived from them', In Coll.: *The Prospect Of Modern Chemistry. Synthesis And Modification Of Polymers*, Moscow, Khimia, 2003, 356 p. (Rus)

[5] Lenin A.S. and Kargin Yu.N., 'On radical copolymerization of methacrylates containing tri- and pentavalent phosphorus', In Coll.: *Chemistry And Technology Of Elementorganic Intermediates And Polymers*, VolgTGU, Volgograd, 1981, pp. 154 – 158. (Rus)

[6] Khardin A.P., Kargin Yu.N., Bakhtina G.D., Lenin A.S., and Gaidukov V.A., 'On polymerization of $\beta$-mathacryloyl-$\alpha$-chloromethyletoxyphenoxymethylphosphonate', *Vysokomol. Soed.*, 1981, vol. **B23**(9), pp. 678 – 680. (Rus)

[7] Novakov I.A., Litinsky A.O., Grekov L.I., and Bakhtina G.D., 'The features of intermolecular interaction in the sequence of phosphorus-containing methacrylates', *Volgograd State University Mercury, Ser. 1, Mathematics. Physics, Iss.* 7, 2002, pp. 103 – 110. (Rus)

[8] Kryukov N.V., Bakhtina G.D., and Kochnov A.B., '*On reactivity of methacrylic derivatives of phosphoric acids in homo- and heterophase copolymerization*', 1986, Cherkassy, 5 p., Dep. to ONIITEkhim, No. 570-hp. (Rus)

[9] Khardin A.P., Kargin Yu.N., BakhtinaG.D., and Gaidukov V.A., 'Complex radical copolymerization of phosphorus-containing methacrylates', In: *Proc. VII All-Union Conference On Chemistry of Organophosphorous Compounds*, Leningrad, 1983, p. 245. (Rus)

[10] Korolev G.V., Mogilevich M.M., and Golikov I.V., *Network Polyacrylates. Microheterogeneous Structures, Physical Networks, Deformation And Strength Properties*, Moscow, Khimia, 1995, 274 p. (Rus)

[11] Bakhtina G.D. Tuzhikova O.I., and Kryukov N.V., 'Copolymerization of unsaturated polyesters with phosphorus-containing monomers', *Plasticheskie Massy*, 1987, No. 12, pp. 15 – 17. (Rus)

In: Chemical Physics of Pyrolysis, Combustion and Oxidation ISBN 1-59454-163-9
Editors: A. A. Berlin et al., pp.111-118 

*Chapter 12*

# VARIATION OF INHIBITED OXIDATION PARAMETERS AT MELTING OF POLYETHYLENE

***Yu. A. Shlyapnikov and N. N. Kolesnikova***
Institute of Biochemical Physics, Russia Academy of Sciences,
119991, Moscow, Russia

## ABSTRACT

Oxidation of low-density polyethylene by molecular oxygen has been studied in the temperature range of 80° to 210° C. The curves of temperature dependencies of several parameters of inhibited oxidation the breaks were observed at the melting temperature, near 110° C. At this temperature the pre-exponential factor calculated from temperature dependence of the rate constant of inhibitor consumption changes $10^5$ times, sharply increasing at transition from the polymer melt to "solid" polymer. This indicates that the essential difference exists between properties of intercrystalline amorphous polymeric substance and completely molten one.

**Keywords:** antioxidant, melting, oxidation, polymer, polyethylene

Polymeric substance differs from its low-molecular analogs in several characteristics, the most essential for investigation of the polymer oxidation processes are virtually zero pressure of the polymer vapors and combination of extremely low transition mobility of its macromolecules with high vibrational mobility. The polymer melting proceeds in relatively wide temperature range in which amorphous substance coexists with the crystalline one. It was interesting to compare this intercrystalline amorphous substance with completely molten polymeric one.

In this study we used the simplest polymer – low density polyethylene. As the method of investigation we used inhibited oxidation of the polymer. In the temperature range of 80° to 210° C which includes the melting temperature of this polymer (110° C). Antioxidant was 2,2'-methylene-bis(4-methyl-6-methylcyclohexylphenol) (MCBP). Relatively high molecular mass of this antioxidant decreased the role of the antioxidant evaporation in the course of oxidation. On the other side, difficulties of quantitative extraction of this heavy antioxidant caused us to use the indirect method: to calculate the rate constants of antioxidant consumption from the concentration dependence of induction period of inhibited oxidation.

In polymeric substance substantially increases the role of cage effect compared to its low-molecular analogs which results in decrease of the yield of hydroperoxides from approximately 100% in hydrocarbons to 5 – 30% in polyolefins [1, 2]. The irregularities in polymer chain arrangement in polymers are stable and may be considered as the second component dissolved in the polymeric substance [3]. In crystalline polymers below its melting temperature the crystalline structures appear? which content increases with decreasing temperature. In some cases these structures behave as free radical traps (more correctly – the traps for free valences). It was interesting to understand whether intercrystalline amorphous substance differs in its properties from the same substance in completely molten polymer.

The main steps of hydrocarbon polymer oxidation are analogous to those of low-molecular hydrocarbons::

$$R\bullet + O_2 \rightarrow RO_2\bullet \quad (1)$$

But in the second step $RO_2\bullet + RH$ we see the marked difference. When in the liquid state the molecular motion rapidly separates the primary products of this reaction and nearly all hydroperoxide groups formed are preserved. In polymers these products are either neighboring groups of the same macromolecule or are present on the neighboring molecules. In both cases these groups are for a long time situated aside and interact with high probability, for example as:

$$ROOH + R\bullet \rightarrow RO\bullet + ROH \quad (2a)$$

As the result the yield of hydroperoxide per oxygen consumed decreases from 1 (100%) to much smaller value and depends on oxygen concentration because reacting with $R\bullet$ oxygen transforms these radicals into $RO_2\bullet$ which cannot decompose hydroperoxide groups [1, 2]. It is more convenient to describe the process by scheme:

$$RO_2\bullet + RH \rightarrow \alpha\, ROOH + R\bullet \quad (2)$$

Usual antioxidants, alkylated phenols and aromatic amines, more rapidly react with molecular oxygen than hydrocarbons and monomeric units of the polymer [1, 4], and the initiating step in the oxidation process becomes the interaction of antioxidant with oxygen. Due to the cage effect the most free radicals formed in this process are disappearing in reactions with the same antioxidant, the process may be written as [5]:

$$IH + O_2 \rightarrow I\bullet + HO_2\bullet \text{ (the rate constant } k_o) \quad (3)$$

With subsequent interaction of $HO_2\bullet$ with monomeric unit RH:

$$HO_2\bullet + RH \rightarrow H_2O_2 + R\bullet \quad (4)$$

Initiating the polymer oxidation, and with antioxidant IH:

$$HO_2\bullet + IH \rightarrow H_2O_2 + I\bullet \tag{5}$$

Thus one elementary step of the reaction of antioxidant with molecular oxygen results in consumption of several f > 1 antioxidant molecules. A small part of radicals involved into this process react with the monomeric units RH initiating the polymer oxidation:

$$HO_2\bullet + RH \rightarrow H_2O_2 + R\bullet, \tag{6}$$

The kinetical chains initiated by these radicals are terminated by the same antioxidant:

$$RO_2\bullet + IH \xrightarrow{k_t} ROOH + I\bullet \tag{7}$$

Thus reaction (6) does not change the value of f. The most radicals of antioxidant I• transform into the product of their condensation which can retard the oxidation in the absence of initial form of antioxidant [1, 6].

Chain branching proceeds at interaction of hydroperoxide groups with adjacent monomeric units:

$$ROOH + RH \rightarrow RO\bullet + R\bullet + H_2O \rightarrow \ldots \rightarrow \sigma R\bullet + \text{products} \tag{8}$$

In developed chain reaction the main source of free radical is not chain initiation, but chain branching. If the rate of chain branching exceeds that of chain termination the process remain self-accelerated, non-stationary, If the rate of chain termination is equal or greater than that of chain branching the process remains stationary, and its rate after initial period is the unique function of reaction conditions: concentrations of antioxidant, oxygen and monomeric units. The border between non-stationary and stationary oxidation corresponds to the critical antioxidant concentration. According to [1, 7] its meaning is equal to:

$$[IH]_{cr} = \frac{\alpha\sigma k_2[RH]}{(1-\sigma)k_t} \tag{9}$$

As seen from (9), the expression for $[IH]_{cr}$ contains the ratio of two rate constants and the product of two coefficients each of which may depend on temperature.

If above critical concentration antioxidant is consumed mainly in reactions of direct oxidation (3) – (5), which follow the first kinetical order according to antioxidant, we may write:

$$-\frac{d[IH]}{dt} = k_{eff}[IH], \tag{10}$$

where the constant $k_{eff}$ is equal to the rate constant of the reaction $IH + O_2$ ($k_0$) multiplied by f – the number of antioxidant molecules consumed in reactions (4) – (6), following (3) [1, 3].

When in the course of oxidation the current antioxidant concentration decreases to its critical meaning the reaction rate rapidly increases and remaining antioxidant is consumed in the short time period equal to $[IH]_{cr}$. Calculated according to this assumption dependence of induction period on initial antioxidant concentration $[IH]_o$ will be [1, 7]:

$$\tau = \tau_o + \frac{1}{k_{eff}} \ln \frac{[IH]_o}{[IH]_{cr}} . \quad (11)$$

This formula (11) may be used for calculation of the oxidation parameters of inhibited oxidation, namely critical antioxidant concentration and apparent (effective) rate constant of antioxidant consumption. So rate constant $k_{eff}$ is:

$$k_{eff} = \frac{\Delta \ln [IH]_o}{\Delta \tau} , \quad (12)$$

and critical concentration corresponds to the sharp bend in the curve of induction period as function of initial antioxidant concentration. This constant may be written as $k_{eff} = A_o exp(-E_a/RT)$, where $A_o$ is pre-exponential factor.

Taking into account that specific structure of polymeric compound above and below the range of the polymer melting essentially differ, the goal of our work was to study the changes of inhibited oxidation parameters at transition from the completely molten polymer to partially crystallized one. It was especially interesting to elucidate how changes the rate of oxidation of antioxidant dissolved in the polymer.

## Experimental

Low density polyethylene, $M_W = 55700$, and antioxidant 2,2'-methylene-bis(4-methyl-6-methylcyclohexylphenol) (MCHP) were used. Polyethylene has been precipitated from m-xylene and washed with alcohol, MCPH was precipitated from alcohol solution. The samples were prepared by mixing the powders with subsequent adding small amounts of alcohol, which was then evaporated. The samples 0.05 grams by weight were oxidized in static vacuum device [1] at oxygen pressure 150 mm Hg (which approximately corresponds to its partial pressure in atmosphere).

## Results and Discussion

At Fig. 1 the curves of oxygen consumption and accumulation of groups –OH and >CO at oxidation of the sample containing $1.25 \cdot 10^{-4}$ mol/kg MCHP at 100° C are shown. Oxidation of this sample proceeds with marked induction period, and the time corresponding to increase of all these values coincide. According to the theory [1, 7] this time corresponds to the moment at which the antioxidant concentration, decreasing in the course of oxidation, reaches its critical meaning.

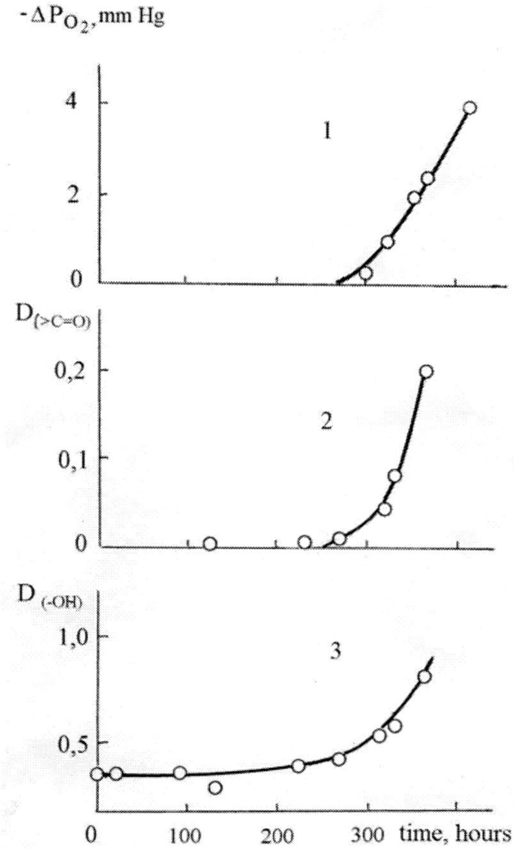


**Fig. 1.** Oxygen consumption (1), accumulation of carbonyl (2) and hydroxyl groups (3) ()in units of optical density at 1700 $cm^{-1}$ for =CO and 3600 $cm^{-1}$ for -OH) during oxidation of polyethylene, containing MCHP, $1.25\times10^{-4}$ mol/kg. 100° C, oxygen, 150 mm Hg.

According to (11), the induction periods of polyethylene oxidation vary with initial antioxidant concentration according to the logarithmic law (Fig.2). This enabled us to calculate dependence of the rate constant $k_{eff}$ on the oxidation temperature. As seen from Fig. 3, in both ranges, below and above the polyethylene melting point (XXX) with the activation energy 36.5 kcal/mol (153 kJ/mol) below and 30.0 kcal/mol (126 kJ/mol) above the melting point. The most interesting was considerable variation of pre-exponent, the decimal logarithm of which decreases from 15.6 to 10.0 ($c^{-1}$) at the polymer melting.

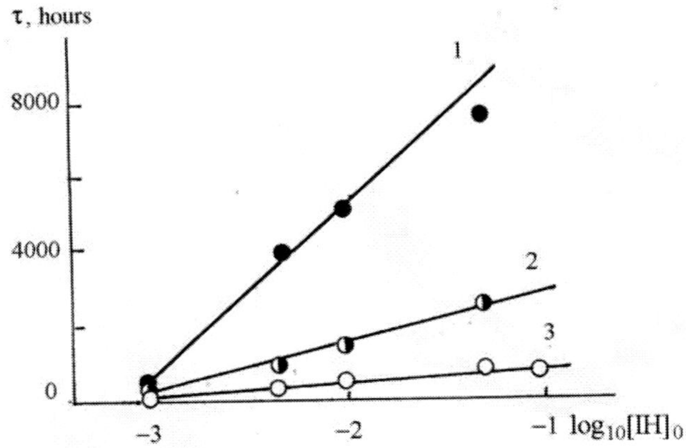


**Fig. 2.** Induction period of polyethylene oxidation as function of logarithm of initial antioxidant concentration at 80° (1), 90° (2), and 100° C (3).

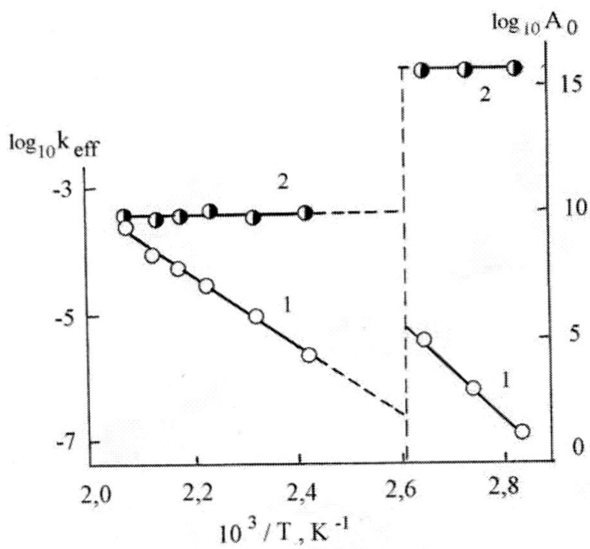


**Fig. 3.** Dependence of apparent rate constant of antioxidant consumption (1) and pre-exponential factor of this constant (2) on temperature in coordinates of Arrhenius equation ($\log_{10} k_{eff}$ vs 1/T). Dotted line shows the polymer melting temperature.

Similarly changes at transition over the melting temperature the critical concentration of the antioxidant which we determined from transformation of the curves of dependence of induction period on initial antioxidant concentration. Opposite to $k_{eff}$, the apparent activation energy calculated for $[IH]_{cr}$ in the solid polymer was lower than in the polymer melt: 3.0 kcal/mol (12.6 kJ/mol) and 11.5 kcal/mol (48.2 kJ/mol) correspondingly (Fig. 4). The expression for critical antioxidant concentration contains several rate constants and we cannot attribute observed change to any of these.

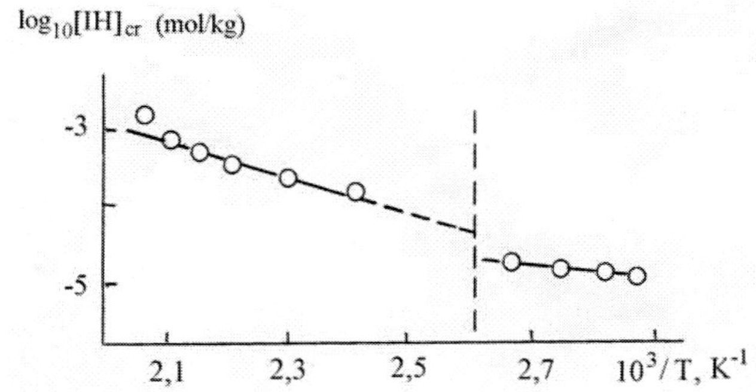


**Fig. 4.** The temperature dependence of antioxidant critical concentration on polyethylene in coordinates of Arrhenius equation.

According to (3) - (6) the reaction of interaction of antioxidant with oxygen is followed with several secondary reactions. Taking into account these reaction we must write

$$k_{eff} = k_o f [O_2] \tag{12}$$

where f is the number of antioxidant molecules consumed in primary and secondary reactions. According to (3) – (6), f = 3. Change of the local IH concentration at transition over the melting temperature can not change the rate constant $k_{eff}$ which is first order in antioxidant concentration. The value of f also cannot markedly change even if the consequence of secondary reactions differs from our assumption.

The greater activation energy of antioxidant oxidation below melting temperature of the polymer may be explained taking into account the sorption mechanism proposed in [3, 8]. The molecules of antioxidant dissolved in a polymer are mainly present in the sorption centers $Z_a$ , the zones of short range violation containing free volume which fits in size and shape for the antioxidant molecule, i.e. PCHA. Below the melting temperature of the polymer in its amorphous phase crystalline formations appear which bind the polymer chains and rearrangement of the center necessary for formation of the reaction complex needs some additional energy,. It is much more difficult to explain the great (on fife decimal orders) increase of pre-exponential factor at transition from the polymer melt to partially crystallized gel-like substance. The authors think that discussion of this phenomenon is now premature.

With appearance of elements of crystalline phase below the melting temperature non-crystallized part of the polymer transforms from the homogenous melt into gel-like substance.

## REFERENCES

[1] Shlyapnikov Yu.A., Kiryushkin S.G., Mar'in A.P. , *Antioxidative stabilization of polymers*. London, Taylor & Francis, 1996

[2] Monakhpova T.V., Bogaevskaya T.V., Gromov B.A., Shlyapnikov Yu.A., *Vysokomolek. Soedin*., 1974, v. B16, p. 91.

[3] Shlyapnikov Yu.A., *Russian Chemical Reviews*, 1997. T. 66. p. 963.

[4] Shlyapnikov Yu.A., Miller V.B., Torsueva E.S., *Izvestiya Akademii Nauk SSSR*, 1961, p. 1966.

[5] Tyuleneva N.K., Shlyapnikov Yu.A.// *Polymer Degrad. Stability*. 1995. v. 47, p. 257.
[6] Shlyapnikov Yu.A., Miller V.B., Neyman M.B., Torsueva E.S., *Vysokomolek. Soedin.*, 1963, v. 5, p. 1507..
[7] Shlyapnikov Yu.A. in *Developments in Polymer Stabilisation*, Ed. G. Scott, London, Applied Science Pub., 1982, v.5, p. 1.
[8] Shlyapnikov Yu.A., *Kinetika i kataliz*, 1978, v. 19, p. 503..

In: Chemical Physics of Pyrolysis, Combustion and Oxidation ISBN 1-59454-163-9
Editors: A. A. Berlin et al., pp.119-124 

*Chapter 13*

# PHYSICAL-MECHANICAL PROPERTIES CHANGE IN A POLYAMIDE FILAMENT UNDER OZONE INFLUENCE

***O. V. Alexeeva*[1*]*, M. L. Konstantinova*[1]*, S. D. Razumovskii*[1]*,***
***V. I. Solodilov*[2]*, A. M. Kuperman*[2] *and G. E. Zaikov*[1]**
[1]Institute of Biochemical Physics of Russian Academy of Sciences,
119991 Moscow, str. Kosigina, 4,Russia
[2]Institute of Chemical Physics of Russian Academy of Sciences,
119991 Moscow, Leniskyi Avenue, 38, Russia

## ABSTRACT

At contact of ozone with the polyamide filaments (PA) the ozone reacts with it rather fast. Study the topochemistry of reaction showed that it proceeds in surface nanolayer ($10^{-9}$м). After treatment only small amount of PA was changed, but many properties of a filaments essentially vary. Physicochemical properties of PA filaments were measured during its reaction with ozone-oxygen mixture. The reaction is accompanied by considerable increase of hardness (up to 13 %), modulus of elasticity (up to 25 %), increase of moisture capacity and inching of elasticity.

**Keywords:** polyamides, polyamide fibers, nanolayers, hardness

## INTRODUCTION

The polyamides are one of the best constructional thermoplastics [1]. They have high mechanical hardness and elasticity in wide temperature range, high fatigue resistance and small creep. The resistance of polyamides to alkalis allows them to work in contact with electrolytes, the stability to oils and benzene provides wide application them in auto an

---

* Correspondence to: Alexeeva Olga, Institute of Biochemical Physics of Russian Academy of Sciences, 117334 Moscow, str. Kosigina, 4,Russia. E-mail: *alexol@yandex.ru*
Contract/grant sponsor: Moscow Government, Department of Science and Industrial policy, Public Corporation "Moscow committee by science and technology "; contract/grant number: ГА 150-03

aircraft construction, to carbons and petroleum - in an oil refining and oil-extracting industry. With each year the volume of production of polyamides is increased. The wide application was found by polyamides PA-6, PA-6,6, PA-12 and PA-610, on the basis of which the number of compositions with various properties intended for usage in many areas of a world industry. Already more than 60 years the filaments from a polyamide - 6,6 are applied in a textile industry. Polyamide - 6,6 and polyamide - 6 discover wide application for manufacture of sports knitted products, of underwear and other items of clothes [2]. However so wide area of usage of polyamides is stipulated by their preliminary modification by chemical, physical or their combined methods [3,4], for improvement of consumer properties of polyamide filaments or films in different range of application. The processes occurring in a material under various modifying agents influence are caused with the deserved interest.

## EXPERIMENTAL

As objects of research the complex polyamide filament PA-6 was selected. The strand consists of 50 filaments by width 30 microns. The line density of a filament was 100 Tex. The filaments were treated by ozone-oxygen mixture. Ozone was obtained by passing the stream of oxygen through a zone of electric discharge (4 kV) with the rate 10-100 ml/min. Concentration of ozone measured spectrophotometrically by change of optical density of a gas stream at $\lambda$=254nm [5]. The special reactor was used where samples of filaments were anchored on a spinned console and treated by ozone-oxygen mixture at constant rotation speed. Uniform effect of ozone on a surface of a polymeric filament thus was provided. The depth and rate of the reaction judged by difference in concentrations of ozone at inlet and outlet of the reactor. Time of ozonation of each series of samples was: 3, 6 9, 12, 20, 24, 36 min, by 6 samples in each series, concentration of ozone - $1{,}44 \cdot 10^{-3}$mol/l.

Width of a bundle and separate filaments during ozone treating controlled through a microscope “POLAM S - 111”. For this purpose were measured the width of initial untreated samples, then them seated in a reactor and in accordance with ozone processing samples were extracted from a reactor with ozone-oxygen mixture and were measured of their width. The deep changes in a composition of filaments surface layer under ozone action have influenced on their waters sorption. Change in water sorption by polyamide filaments, depending on an amount of absorbed ozone, determined on a difference between wetted in water filaments weight, both initial bundle, and treated by ozone bundle at room temperature. The effect of ozone could influence on mechanical characteristics of polyamide filaments. For studying it influence we produced a number of experiment on expansion of polyamide filaments [3]. Measuring of mechanical characteristics of filaments was carried out at room temperature on a testing machine INSTRON with the digital recording of the diagram of a stretching. The stretching rates at any kind of trials remained stationary values were $5 \cdot 10^{-4}$ m/s. The treated and untreated filaments were cut on samples of length 0,17 m. Length of a working part was 0,15 m. The strength of researched filaments at tearing $\delta_P$, their initial modulus of elasticity E and limiting deformations $\varepsilon_K$, settled up from the diagrams of a stretching [6,7]. The strength at tearing up $\delta$ settled up under the formula: $\delta$ = P:S, where P - strength, S - cross sectional area of a filaments bundle; an modulus of elasticity E defined by a linear part of the diagram of expansion of a filaments bundle: $\delta$=f ($\varepsilon$).

## RESULTS AND DISCUSSION

In Tab. 1 you can see the data of width change of a bundle and separate filaments under ozone influence. Visible, that during the reaction of a material with ozone it undergoes essential changes. While the width of a separate filament remains to stationary value, the width of a bundle in the beginning grows, and then starts to decrease. The most reasonable explanation to this fact is, that the relative affinity of filaments in a bundle in the beginning is reduced, and then grows. In the long term it could allow to control by the process of modification and to influence quality of products from such filament. The similar method was used for modification of a wool surface [8].

**Table 1 Changes of a bundle width and separate filaments under ozone action.**

| | | | | | | |
|---|---|---|---|---|---|---|
| Width of a filament bundle (micron) | 940 | 970 | 1070 | 980 | 960 | 880 |
| Width of a separate filament (micron) | 40 | 30 | 30 | 30 | 30 | 30 |
| Time of ozonization (min) | 0 | 45 | 90 | 105 | 120 | 135 |

Earlier was shown [9], that measure of absorbed ozone quantity during PA filaments and films and on the basis of the previous researches in the field of absorption of ozone by polymers [10], it is possible to judge that only surface nanolayer by width 4-6 monolayers ($1{,}2\text{-}1{,}4*10^{-9}$m) take part in the reaction. The stopping of reaction is observed after expending reactive groups in this surface layer, an amount of the connected ozone during the reaction does not vary a lot. As a conclusion it is possible to say that the process goes on a surface of a sample in main.

The influence of ozone on a surface of polymeric filaments and films is accompanied by change of their sorption characteristics [11-14]. Polyamides are not exception. In a fig. 1 shown the change data of water sorption by a polyamide filament, as a function from an amount of absorbed ozone. One can see that at the beginning the rise of sorption, and then the sharp drop is observed. Apparently, at a small amount of absorbed ozone the function groups, generated in surface layer, have a lyophilic nature, at further ozone processing they turn to more hydrophobic.

Strength properties of PA filaments have been researched (table №2, 3 and 4). The obtained outcomes deserve the special attention. So at depth of penetration for about some nanosize layers ($10^{-9}$) and small expenditures of ozone the noticeable effects in change of physical-mechanical properties are received. From the given tables follows, that strength characteristic not monotonically vary. At first the hardness of filaments accrues up to a particular limit and then occurs it droop as a corollary of penetrating changes in material nature. In some cases the change of main mechanical characteristics of a sample material under ozone influence is observed. The similar examples are given in works [15-18], where the increase of strength of the carbon filament, marked by the authors after ozone processing was circumscribed. The ability of a filament to extend (Tab. 2), long enough remains without changes and only to the end of the process is raising by ~10 %. On a background of changes of other properties of a filament this quality seems rather stable.

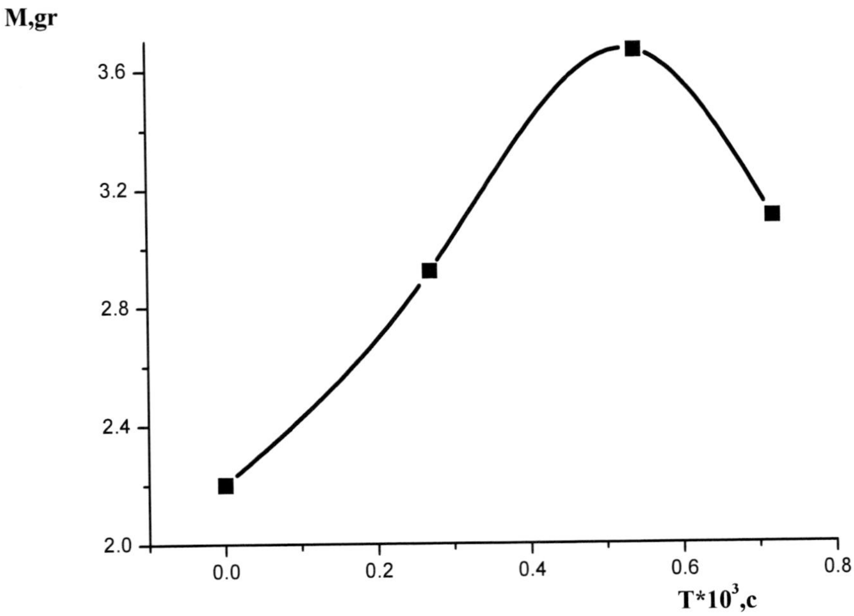


**Fig. 1.** Change of sorption of water (M, gr.) by a polyamide filament, as the function of an amount of absorbed ozone in time.

**Table 2 Change of elongation of a sample in a dependence from time of processing.**

| Processing Time (min) | 3 | 6 | 9 | 12 | 20 | 24 | 36 |
|---|---|---|---|---|---|---|---|
| Elongation (%) | 22,95 | 23,2 | 20,95 | 22,1 | 24,7 | 25,95 | 26,2 |

**Table 3 Change of a modulus of elasticity of a filament from time of processing.**

| Processing Time (min) | 0 | 3 | 6 | 9 | 12 | 20 | 24 | 36 |
|---|---|---|---|---|---|---|---|---|
| Modulus of elasticity (GPa) | 6,6 | 6,75 | 8 | 8,6 | 8,85 | 8,35 | 8 | 6,6 |

**Table 4. Dynamic of hardness change**

| Processing Time (min) | 0 | 3 | 6 | 9 | 12 | 20 | 24 | 36 |
|---|---|---|---|---|---|---|---|---|
| Strength (GPa) | 0,72 | 0,75 | 0,78 | 0,8016 | 0,83 | 0,81 | 0.76 | 0,77 |

The modulus of elasticity (Tab. 3) calculated as the ratio of mechanical strain to elongation, undergoes major changes. Thus, the dynamic of change carries clearly expressed extreme character. The maximal value of an elastic modulus (extremum) is reached at 12 min. effects with the subsequent smooth droop. In comparison with other physical-chemical characteristics the elastic modulus varies to the greatest degree during ozone processing (~19 %).

Dynamic of hardness change (Tab. 4) on the character cymbate to change of an elastic modulus as functions of processing time, though it scale is a little bit less (~15 %), but really visible and important for an evaluation of technical usefulness of the obtained data.

## CONCLUSIONS

As follows from the described experimental data the ozone has occurred an effective modifying agent for polyamide materials. Unlike the majority of other polymeric materials [19] the complex of properties of polyamides under it effect is improved. Its influence is not accompanied by noticeable destruction of a material. It is interesting to mention, that at inappreciable depth of penetration (nanosize layer) and small expenditures of ozone the large effects in change of physical-mechanical properties were obtained. The initial value of strength of filaments during the reaction accrues up to 15 %, with further drop; the modulus of elasticity is increased by 19 %; and elongation - on 9 %.

## REFERENCES:


[1] KorshakV.V., FrunzeT.M., "*Synthetic geterocycle polyamides*", Moscow, Chemistry (1962).

[2] Piller B., Travnichek Z. "*Synthetic fibers and specialties of their processing in textile industry*", Moscow, Published by Foreign Literature (1960).

[3] Anbarasan R, Jayaseharan J., Sudha M., Gopalan A., *J. Appl. Pol. Sci.* (2002), v.86, №12, p. 3108.

[4] Wa-Oh K., Kim S.H., Kim E.A. // *J. Appl. Pol. Sci.* (2001), v.81, №3, p. 684.

[5] Razumovskii S.D., Zaikov G.E., *Polymer Science* (1981), v.23, №3, c. 513.

[6] Grebennikov S.F., Andrianova I.I., Smotrina T.V., Suhovarova O.A. "Intermolecular interactions in water-biopolymer system", *Chemistry and computer design, Butler reports* (2001) № 4; VII All-Russian Conference "Structure and dynamic of molecular systems" reports.

[7] Jang, B. Z., *Advanced Polymer Composites: Principles and Applications*, ASM International, Materials Park, OH, (1994).

[8] W. J. Thorsen, W. H. Ward, M. M. Millard, "Wool shrinkage control and surface modification by ozone", *Journal of Applied Polymer Science* Volume 24, Issue 2, (1979), Pages: 523-546

[9] Alexeeva O.V., KonstanttinovaM.L.,Razumovskii S.D., "Dynamic of ozone sorption by fibers and films from polyamide", *Polymer Science*, 2004 (submitted)

[10] Alexeeva O., Razumovsky S.D., "Oxidizing by ozone of shallow layers of polystyrene marked on a hard substrate" Abstracts, *The First Youth Conference on Chemical Physics*, March 2001, Moscow

[11] Kluev L.E.,Grebennikov S.F., *Journal of physical chemistry,* (1996), V. 70, № 11, c. 2070.

[12] Kluev L.E.,Grebennikov S.F., *Journal of physical chemistry*, 1999,v. 73, № 9. c. 1720. 9.

[13] Loong-Tak Lim, Ian J. Britt, Marvin A., "Tung Sorption and transport of water vapor in nylon 6,6 film", *Journal of Applied Polymer Science*, Volume: 71, Issue: 2, Date: 10 January 1999, Pages: 197-206

[14] G. D. Danilatos, R. Postle , "Dynamic mechanical properties of keratin fibers during water absorption and desorption", *Journal of Applied Polymer Science*, 26 (1981), Pages: 193-200

[15] X. Fu, W. Lu, D.D.L. Chung, "Ozone treatment of carbon fiber for reinforcing cement", *Carbon* 36 (9) (1998) 1337–1345.

[16] Y. Xu, D.D.L. Chung, "Improving the workability and strength of silica fume concrete by using silane treated silica fume", *Cement and Concrete Research* 29 (3) (1999) 451–453.

[17] X. Fu, W. Lu, D.D.L. Chung, "Improving the tensile properties of carbon fiber reinforced cement by ozone treatment of the fiber", *Cement and Concrete Research* 26 (10) (1996) 1485–1488.

[18] Yunsheng Xu, D.D.L. Chung, "Carbon fiber reinforced cement improved by using silane-treated carbon fibers", *Cement and Concrete Research*, 29 (1999) 773–776

[19] H. Chtourou, B. Riedl, B. V. Kokta, A. Adnot, S. Kaliaguine, "Synthetic pulp fiber ozonation: An ESCA and FTIR study", *Journal of Applied Polymer Science* Volume 49, Issue 2, (1993) Pages: 361-373.

In: Chemical Physics of Pyrolysis, Combustion and Oxidation ISBN 1-59454-163-9
Editors: A. A. Berlin et al., pp.125-133 

*Chapter 14*

# TERMINATION REACTION IN THE CHAIN OXIDATION OF CAPROLACTAM

**B. Lanska[a], G. G. Makarov[b] and J. Sebenda[a]**
[a] Institute of Macromolecular Chemistry, Czechoslovak Academy of Sciences, 162 06 Prague 6, Czechoslovakia
[b] Emanuel institute of biochemical physics Russian academy of sciences ul. Kosygina 4, Moscow, 119991 Russia

## ABSTRACT

The oxidation of caprolactam initiated with 0.008-0.82 mol/kg 1,1'-azo-di(cyclo-hexane carbonitrile) was studied in the temperature range 80-98°C. At the conversion lower than 3%, caprolactam hydroperoxide, adipimide, and hydrogen peroxide were determined in the oxidation products. Depending on the initiator concentration, the kinetic chain length of oxidation ranged between 3 and 30. The formation rate of hydrogen peroxide was equal to one half of the rate of initiation reaction, i. e. to one half of the rate of bimolecular termination in the steady state. It may be assumed that hydrogen peroxide is formed by the bimolecular termination reaction. This is in agreement with the fact that hydrogen peroxide was not found in the thermally initiated oxidation of caprolactam which proceeded with the kinetic chain length more than 300. Another product of the bimolecular termination reaction of caprolactam peroxy radicals is adipimide, which is also formed by their monomolecular decomposition.

## INTRODUCTION

In the chain oxidation of caprolactam (L), caprolactam hydroperoxide is formed by a propagation reaction [1]. The termination proceeds bimolecular-ly [2-4] and several possible mechanisms have been suggested [5]. Concerning the quality of caprolactam as a monomer for polyamides and the stability of polyamides, the knowledge of the termination mechanism of oxidation chains is substantial. The kinetic chain length may be relatively short under practical conditions [5,6] so that the products of termination reactions could affect the further course of oxidation. For this reason we have studied the termination mechanism of oxidation chains of caprolactam with the aim to find whether the product of this reaction is a compound able to degenerate the oxidation chains and thus acts as an initiator of oxidation.

In the thermally initiated oxidation of caprolactam [1], at 90-155°C, the kinetic length of oxidation chains is more than 300 so that the products of termination reactions represent only a slight fraction of oxidation products. In order to increase this fraction, we reduced in this study the kinetic chain length of oxidation by a suitable concentration of initiator, l,l'-azo-di(cyclo-hexane carbonitrile), for which the rate constants of molecular decomposition and the efficiency constant in caprolactam melt were published [4'7].

## EXPERIMENTAL

### Materials

Caprolactam (technical grade, Spolana Neratovice, Czechoslovakia) was refined by threefold crystallization (water (twice) and benzene) and dried for 30 h at 50°C/130 Pa.

l,l'-Azo-di(cyclohexane carbonitrile) was twice crystallized from ethanol and dried for 20 h at 25°C/130 Pa. The rate constant of its molecular decomposition in chlorobenzene solution complies with the relationship [7] log $k_d$ = 15.49 - 7 261/T; the efficiency (e) for the initiation of oxidation of caprolactam melt [4] is equal to 0.35.

2,2,6,6-Tetramethyl-4-hydroxypiperidine-l-oxyl was twice crystallized from diethyl ether — hexane (2:1); m.p. 66-70°C. Its concentration was determined by HPLC [4]. The values of efficiency were checked by the thermolysis of initiator in caprolactam melt in the presence of this stable radical.

### Procedures

Caprolactam was oxidized in the described apparatus [4] in the presence of initiator (0.008-0.82 mol/kg) at 80, 90, or 98°C under oxygen pressure of 98 kPa. Under the



given conditions, the reaction rate did not depend on oxygen pressure. The volume of gaseous phase was determined with the accuracy of 0.5 mmol/kg. The thermolyses were followed in argon atmosphere.

### Analyses

The content of oxygen in gaseous phase was determined with an Orsat microappara-tus, the contents of hydrogen, water, carbon monoxide and dioxide, saturated and unsaturated hydrocarbons by means of IR spectroscopy and gas chromatography.

The oxidation products were analyzed as described earlier [1]. Carboxy groups, caprolactam hydroperoxide, and hydrogen peroxide were determined with the accuracy of 0.5 mmol/kg (imide groups: 2 mmol/kg).

# RESULTS AND DISCUSSION

## The Rate of Oxidation

The oxidation of caprolactam was followed at three temperatures in the presence of 1,l'-azo-di(cyclohexane carbonitrile) at concentrations 0.008 - 0.820 mol/kg up to a conversion of caprolactam of max. 3 °7o. Therefore the concentration of caprolactam was considered constant in calculations. Oxidation time was shorter than 2 h.

The thermolysis of initiator in the melt of caprolactam revealed that the rate of nitrogen formation observes the relationship

The volume change of gaseous phase is given by the difference between oxygen consumption and the volume of gaseous products formed. In the studied extent of reaction, the gaseous phase did not contain products of deeper oxidation stages, i.e. $CO_2$, CO, $H_2$, saturated or unsaturated hydrocarbons. Nitrogen formed in the decomposition of initiator (I) was the only gaseous product:

$$I \xrightarrow{k_d} 2R^{\bullet} + N_2 \qquad (1)$$

$$v_{N_2} = 0.35\ k_d\ [I]_0 \qquad (2)$$

where $k_d$ is the published [7] value of rate constant of the molecular decomposition of initiator in chlorobenzene solution and $[I]_o$ is the initial concentration of initiator. The factor 0.35 corresponds to the earlier found efficiency of initiator 1,l'-azo-di(cyclohexane carbonitrile) in the oxidation of caprolactam [4]. This efficiency was confirmed also by the thermolysis of initiator in caprolactam in the presence of stable radicals.

At the given time, the relevant volume of nitrogen corresponds to the rate of initiation reaction in caprolactam oxidation:

$$v_i = 2\ e\ k_d\ [I] = k_i\ [I] \qquad (3)$$

This volume was always found in gaseous products of oxidation within the limits of experimental error. The consumption of oxygen ($\Delta O_2$), given by the sum of absolute values of volume change of the gaseous phase and volume of nitrogen evolved from the initiator, increases linearly in time (Fig. 1). The kinetic chain length of oxidation was calculated from the oxidation rate $v_0 = d(\Delta O_2)/dt$ and the rate of initiation reaction $V_i$ (Eq. (3)) as $v = v_o/v_i$ (Tab. 1).

## Oxidation Products

Caprolactam hydroperoxide, hydrogen peroxide, and adipimide were found in the products of oxidation and their concentration increased linearly from the beginning of reaction in all cases (Tab. 1). Any product did not contain carboxylic acid; the content of water in products was not determined.

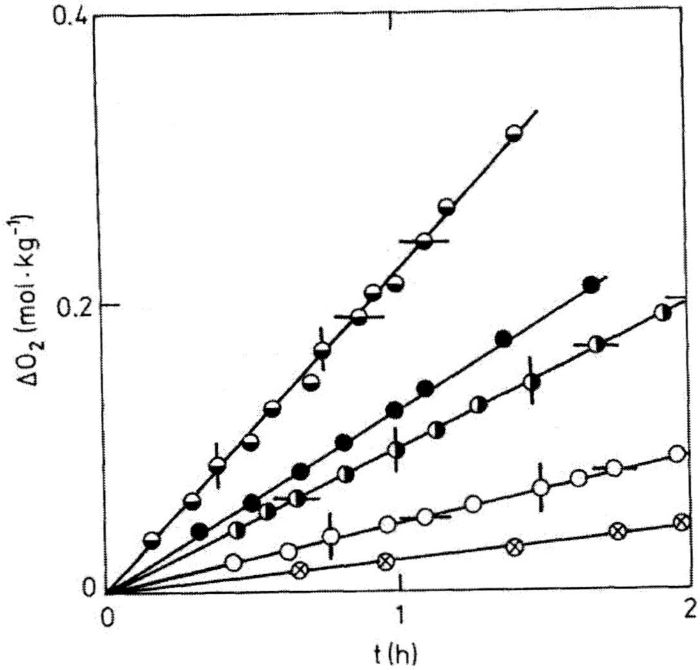


Fig. 1. Oxygen consumption ($\Delta O_2$) in the caprolactam oxidation initiated with 1,1′-azo-di(cyclohexane carbonitrile) (I) at 90°C.
$[I]_0$ (mol/kg): (⊗) 0.008, (○) 0.04, (◑) 0.21, (●) 0.36, (◒) 0.82, (–○–) 0.04 + 0.08 $H_2O$, (○ with vertical bar) 0.04 + 0.04 $H_2O_2$, (–◑–) 0.21 + 0.04 $H_2O$, (◑ with vertical bar) 0.21 + 0.014 $H_2O_2$, (–◒–) 0.82 + 0.052 caprolactam hydroperoxide, (◒ with vertical bar) 0.82 + 0.060 adipimide.

The linear time dependence of oxygen consumption proves that none of the products caused the degeneration of chain reaction in the following stage of caprolactam oxidation. Inactivities of adipimide and water in the oxidation of caprolactam were already observed in the non-initiated thermal oxidation of caprolactam [1,8]. Adipimide as the oxidation product did not undergo successive reaction even at 155°C [1]. Also the initiated oxidation of caprolactam proceeded in the presence of adipimide or water at the same rate and with formation of the same products as the reaction of caprolactam alone (Fig. 1, Tab. 2).

Caprolactam hydroperoxide and hydrogen peroxide are able to undergo homolytic cleavage to radicals [1,9,10]. In spite of this, neither the addition of caprolactam hydroperoxide to the initiated oxidation of caprolactam nor that of hydrogen peroxide influenced the consumption of oxygen and the composition of oxidation products (Fig. 1, Tab. 2). Although caprolactam hydroperoxide proved as autocatalyst in the thermal oxidation of caprolactam [1], it is stable at 90° C in caprolactam melt to such extent that in the absence of initiator its decomposition cannot be perceived by the used experimental methods within 2 h (Tab. 3).

## Termination Reaction

Several mechanisms have been proposed for the bimolecular termination in the oxidation of N-alkylamides [5], but so far none of them has been confirmed, e. g. by a balance of products or by the rate of their formation. One of the proposed routes consists in the reaction of two peroxy radicals giving rise to two molecules of imide and one molecule of hydrogen peroxide:

$$2\ \text{—CO—NH—}\underset{\displaystyle OO^{\bullet}}{\underset{|}{\text{CH}}}\text{—} \longrightarrow 2\ \text{—CO—NH—CO—} + H_2O_2 \qquad (4)$$

As follows from Tab. 1, the formation rate of hydrogen peroxide is equal to one half of the initiation rate. In the steady state, the rate of initiation (Eq. (3)) is equal to the rate of termination ($v_i = v_t$), consequently the termination by reaction according to Eq. (4) is highly probable.

**Table 1. Rates of initiation ($v_i$), oxidation ($v_0$), increase in the concentration of caprolactam hydroperoxide ($v_{HP}$), hydrogen peroxide ($v_{H2O2}$), adipimide ($v_{AI}$), and the kinetic chain length (v) in the oxidation of caprolactam initiated with l,l'-azo-di(cyclohexane carbonitrile).**

| T (°C) | $[I]_0$ ($mol\ kg^{-1}$) | $v_i$ | $v_0$ | $v_{HP}$ | $v_{H2O2}$ | $v_{AI}$ | v |
|---|---|---|---|---|---|---|---|
| | | ($10^{-6}\ mol\ kg^{-1}\ s^{-1}$) | | | | | |
| 80 | 0.080 | 0.48 | 4.8 | 3.65 | – | 1.15 | 10 |
| 90 | 0.008 | 0.18 | 6.0 | 5.1 | – | 0.9 | 33 |
| 90 | 0.040 | 0.9 | 12.3 | 9.8 | 0.45 | 2.5 | 14 |
| 90 | 0.210 | 4.6 | 28.15 | 20.15 | 2.3 | 8.0 | 6 |
| 90 | 0.360 | 7.9 | 37.7 | 25.25 | 3.9 | 12.45 | 5 |
| 90 | 0.820 | 18.0 | 62.0 | 37.3 | 9.0 | 25.0 | 3-4 |
| 98 | 0.040 | 2.4 | 26.2 | 20.15 | 1.2 | 6.05 | 11 |

## Formation of Adipimide

The results in Tab. 3 reveal that, in the absence of oxygen and initiator, caprolactam hydroperoxide is stable at 90° C for at least 2 h. However, it is decomposed in the presence of initiator by a rate close to the rate of initiation giving rise to adipimide by a constant rate. It may be supposed that, in the absence of oxygen, alkyl radicals $R^{\bullet}$ and $L^{\bullet}$ form a peroxy radical from caprolactam hydroperoxide:

$$R^{\bullet} + LOOH \longrightarrow RH + LOO^{\bullet} \qquad (5)$$

$$L^{\bullet} + LOOH \longrightarrow LH + LOO^{\bullet} \qquad (6)$$

Peroxy radicals give adipimide and hydrogen peroxide by the bimolecular termination reaction according to Eq. (4). It cannot be also excluded that a C-radical is formed from hydroperoxide [11],

(7)

which is cleaved to adipimide and a $HO^{\bullet}$ radical analogously as it was considered in the initiated decomposition of tetrahydronaphthalene hydroperoxide [12]:

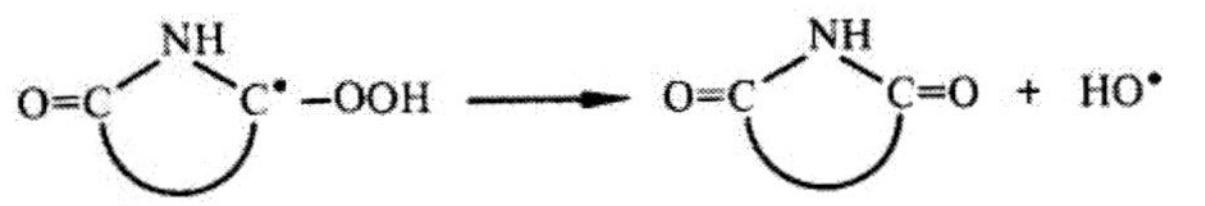
(8)

The fact that, in the initiated oxidation, the formation rate of adipimide ($v_{AI}$) is higher than the rate of termination (lab. 1) indicates that adipimide is formed not only by reaction according to Eq. (4). The constant value of $v_{AI}$ since the very beginning of oxidation indicates that also the second reaction yielding adipimide ensues from peroxy radical, the concentration of which is constant in the steady state. It can be conceived that the peroxy radical may undergo also a monomolecular cleavage:

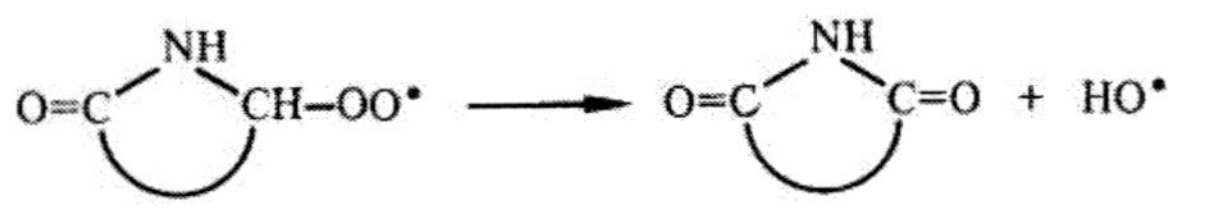
(9)

In the presence of oxygen, the extent of reactions according to Eq. (5) and (6) will be negligible in the initial stage of oxidation. Hence, it can be supposed that the radicals $LOO^{\bullet}$ are formed in the reaction of radical $L^{\bullet}$ with oxygen and that their concentration is constant in the following extent of reaction. The formation rate of adipimide is thus given by the sum of the rate of bimolecular termination (Eq. (4)) and the rate of decomposition of peroxy radicals (Eq. (9)):

$$v_{AI} = v_i + v_m = k_t\,[LOO^{\bullet}]^2 + k_m\,[LOO^{\bullet}] \qquad (10)$$

$$v_{AI} = k_i\,[I] + k_m\,k_t^{-1/2}\,k_i^{1/2}\,[I]^{1/2} \qquad (11)$$

Leading to the relationship:

At 90°C, the values of $k_i$ ($2.2{\bullet}10^{-5}$ $s^{-1}$) and $k_m\,k_t^{-1/2}$ ($1.6{\bullet}10^{-3}$ $mol^{1/2}$ $kg^{-1/2}$ $s^{-1}$) are obtained from the plot $v_{AI}/[I]_0^{1/2}$ vs. $[I]_0^{1/2}$ (Fig. 2). The slope of this dependence agrees with the value $k_i$ in Eq. (3).

**Table 2. The effect of caprolactam hydroperoxide (HP), adipimide (AI), water or hydrogen peroxide on the oxygen consumption ($\Delta O_2$) and formation of peroxides (caprolactam hydroperoxide + hydrogen peroxide, $\Delta(HP + H_2O_2)$) and adipimide ($\Delta AI$) in the initiated oxidation of caprolactam at 90°C.**

| $[I]_0$ (mol kg$^{-1}$) | Additive | [Additive] (mmol kg$^{-1}$) | Time (h) | $\Delta O_2$ | $\Delta(HP+H_2O_2)$ (mmol kg$^{-1}$) | $\Delta AI$ |
|---|---|---|---|---|---|---|
| 0.82 | – | | 1 | 218 | 163 | 86 |
| 0.82 | HP | 52 | 1 | 210 | 150 | 88 |
| 0.82 | AI | 60 | 1 | 216 | 162 | 88 |
| 0.22 | – | | 0.5 | 51 | 40 | 15 |
| 0.21 | $H_2O$ | 40 | 0.5 | 50 | 40 | 15 |
| 0.21 | $H_2O_2$ | 14 | 0.5 | 52 | 41 | 15 |
| 0.04 | | | 1 | 43 | 34 | 9 |
| 0.04 | $H_2O$ | 80 | 1 | 42 | 33 | 8 |
| 0.04 | $H_2O_2$ | 40 | 1 | 43 | 35 | 9 |

**Table 3. The decrease in caprolactam hydroperoxide (–ΔAHP) and formation of adipimide (ΔAI) during thermolysis of caprolactam hydroperoxide in caprolactam at 90°C.**

| $[HP]_0$ (mmol kg$^{-1}$) | 120 | | 120[a] | | 115[b] | 126[c] |
|---|---|---|---|---|---|---|
| Time (h) | –ΔHP (mol kg$^{-1}$) | ΔAI | –ΔHP | ΔAI | ΔAI | ΔAI |
| 0.25 | 0 | 0 | 0 | 0 | – | 4 |
| 0.50 | 0 | 0 | 0 | 0 | 8 | 8 |
| 1.00 | 0 | 0 | 0 | 0 | – | 19 |
| 1.50 | – | – | – | – | 24 | 25 |
| 2.00 | 1 | 1 | 1 | 1 | – | 32 |
| 3.00 | 1-2 | 1-2 | 2 | 1-2 | 51 | 50 |

[a] The initial reaction mixture contained 60 mmol kg$^{-1}$AI.
[b] The initial reaction mixture contained 0.36 mol kg$^{-1}$ 1,1'-azo-di(cyclohexane carbonitrile).
[c] The reaction mixture was the product after 1.5 h of initiated ($[I]_0$ = 0.60 mol kg$^{-1}$) oxidation of caprolactam at 90°C, which contained 68 mmol kg$^{-1}$ AI and the residual initiator ($[I]$ = 0.36 mol kg$^{-1}$).

The rate of propagation reaction in the steady state $v_p = k_p k_t^{-1/2} [L] v_i^{1/2}$ can be calculated from the rate of oxidation $v_0 = v_p + v_t$. The parameter of oxidability $k_p k_t^{-1/2}$ determinded in this way agreed with the earlier found [4] relationship $\log (k_p k_t^{-1/2}) = 5.385 + 3030/T$. The rate of propagation reaction at 90°C was $v_p = 1.5 \cdot 10^{-6}$ mol kg$^{-1}$ s$^{-1}$ + $4.7 \cdot 10^{-5}$ mol$^{1/2}$ kg$^{-1/2}$ s$^{-1/2}$ $[I]_0^{1/2}$. The value $1.5 \cdot 10^{-6}$ mol kg$^{-1}$ s$^{-1}$ expressed the contribution of autoxidation at 90° C in the stage of maximum reaction velocity [1]. From the above given values, there was found the ratio $k_p/k_m = 1.03 \cdot 10^{-3}$ s$^{-1}$ / $1.6 \cdot 10^{-3}$ mol kg$^{-1}$ s$^{-1}$, i.e. 0.64 kg mol$^{-1}$ for 90°C.

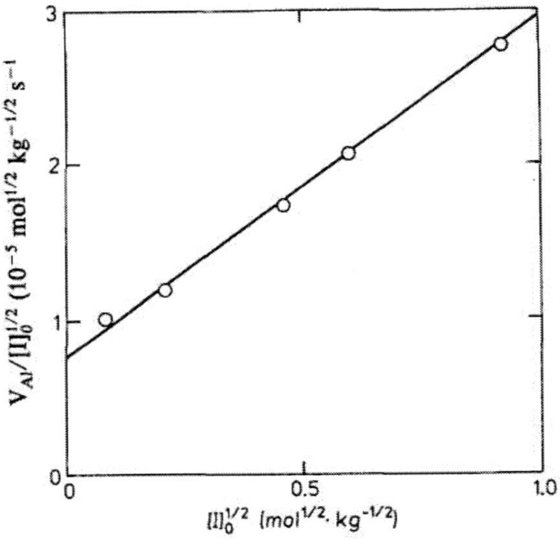


**Fig. 2.** Initiated oxidation of caprolactam at 90°C.

## Conclusion

In the initial stage of radical initiated oxidation of caprolactam at 80 – 98 °C under the conditions when the rate does not depend on oxygen pressure, caprolactam hydroperoxide, adipimide, hydrogen peroxide, and water are formed. The kinetic chain length of oxidation corresponds to the concentration of initiator used. The content of hydrogen peroxide in the products of oxidation gives the evidence that the bimolecular termination reaction of peroxy radicals gives rise to two molecules of adipimide and one molecule of hydrogen peroxide. Adipimide is also formed by the monomolecular decomposition of peroxy radicals.

The formation of hydrogen peroxide in the oxidation of N-alkylamides is of consequence to the durability of polyamide articles. Hydrogen peroxide is capable of homolytic cleavage to radicals under exposure to heat and light and may cause deeper degradation changes.

## Acknowledgements

The authors thank Dr. E. Knizakova and Mrs. N. Plecita, Institute of Macromolecular Chemistry, Prague, for careful titrimetric determinations.

## References

[1] B. Lanska, J. Sebenda, *Angew. Makromol. Chem.* **164** (1988) 181
[2] B. F. Sagar, *J. Chem. Soc* **B 1967,** 1047
[3] A. L. Aleksandrov, *Izv. Akad. Nauk SSSR, Ser. Khim.* **1982,** 2223; *C A.* **98** (1983) 71301 h

[4] B. Lanska, L. M. Postnikov, A. L. Aleksandrov, J. Sebenda, *Collect. Czech. Chem. Commun.* **46** (1981) 2650
[5] B. F. Sagar, *J. Chem. Soc B* **1967,** 428
[6] A. L. Margolin, I. A. Kabanova, L. M. Postnikov, V. Ya. Shlyapintokh, *Vysokomol. Soedin., Ser. A.* **18** (1976) 1094; *C.A.* 85 (1976) 63574r
[7] S. Wu Chin-Hua, G. S. Hammond, J. M. Wright, *J. Am. Chem. Soc.* **82** (1960) 5386
[8] J. Sebenda, L. M. Postnikov, B. Lanska, *Collect. Czech. Chem. Commun.* 40 (1975) 1877
[9] C. H. Bamford, R. P. Wayne, *Photochem. React. Kinet.* **1967,** 26
[10] N. M. Emanuel, E. T. Denisov, Z. K. Maizus, Tsepnyie reakcii okisleniya uglevodorodov v zhidkoi faze *(Russ.J,* Nauka, Moscow 1965, p. 92
[11] Ref.10, p. 149
[12] A. Robertson, W. A. Waters, *J. Chem. Soc* **1948,** 1578

In: Chemical Physics of Pyrolysis, Combustion and Oxidation ISBN 1-59454-163-9
Editors: A. A. Berlin et al., pp.135-144 

*Chapter 15*

# INTERACTION OF NITROGEN OXIDES WITH POLY(VINYLPYRROLIDONE) AND POLY(CAPROAMIDE): MOLECULAR AND FREE-RADICAL PRODUCTS[1]

***I. S. Gaponova, E. Ya. Davydov, G. G. Makarov, G. B. Pariiskii and V. P. Pustoshnyi***
Emanuel Institute of Biochemical Physics, Russian Academy of Sciences, ul. Kosygina4, Moscow, 117977 Russia

## ABSTRACT

The mechanism for the reaction of nitrogen dioxide and poly(vinylpyrrolidone) or poly(capro-amide) is discussed. As evidenced by UV and IR spectroscopy, the reaction with poly(vinylpyrrolidone) yielded mainly nitrites and nitrosoamides, while, in the case of poly(caproamide), nitrosoamides were the major products. Stable acylalkylnitroxyl and iminoxyl radicals were registered for the reaction with poly(vinylpyrroli-done), while acylalkylnitroxyl radicals were detected in poly(caproamide). The formation of molecular products and stable radicals was initiated by the selective H abstraction from either N-H or C-H bonds taking a-positions with respect to the amide groups of monomer units.

Nitrogen oxides are toxic and corrosive atmospheric components. To develop methods for protecting various materials, among them polymeric materials, it is necessary to elucidate the mechanism of the reactions involving nitrogen oxides. There is extensive literature on the reactions of nitrogen oxides with various organic compounds in liquid phase [1]. However, the data on the mechanism for the reactions of nitrogen oxides and solid polymers are scarce [2]. In order to clarify the mechanism of reactions in polymers, we need to identify the molecular and free-radical products, and to establish the reactive chemical bonds in monomer units, which are responsible for the initiation of the processes of polymer aging in the presence of nitrogen oxides. In this work, we performed a comparative study of the reaction products and the mechanism for the reaction of nitrogen dioxide with poly(caproamide)

(PCA) and poly(vinylpyrrolidone) (PVP), which contain amide groups in the backbone and side fragments.

## EXPERIMENTAL

Nitrogen dioxide was obtained by the thermal decomposition of lead nitrate. Nitrogen oxide, which was used in some experiments, was prepared by the reaction of 30% nitric acid with copper [3]. The experiments were performed with powdery PCA and PVP with $M_\eta = 2.3 \times 10^4$ and $3.0 \times 10^5$, respectively. The polymer samples (50-100 mg) were placed in quartz ampules used to measure the ESR spectra and equipped with vacuum stopcocks. The ampules were connected to a 1-l glass vessel. After the system was evacuated to a residual pressure of 0.133 Pa, the stopcock was shut down, and the vessel was filled with $NO_2$ until the $NO_2$ concentration range $1 \times 10^{-4}$-$8.5 \times 10^{-4}$ mol/l was achieved. When $NO_2$ was introduced into an ampule, the registration of the ESR spectra with an ESR-V (Institute of Chemical Physics, Moscow) radiospec-trometer at room temperature was started. When the radicals were generated in the presence of $NO_2$ and NO, the glass vessel was initially filled with a mixture of these gases. The kinetics of radical generation was measured in relative units by comparing the intensities of their components in the ESR spectra with the signal of the external copper chloride reference. The molecular products obtained by the reactions of PVP and PCA with $NO_2$ were identified by UV and IR spectroscopy using Specord UV-VIS and Specord IR-75 spectrometers. For PCA, the absorption spectra in the UV and visible regions were measured using samples prepared as l-2-μm films cast from a polymer solution in concentrated formic acid. In the case of PCA, the IR spectra were recorded with l-5-μm films supported on KRS-5 plates. Some measurements were carried out with an IFS-45 Beckman IR spectrometer.[2] Before recording the IR spectra, the PCA films exposed to $NO_2$ were evacuated for 1 h. For PVP, the reaction products were identified by the UV spectra of the polymer chloroform solutions subjected to the action of $NO_2$. The concentration of $NO_2$ in the gas phase was determined from the optical density at 410 nm [4].

## RESULTS AND DISCUSSION

The products of the reactions between nitrogen oxide, PCA, and PVP show similar and different absorption bands in the UV spectra. Thus, the spectra (Fig. 1) display the bands in the range of 390-435 nm, which are similar in shape. The spectra of PCA display three maxima at 386, 410, and 430 nm. The spectra of PVP show the distinct maxima at 413 and 435 nm. The characteristic spectral region and the presence of the structure allow us to assign the UV bands to nitroso-amide groups [5].

The transformation of N-H groups of PCA into nitrosoamides is also confirmed by IR spectroscopy. Figure 2 shows the spectra of the unexposed film and the film exposed to $NO_2$. Comparison of the spectra indicates that the intensity of band $\nu = 3293$ cm$^{-1}$, which is associated with the stretching vibrations of H-bonded N-H groups of the polymer, decreases sharply. The intensity of the Amide I ($\nu = 1642$ cm$^{-1}$) and Amide II ($\nu = 1563$ cm$^{-1}$) bands, which are characteristic of PCA, also drops markedly. Instead of these bands, band $\nu = 1730$ cm$^{-1}$, which corresponds to the vibrations of the C=O group of nitrosoamides, and the bands ν

= 1504 and 1387 $cm^{-1}$, which correspond to the asymmetric and symmetric stretching vibrations of N=O groups of nitrosoamides [6], appear in the spectrum of the polymer exposed to $NO_2$. The band v = 1580 $cm^{-1}$, which is typical of the vibrations of $NO_2$ groups in nitroamides, is not observed. In fact, nitrosa-tion through an amide group is the main process of PCA transformation in an $NO_2$ atmosphere. We found that the initial rate for the accumulation of the optical density of the band v = 1730 $cm^{-1}$ in the $NO_2$ concentration range $4 \times 10^{-5}$-$4 \times 10^{-4}$ mol/l is proportional to $[NO_2]^n$, where $n \approx 2$. The initial rate was estimated at the zero point from the tangent to the kinetic curve obtained for the optical density of this band.

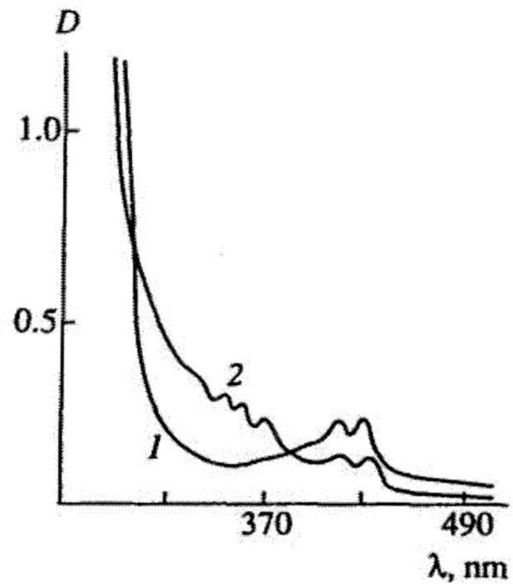


**Fig. 1.** Absorption spectra of (*1*) PCA and (*2*) PVP exposed to $NO_2$: (*1*) film and (*2*) PVP solution in chloroform.

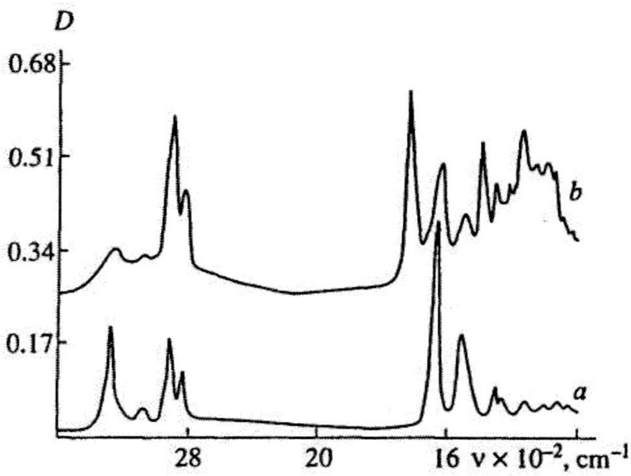


**Fig. 2.** IR spectra of (*a*) the starting PCA films and (*b*) the PCA films exposed to $NO_2$

In addition to the bands of nitrosoamides, the UV spectrum of PVP exhibited the absorption at 320-380 nm (Fig. 1). This absorption is characterized by a set of bands spaced at ~10 nm. This indicates that alkylni-trites [7] are formed by the reaction of PVP with $NO_2$.

Figure 3 shows the ESR spectra recorded after the introduction of $NO_2$ into the ampules with PVP and PCA samples. For PCA, the spectrum displays an anisotropic triplet with $A^{N}_{II}$ = 1.94 mT and $g_{II}$ = 2.003 (radical I), which is characteristic of acylalkylnitroxyl radicals in a solid phase [8, 9]. For PVP, the ESR spectrum represents a superposition of the signals from the radicals I and the iminoxyl radicals П with parameters $A^{N}_{II}$ = 4.33 mT and $A^{N}_{II}$ = 2.44 mT; $g_{\perp}$ = 2.0029 and $g_{II}$ = 2.0053 [10]. Note that the limiting concentrations of the stable nitrogen oxide radicals in PVP and PCA are significantly smaller than the concentration of the molecular products. As evidenced by the approximate estimation on the basis of the known extinction coefficients [5, 7], 90% $NO_2$ is converted to nitrosoamides and nitrites.

The examination of the UV and ESR spectra allow us to suggest that the mechanism of PVP transformations initiated by nitrogen dioxide is as follows. It is probable that C-H bonds at α-positions with respect to amide groups show the highest activity in the H-abstraction reaction. PVP has two types of C-H bonds with similar bond dissociation energies. Hence, one might expect that two macroradicals, in which a free valence is localized in the backbone and in side ring, will appear upon the primary reaction with $NO_2$

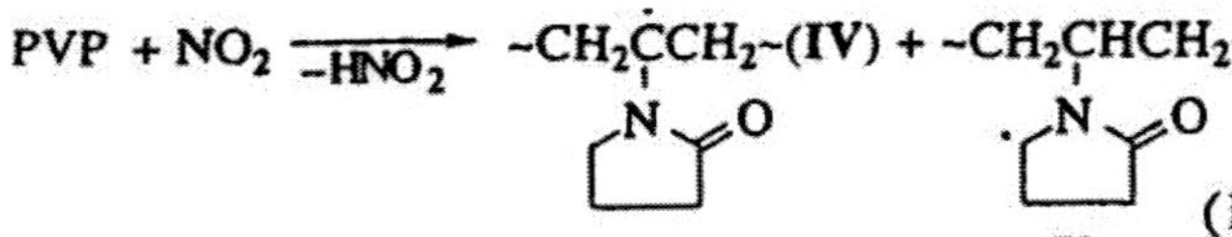

V (1)

Recombination of IV and V radicals with $NO_2$ molecules gives rise to nitrites, which are detected by the UV spectra

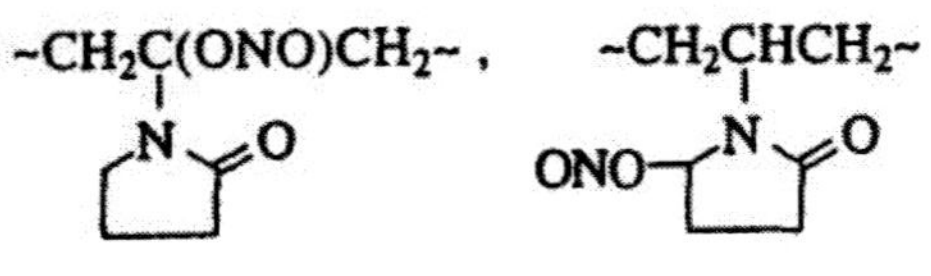


β-Decomposition of radicals V yields nitroso-amides. A cyclic fragment is initially splitted off the backbone chain

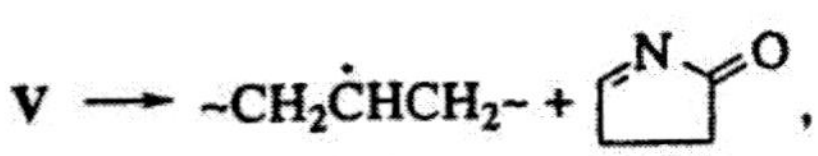

VI (2)

Then, $NO_2$ reacts through double bonds to form the final product [11]

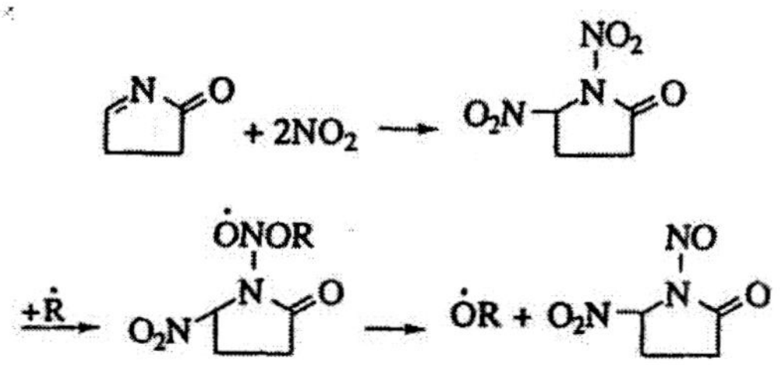

(3)

The mechanism for the formation of nitrogen oxide radicals is also related to the reactions of primary radicals FV and V. It appears that nitrous acid can function as a source of nitrogen oxide in the system under consideration

$$2HNO_2 \longrightarrow H_2O + NO_2 + NO \qquad (4)$$

Recombination of nitrogen oxide with radicals V initiates the formation of iminoxyl radicals II detected by ESR spectroscopy. This process includes the following consecutive reactions: the formation of nitroso compounds, their isomerization to oxime [1], and the splitting off a hydrogen atom from oxime by $NO_2$

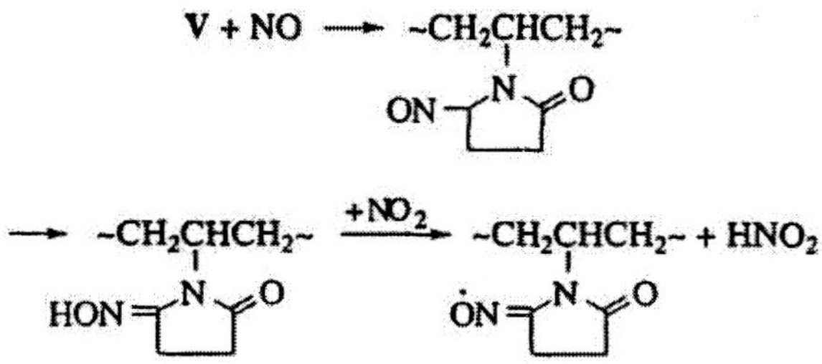

(5)

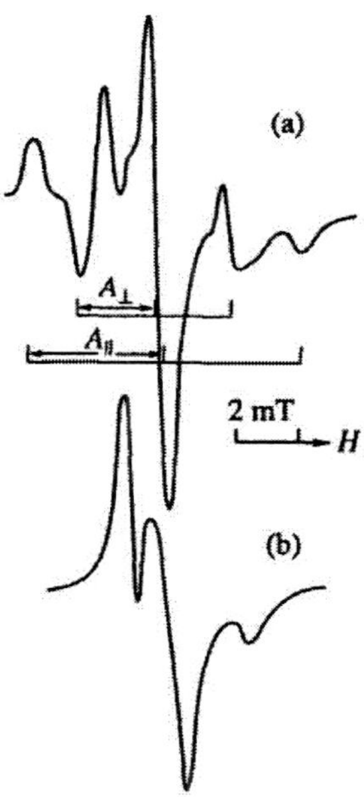


**Fig. 3.** ESR spectra of radicals in (a) PVP and (b) PCA generated by $NO_2$.

The formation of nitrogen oxide via reaction (4) can also explain the generation of the acylalkylnitroxyl radicals I in PVP. Allowances should be made that macro-radicals IV can decompose with the opening of a pyr-rolidone ring in similar fashion to their photodissociation [12]

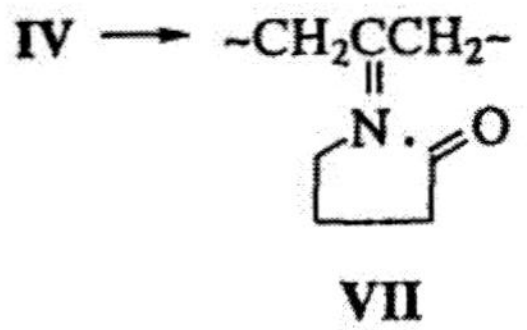


(6)

Then, the reaction of double radical accentuation involving NO first yields acylnitroso compounds—the effective spin scavengers [13]—and, then, radicals I

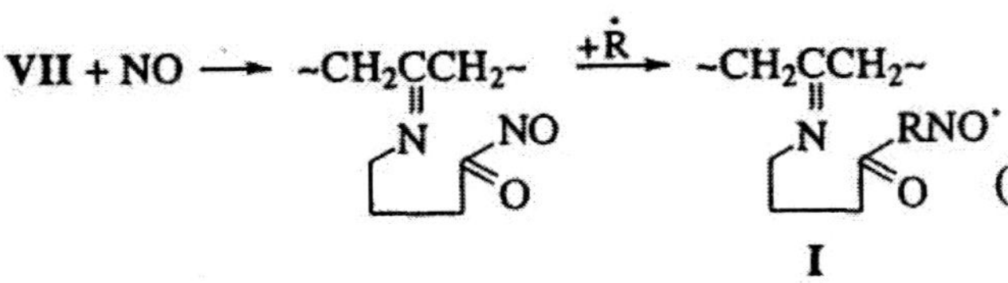


(7)

Note that any free radical, which is not stabilized at ordinary temperatures, for example, IV, V, and VI can function as R.

The accumulation kinetics of the stable nitroxyl radicals in PVP reflects the mechanism of their formation. In the kinetic experiments, we varied the concentration of $NO_2$ in the gas phase and used a $NO_2$ + NO mixture.

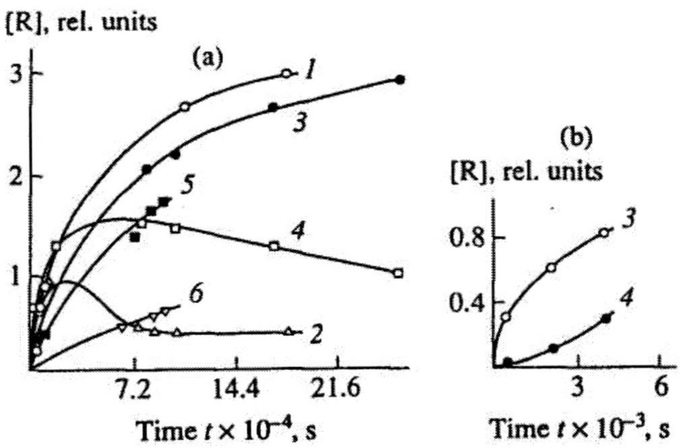


**Fig. 4.** (a) Kinetics of accumulation of (*1, 3, 5*) radicals I and (*2, 4, 6*) radicals II in PVP and (b) initial portion of curves *3* and *4*. The content of $NO_2$: (1, 2) 8.55 x $10^{-4}$ and (*3, 4*) 4.0 x $10^{-4}$ mol/l; (*5, 6*) is the mixture of $NO_2$ ($10^{-4}$ mol/l) + NO (3.0 x $10^{-4}$ mol/l).

As follows from Fig. 4a, the accumulation kinetics for radicals I and II differ qualitatively. The concentration of radicals I achieves the limiting value, whereas, in the case of radicals II, this value passes through a maximum. The difference in the accumulation kinetics is pronounced, even at the initial stages (Fig. 4b). Contrary to radicals Π, radicals I are formed with marked acceleration.

These results are consistent with the reactions above (5). Actually, the generation rate of radicals VII should increase with the accumulation of nitroso groups, which are the scavengers of macroradicals, until a certain steady-state concentration is achieved. Apparently, an approach of a nitro group to a macro-radical R is the limiting stage in the formation of radicals VII. In a solid polymer, this process can be hampered markedly due to the low molecular mobility of the species.

The absence of a pronounced induction period at the initial stages of the generation of radicals II can be explained by the pseudocage character of the process. Taking into account that the limiting yield of the stable radicals is low compared to that of the major molecular products of PVP transformation, it is conceivable that the generation of radicals Π is associated with the reaction of the dimer form $N_2O_4$, which occurs as an admixture in nitrogen dioxide. Taking into account the equilibrium $N_2O_4 \leftrightarrow 2NO_2$, the hydrogen atom abstraction by the dimer yields two molecules of nitrous acid, which are in close proximity. Their transformation via reaction (4) leads to the situation where NO and $NO_2$ turn out to be close to radicals V. Therefore, it is conceivable that the sequential reactions (5) virtually proceed in a cage and are not limited by the mutual diffusion of the reagents. It is possible that the passage of the concentration of radicals II through the maximum is associated with their decay by the subsequent recombination with NO, which accumulates during the polymer nitration

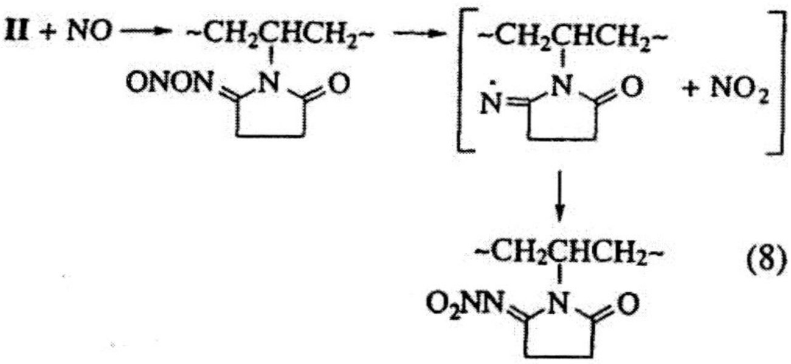


(8)

This decay mechanism of iminoxyl radicals was discussed in [14]. The higher the initial concentration of $NO_2$, the shorter the time of attaining the maximum concentration of II and the lower the value of this concentration (Fig. 4a).

Figure 4a shows that the accumulation rate of radicals I and their limiting concentration remain almost unchanged as the initial $NO_2$ concentration varies twofold. This leads us to draw a conclusion about the limiting stage of their formation. The rate of $NO_2$ absorption by the polymer is proportional to the concentration of $NO_2$ in the gas phase. This simplified scheme of the reactions of PVP nitration can be presented as follows:

$$RH + NO_2 \xrightarrow{k_1} HNO_2 + \dot{R} \qquad (9)$$

$$\dot{R} + NO_2 \xrightarrow{k_2} RONO \qquad (10)$$

$$\dot{R} + NO \xrightarrow{k_3} RNO \xrightarrow{NO_2} II \qquad (11)$$

$$\dot{R} \xrightarrow{k_4} \dot{R}_{VII} \qquad (12)$$

$$\dot{R}_{VII} + NO \xrightarrow{k_5} R_{VII}NO \qquad (13)$$

$$R_{VII}NO + \dot{R} \xrightarrow{k_6} I \qquad (14)$$

$$\dot{R} \xrightarrow{k_7} \dot{R}_{VI}, \qquad (15)$$

where R are the macroradicals generated via the primary reaction of PVP with $NO_2$, and $R_{IV}$ and $R_{VII}$ are the free-radical products of their decomposition. Ignoring the initial stages of this process (Fig. 4b) and assuming that R, $R_{VII}$, and $R_{VII}NO$ concentrations are steady-state, we get

$$[\dot{R}] = \frac{k_1[RH][NO_2]}{k_2[NO_2] + k_3[NO] + k_4 + k_7} \qquad (16)$$

$$[R_{VII}NO] = \frac{k_4}{k_6} \qquad (17)$$

As $[NO_2] > [NO]$ and $k_2 \sim k_3$, we assume that $k_2[NO_2] > k_3[NO]$. If $k_2[NO_2] > k_4 + k_7$, then, $[R] \sim k_1[RH]/k_2$, and $d[I]/dt \sim (k_4k_1[RH])/k_2$. Thus, the fact that the rate of accumulation of acylalkylnitroxyl radicals is independent of the $NO_2$ concentration in the gas phase indicates that the transformation of macroradicals R via reactions (12) and (15) is the limiting stage of their formation. A piece of indirect evidence for this conclusion is provided by the accumulation kinetics of radicals I and II for the reaction of PVP with the $NO_2$ + NO mixture (Fig. 4a, curves *5* and *6*). In this case, the $NO_2$ concentration is approximately eight times lower than that in the experiments represented by curves *1* and 2; however, the initial accumulation rate of radicals I decreases only by a factor of two. Under these conditions, the rate of generation of iminoxyl radicals II drops markedly. This is consistent with the mechanism of their decay via reaction (8) [14].

As evidenced by the kinetic curves *1* and *3* (Fig. 4a), the concentration of $NO_2$ in the gas phase decreased only insignificantly (~10%) during the entire period of accumulation of radicals I. Therefore, the decrease in the generation rate of radicals I during exposure of PVP apparently reflects the consumption rate only of the reactive pyrrolidone rings involved in reactions (12)-(14). The main fraction of the absorbed $NO_2$ converts to alkylnitrites and nitrosoamides.

The formation of nitrosoamide groups in PCA can be explained in terms of the notions about the reaction of $NO_2$ with N-H bonds [6]. Taking into account equilibrium

$$NO^+NO_3^- \rightleftharpoons N_2O_4 \rightleftharpoons 2NO_2, \qquad (18)$$

the formation of nitrosoamide can be presented as follows:

$$\begin{aligned} &\sim CONHCH_2\sim + N_2O_4 \\ &\longrightarrow \sim CON(NO)CH_2\sim + HNO_3 \end{aligned} \qquad (19)$$

The occurrence of reaction (19) is supported above by the fact that the initial rate of $NO_2$ absorption in PCA is proportional to $[NO_2]^n$, where $n \approx 2$. Probably, the reaction of $NO_2$ with N-H bonds proceeds according to the radical mechanism to yield acylalkylnitroxyl radicals III (Fig. 3b). The reaction initially generates macroradicals that undergo recombination with $NO_2$ to form an unstable intermediate compound

$$\begin{aligned} &\sim CONHCH_2\sim + NO_2 \\ &\xrightarrow[-HNO_2]{} \sim CO\dot{N}CH_2\sim \xrightarrow{NO_2} \sim CON(ONO)CH_2\sim \end{aligned} \qquad (20)$$

The latter generates the detectable radicals **III** through the equilibrium dissociation

$$\sim CON(ONO)CH_2\sim \rightleftharpoons \underset{\textbf{III}}{\sim CON(\dot{O})CH_2\sim} + NO \quad (21)$$

The occurrence of equilibrium (21) is confirmed by the fact that the evacuation of the PCA samples, which were subjected to the action of $NO_2$, causes the intensity of the ESR spectrum to increase appreciably, because the equilibrium shifts to form acylalkylnitroxyl radicals of PCA.

Thus, nitrosoamides and nitrites are the major molecular products of PVP and PCA nitration. Stable nitroxyl radicals are also formed. Their structure and the mechanism of their formation provide evidence for the selectivity of the primary reaction of hydrogen atom abstraction from macromolecules. Either N-H bonds (PCA) or C-H bonds at oc-position with respect to the amide group (PVP) are involved in this reaction.

## REFERENCES

[1] *The Chemistry of the Nitro and Nitroso Groups,* Feuer, H., Ed., New York: Wiley, 1969.

[2] Jelinek, H.H.G., *Aspects of Degradation and Stabiliza tion of Polymers,* Amsterdam: Elsevier, 1978, ch. 9.

[3] Karyakin, Yu.V., *Chistye khimicheskie reaktivy* (Pure Chemicals), Moscow: Goskhimizdat, 1947, pp. 15, 18.
[4] Hall, T.C. and Blacet, F.E., *J. Phys. Chem.,* 1952, vol. 20, no. 11, p. 1745.
[5] Djerassi, C, Lund, L., Bunnenberg, E., and Sjoberg, B., *J. Am. Chem. Soc,* 1961, vol. 83, no. 10, p. 2307.
[6] White, E.H., /. *Am. Chem. Soc,* 1955, vol. 77, no. 22, p. 6008.
[7] Ungrande, H.E. and Smiley, R.A., *J. Org. Chem.,* 1956, vol. 21, no. 9, p. 993.
[8] Buchachenko, A.L. and Vasserman, A.M., *Stabil'nye radikaly* (Stable Radicals), Moscow: Khimiya, 1973, p. 199.
[9] Gaponova, I.S., Pariiskii, G.B., and Toptygin, D.Ya., *Vysokomol. Soedin., Sen A,* 1988, vol. 30, no. 2, p. 262.
[10] Lakkaraju, S., Zhang, J., and Roth, H.D., *J. Chem. Soc, Perkin Trans. 2,* 1993, no. 11, p. 2319.
[11] Arens, W. and Bernatt, A., *Tetrahedron Lett.,*1973, no. 43, p. 4281.
[12] Mel'nikov, M.Ya. and Bochkareva, L.I., *Dokl. Akad NaukSSSR,* 1978, vol. 239, no. 1, p. 132.
[13] Maruthamathu, P. and Scaiano, J.C., *J. Phxs. Chem.,* 1978, vol. 82, no. 14, p. 1588.
[14] Mandelhall, G.D. and Ingold, K.U., *J. Am. Chem. Soc,* 1973, vol. 95, p. 2963.

In: Chemical Physics of Pyrolysis, Combustion and Oxidation ISBN 1-59454-163-9
Editors: A. A. Berlin et al., pp.145-174 

*Chapter 16*

# FILLER AND DECREASING OF FLAMMABILITY OF POLYMER MATERIALS

***M. A. Shashkina, A. A. Donskoi and G. E. Zaikov****
Institute of Aviation Materials, 17 Radio st., Moscow 119105, Russia
*N.M.Emanuel Institute of Biochemical Physics, 4 Kosygin st., Moscow 119991, Russia

Fire-resistant polymeric materials represent composites containing many components of various end uses: fillers, plasticizers, combustion decelerators, and others.

A great variety of substances have been serving as combustion decelerators and smoke formation suppressors are suggested, among which the most important are mineral natural and synthetic inorganic compounds: asbestos, talc [1], compounds from bismuth and molybdenum or their mixture [2], aluminum oxide and hydroxide [3], ammonium salts [4], and metal carbides [5].

In many works devoted to reduction of polymeric material combustion, fillers are subdivided into active and inert ones.

Mineral compounds unchanging during thermal influence are classed as inactive ones. However, this is quite relative, because particular temperature conditions of material operation, oxidant presence and other factors should be taken into account. For example, at temperature below 900°C, asbestos fiber in combustible polymeric matrix is a simple inert diluter with low heat conductivity. In the temperature range of 900 – 1,400°C, crystallization water is detached, and this process consumes thermal energy.

Above 1,400°C endothermal processes of asbestos chemical structure reconstruction are observed. Chemical interaction of asbestos components with polymer degradation products is possible. Under high-temperature influence, the effect of inorganic fillers in the polymeric matrix is traditionally reduced to a decrease of combustible component concentration, change of thermophysical properties of the system, and physical transformations of the filler.

Solid surface displays various effects on elastomers. Rubber chains are oriented in the border layer in the presence of the dispersed phase. It depends on the dispersion of particles and their affinity to rubber. Vulcanizing agents are sorbed by the surface of the dispersed phase particles. This process can be both of physical and chemical type. The filler surface can cause a catalytic effect on reactions of vulcanizing group components with one another, rubber and the surface. Application of polar groups reduces orientation effects of rubber molecules at the surface. Injection of fillers affects variation of both strength properties and thermal degradation kinetics [6 - 8]. One of the methods of reducing combustibility and increasing fire and heat shield properties is injection of incombustible fillers diluting the combustible phase during high temperature effect on the material or, passing through a series of physical and chemical transformations, changing the heat balance in both condensed and the gas phase.

The effect of fillers reducing polymer combustibility can be studied help the help of the scheme shown in Figure 1 presenting the elementary model of polymeric material combustion. Analysis of the given model shows that the material combustion is decreased in three main directions:

- Change of thermophysical properties of the material;
- Target change of the state and properties of the condensed phase;
- Reduction of combustibility of the gas phase formed.

Thermophysical properties of the material, described by absorption, reflection, irradiation and transmission factors, as well as by heat conductivity, heat capacity and temperature conductivity, depend on the compounding and interaction of components during production and under the effect of high temperature. Organic compounds, to which rubber belongs, possess similar thermophysical properties. That is why their effect shows up mostly on two other directions. Thermophysical properties of the material can be target changes by injecting mineral fillers.

However, mineral fillers injected into a composite material may serve a dual role: some inorganic compounds used as combustion decelerators are also catalysts of polymerization and polycondensation during polymer synthesis [9]. As a result, these compounds are capable of displaying catalytic functions in relation to degradation products during thermal degradation, accelerating or aiming structuring and coke formation.

Of great interest are reactions of hydrocarbon dehydrocyclization and aromatization in the presence of catalysts, among which titanium, cobalt, aluminum oxides and aluminum phosphates are active. The mechanism of highly effective combustion decelerators of the intumescent type is associated with catalysis of the coke formation reactions proceeding with participation of either a polymer or a coal-forming component, specially injected into the intumescent system [10].

Hence, it can be concluded that inorganic fillers participate in realization of all three discussed mechanisms.

Organic fillers mostly affect transformations proceeding in the gas and condensed phases, therewith absorbing the heat energy.

Table 1 shows characteristics of the fillers reducing combustibility of rubbers [1].

Analysis of data shown in Table 1 show that aluminum, magnesium and titanium oxides remain unchanged below test temperatures of fire and heat shield material. Aluminum chloride, antimony compounds, pentaerythritol and benzguanidine melt and decompose at lower temperatures, which increases fire shield capabilities of the material. Antimony chlorides cannot be used for filling high-molecular elastomers because of low decomposition temperature, which is lower than processing temperature of these polymers.

**Table 1. Properties of compounds used as fillers in rubbers**

| Compound | Density, $g/cm^3$ | Melting temperature, °C | Boiling point, °C |
|---|---|---|---|
| Aluminum oxide | 3.5 | 2,050 | 2,980 |
| Aluminum chloride | 2.47 | 190 | 180.7 (ignit.) |
| Aluminum hydrochloride | 2.4 | Decomposes | - |
| Magnesium oxide | 3.5 | About 2,800 | - |
| Antimony trioxide | 5.2 | 656 | About 1,500 |
| Antimony trichloride | 3.14 | 73 | 223 |
| Antimony pentachloride | 2.33 | 4 | 140 (decomp.) |
| Silicon dioxide | 2.20 | 1,725 | |
| Titanium dioxide | | | 2,590 |
| Anatase | 3.84 | 1,840 | - |
| Rutile | 4.26 | 1,825 | - |
| Pentaerythritol | About 1.0 | 260 | - |
| Melamine | 1.573 | 354 | Ignit., decomp. |
| Benzguanidine | - | 225 | 333 (decomp.) |

## 1. Inorganic Fillers and their Function in Decreasing Combustibility

A study of silicon, aluminum and antimony oxides, aluminum chloride and hydrochloride, boron nitride, and calcium carbonate, as well as mixtures of fillers shows that fire and heat shield properties depend on the filler type (Table 2) and its amount in the composite (Table 3). The Table shows tests results on fire and heat shield properties of samples of composites with the filler concentration of 50 wt. parts per 100 wt. parts of sulfochlorinated polyethylene.

Table 2 shows that silicon dioxide and aluminum hydrochloride are the most efficient under the conditions of heat flux (1,100°C) effect on the sample surface. Table 3 shows the dependence of fire shield properties on the filler amount according to estimation of shielded surface temperature change under the effect of 1,100°C temperature on the material.

**Table 2. Change of protected surface temperature with time**

| Filler | Test duration, min | | | | | | | | | | Time taken to reach 400°C, min |
|---|---|---|---|---|---|---|---|---|---|---|---|
| | 1 | 2 | 3 | 4 | 5 | 6 | 7 | 8 | 9 | 10 | |
| No filler | 25 | 60 | 100 | 170 | 220 | 310 | 460 | - | - | - | 6.5 |
| Aluminum oxide | 20 | 50 | 90 | 120 | 140 | 165 | 200 | 230 | 300 | 380 | 11 |
| Aluminum hydroxide | 25 | 45 | 90 | 120 | 145 | 170 | 200 | 220 | 235 | 250 | 14 |
| Aluminum hydrochloride | 42 | 60 | 95 | 110 | 120 | 130 | 140 | 153 | 159 | 165 | 15 |
| Antimony trioxide | 30 | 55 | 100 | 135 | 170 | 200 | 220 | 250 | 300 | 360 | 11 |
| Calcium carbonate | 35 | 70 | 120 | 180 | 215 | 240 | 270 | 300 | 350 | 385 | 11 |
| Silicon dioxide | 30 | 50 | 80 | 120 | 145 | 175 | 200 | 220 | 235 | 250 | 17 |
| Boron nitride | 25 | 45 | 120 | 140 | 170 | 200 | 230 | 250 | 270 | 310 | 12 |

**Table 3. Influence of filler concentration on fire and heat shield properties of the material**

| Filler | Concentration, wt. parts | Protected surface temperature at test duration (min), °C | | | | | | | | | | | | | | |
|---|---|---|---|---|---|---|---|---|---|---|---|---|---|---|---|---|
| | | 1 | 2 | 3 | 4 | 5 | 6 | 7 | 8 | 9 | 10 | 11 | 12 | 13 | 14 | 15 |
| Aluminum oxide | 10 | 22 | 58 | 99 | 142 | 180 | 216 | 247 | 270 | 292 | 310 | 333 | 351 | 373 | 390 | 405 |
| | 30 | 27 | 49 | 76 | 103 | 126 | 153 | 169 | 183 | 211 | 234 | 261 | 297 | 319 | 339 | 360 |
| | 50 | 30 | 50 | 80 | 120 | 145 | 175 | 200 | 220 | 235 | 250 | 265 | 285 | 330 | 350 | 370 |
| | 60 | 22 | 54 | 93 | 126 | 153 | 180 | 202 | 225 | 243 | 280 | 287 | 313 | 328 | 351 | 370 |
| | 70 | 27 | 50 | 90 | 126 | 159 | 180 | 211 | 220 | 238 | 252 | 279 | 310 | 333 | 351 | 368 |
| Aluminum hydroxide | 30 | 25 | 50 | 105 | 140 | 170 | 180 | 210 | 280 | 520 | - | - | - | - | - | - |
| | 50 | 40 | 50 | 95 | 110 | 120 | 130 | 140 | 145 | 150 | 160 | 180 | 240 | - | - | - |
| | 80 | 35 | 70 | 85 | 95 | 110 | 120 | 125 | 135 | 140 | 150 | 155 | 160 | 170 | 190 | 255 |
| | 100 | 42 | 54 | 78 | 93 | 111 | 126 | 126 | 141 | 147 | 153 | 162 | 168 | 176 | 185 | 200 |
| | 120 | 36 | 51 | 69 | 87 | 105 | 120 | 130 | 144 | 153 | 162 | 168 | 174 | 183 | 185 | 204 |

Figure 1 plots the curve of time dependence of protected surface heating up to 400°C.

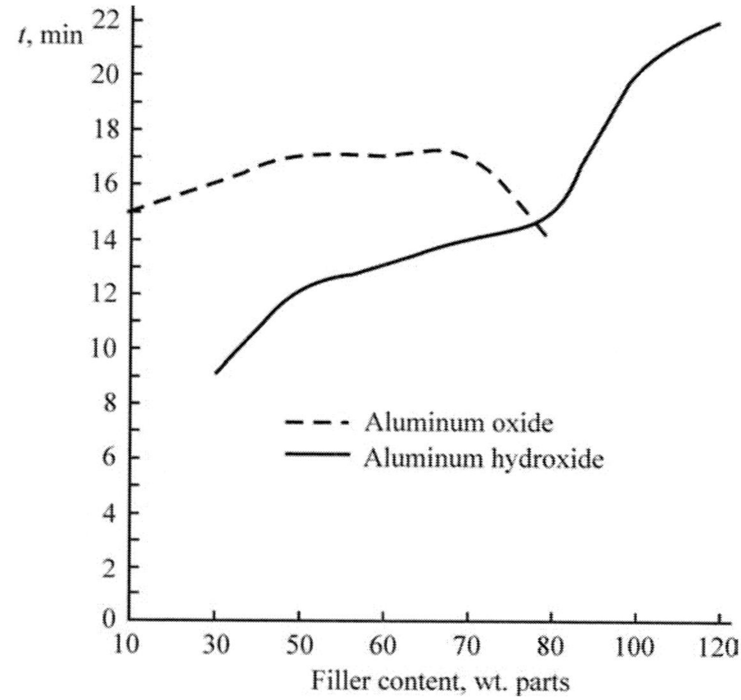


**Figure 1.** Influence of filler concentration of time taken to reach 400°C

Analysis of data from Table 3 and Figure 1 indicates that silicon dioxide possesses the optimal filling degree equal to 30 – 50 wt. parts. Filling degree increasing does not improve fire and heat shield properties. Such dependence is typical of hot condensed phase diluters. In the case of aluminum hydrochloride, fire and heat shield properties are improved with increase in the filler concentration, which is associated with thermal transformations of this compound. Thermal influence on aluminum hydrochloride and materials filled with it induces crystallization water detachment first initially, thereafter chlorine is substituted by oxygen. Further on, aluminum oxide obtained in thermal transformations serves as the fire shield component. However, despite good fire shield properties, application of materials filled with aluminum hydrochloride is limited due to high corrosion activity of this compound.

To reduce corrosion activity, the possibility of substituting a part of aluminum hydroxide by silicon and aluminum oxides has been studied. Table 4 shows results of tests on fire and heat shield properties of materials filled with silicon dioxide and aluminum hydrochloride mixture.

**Table 4. Dependence of fire and heat shield materials on content of fillers**

| Concentration, wt. parts | | Temperature of shielded surface by minutes of tests, °C | | | | | | | | | |
|---|---|---|---|---|---|---|---|---|---|---|---|
| Silicon oxide | Aluminum hydrochloride | 1 | 2 | 3 | 4 | 5 | 6 | 7 | 8 | 9 | 10 |
| 80 | - | 35 | 70 | 80 | 98 | 120 | 140 | 162 | 180 | 204 | 228 |
| 60 | 20 | 30 | 51 | 84 | 108 | 128 | 147 | 159 | 171 | 180 | 186 |
| 50 | 30 | 33 | 60 | 114 | 135 | 150 | 162 | 174 | 183 | 189 | 192 |
| 30 | 50 | 30 | 57 | 81 | 104 | 123 | 141 | 153 | 162 | 168 | 174 |
| 10 | 70 | 33 | 51 | 75 | 102 | 120 | 135 | 150 | 159 | 165 | 172 |
| - | 80 | 35 | 70 | 85 | 95 | 110 | 120 | 125 | 135 | 140 | 150 |

Analysis of data from Table 4 shows that the optimal ratio of silicon dioxide and aluminum hydrochloride is 60:20 and 50:30, respectively. In the case of these ratios, the time taken to reach 400°C on the shielded surface is 23 – 25 minutes. However, corrosion activity is high even at this concentration of aluminum hydrochloride, which limits the application field of these materials.

Oxygen factor is one of the indices of fire resistance of materials that allows prompt estimation of efficiency of fillers used at minimal consumption of the material. Table 5 shows composition and oxygen index of composites with different mineral fillers.

Analysis of data from Table 5 shows that the maximum oxygen index is obtained by applying a complex of fillers. The same composite possesses the best fire and heat shield properties.

Material combustibility is also essentially affected by the amount of filler injected into the composite. This is described well by data shown in Table 6.

Table 6 illustrates the influence of metal oxide nature and concentration, as well as test conditions, on the oxygen index value. Figure 2 shows the effect of metal oxide concentration on variation of the material oxygen index. The flame spreading rate by the sample surface and the oxygen index depend on test conditions. Changing test conditions to candle combustion, horizontal or vertical from bottom to top ignition, and the presence of heat-insulating support show up noticeably on test results, which is clearly observed from Table 6. Sizes of samples and the reactor, and the flow rate of nitrogen-oxygen mixture correspond to standard conditions of the oxygen index determination.

**Table 5. Composition and oxygen index of composites with mineral fillers**

| Filler | Filler concentration, wt. parts | | | | | | | | | | | | | |
|---|---|---|---|---|---|---|---|---|---|---|---|---|---|---|
| Aluminum oxide | - | 80 | 48.1 | 45 | 35 | 25 | 25 | 7 | 7.5 | - | - | - | - | - |
| Zinc oxide | - | - | - | 5 | 5 | - | - | - | - | 48 | - | - | - | 50 |
| Silicon dioxide | - | - | - | - | - | - | - | 43 | 42.5 | - | - | - | 50 | - |
| Titanium dioxide | - | - | - | - | - | 25 | - | - | - | - | - | 25 | - | - |
| Antimony oxide | - | - | - | - | - | - | 25 | - | - | - | - | - | 50 | 50 |
| Chromium oxide | - | - | 1.9 | - | - | - | - | - | 0.0015 | 2 | - | - | - | - |
| Calcium carbonate | - | - | - | - | 10 | - | - | - | - | - | - | 25 | - | - |
| Aluminum hydrochloride | - | - | - | - | - | - | - | - | - | - | 50 | - | - | - |
| Chloroparaffin | - | - | - | - | - | - | - | - | - | - | - | - | 20 | 20 |
| Oxygen index | 24 | 27 | 26 | 28 | 28 | 26 | 34 | 30 | 30 | 25 | 32 | 25 | 42 | 33 |

**Table 6. Influence of the amount of metal oxides on combustibility of sulfochlorinated polyethylene vulcanizates**

| Combustibility indices and flame spreading direction | Metal oxide concentration per 100 wt. parts of polymer, wt. parts | | | | | | | |
|---|---|---|---|---|---|---|---|---|
| | Antimony | | | | Silicon | | | |
| | 0 | 30 | 50 | 70 | 0 | 30 | 50 | 70 |
| Oxygen index at combustion from the top down, % | 27.8 | 37.8 | 40.0 | 42.7 | 27.8 | 34.8 | 39.5 | 42.5 |
| Oxygen concentration, % | 28 | 38 | 40 | 43 | 28 | 36 | 40 | 43 |
| Flame spreading rate from bottom to top, 10 mm/s | 2.01 | 0.56 | 0.52 | 0.85 | 2.01 | 0.7 | 1.18 | 0.73 |
| Oxygen index at combustion from bottom to top, % | 20.5 | 34.5 | 38.5 | 37.8 | 20.5 | 27.5 | 29.5 | 30.5 |
| Oxygen index* at horizontal combustion, % | 26.8 | 38 | 40.5 | 43 | 26.8 | 33.8 | 38 | 40 |
| Oxygen concentration, % | 27 | 38 | 41 | 43 | 27 | 34 | 38 | 40 |
| Flame spreading rate* at horizontal combustion, 10 mm/s | 0.66 | 0.5 | 0.7 | 0.53 | 0.66 | 0.90 | 0.84 | 0.45 |
| Oxygen index** at horizontal combustion, % | 31 | 43.5 | 45.5 | 48 | 31 | 41 | 48.5 | 52.8 |
| Oxygen concentration, % | 31 | - | 46 | 48 | 31 | 41 | 49 | 53 |
| Flame spreading rate** at horizontal combustion, 10 mm/s | 1.64 | - | 1.82 | 0.68 | 1.64 | 1.75 | 2.5 | 1.35 |

* tests performed without heat insulating support;

** tests performed with asbestos support.

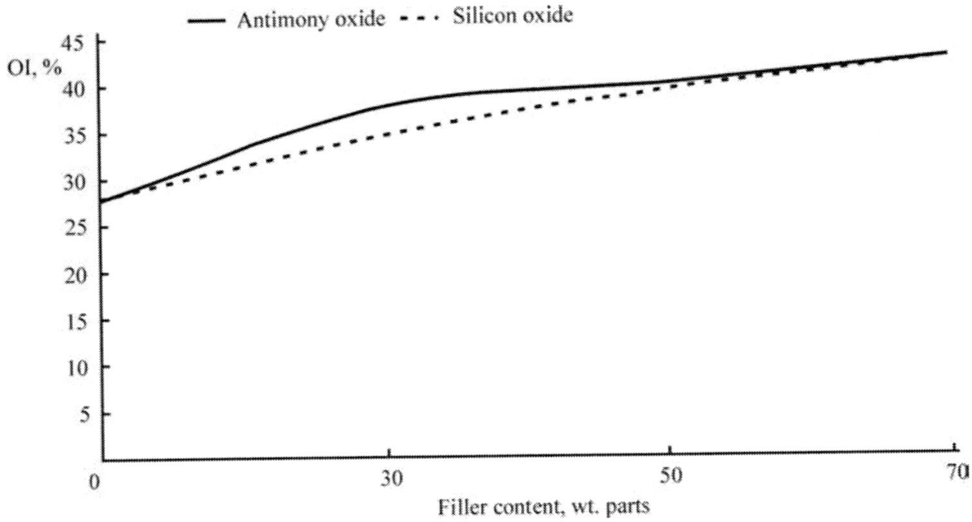


**Figure 2.** Influence of amounts of antimony and silicon oxides on the oxygen index of materials

Orientation of samples in space, flame spreading direction, and the support presence show up on combustibility indices. Values of the oxygen index at candle combustion and horizontal flame spreading by the sample surface in the absence of support are close. The presence of support leads to a noticeable rise of the oxygen index but to increase of the spreading rate of flame by the surface, too. This shows that a significant amount of heat is released from the combustion zone through the condensed phase and the support. The oxygen index varies most significantly at transition from the candle combustion to the one from bottom to top (Table 6). Therewith, the rate of flame spreading by the sample surface is variable, increasing as the combustion front moves. Under strict combustion conditions, unsaturated sulfochlorinated polyethylene vulcanizate is capable of igniting and combusting in the air medium (OI = 20.5).

As metal oxides are injected, combustibility of materials decreases. At the filler concentration of 30 wt. parts, antimony trioxide is more effective than silicon dioxide. However, as the concentration of metal oxides is increased up to 50 – 70 wt. parts, the influence of the filler nature on the oxygen index is leveled in tests under standard conditions. The flame spreading rate at the combustion limit of sulfochlorinated polyethylene saturated vulcanizates is by 2 – 4 times lower than that of unsaturated ones. Compared with silicon dioxide, high efficiency of antimony oxide is much clearer at combustion from bottom to top.

Ref. [12] shows results of studying composites based on CPE with antimony trioxide concentration of 10, 20 and 30 wt. parts. The oxygen index increases with chlorine concentration in the polymer, and this effect is intensified by adding antimony trioxide.

The mechanism of thermal degradation of PE and CPE without fillers and in the presence of antimony trioxide was studied by TGA and DTA methods. The thermogravimetric method shows that PE does not degrade below 400°C, above which intensive degradation proceeds terminated up to 500°C. CPE degrades in two stages (in analogous fashion to poly(vinyl chloride)). In the first stage, it degrades to some 41 – 43% at 250 – 400°C; the second stage of complete degradation proceeds above 400°C.

Behavior of CPE mixed with antimony trioxide is more complicated: high degradation rate in the first stage and lower one in the second stage. PE study by DTA method displays two endoeffects: at 120°C, which corresponds to crystalline phase melting; and at 450°C, intensive degradation of polymer. CPE displays three peaks, the first and the third of which correspond to those of PE. The second peak at 340°C corresponds to dehydrochlorination (the analogous peak is observed for poly(vinyl chloride)). Antimony trioxide injection into CPE reduces the first peak; the second peak is preceded by somewhat exothermal processes, and the third one is reduced.

Chromatographic analysis of PE degradation products shows that the first peak is not accompanied by release of volatiles; the second peak corresponds to release of ethylene, ethane, propylene, propane, butene, and butane. Analysis of CPE chromatograms displays the absence of gas release at the first peak. Hydrogen chloride and benzene release corresponds to the second peak, and the third peak displays release of the same products, but in smaller amounts. The presence of antimony trioxide in PE does not change the gas phase composition. Benzene and ethane yields increase in CPE with the filler concentration.

The following conclusions were made on the basis of the data obtained:

1. Reduction of the polymer combustibility with increase of chlorination degree is associated with dilution of the gas phase by hydrogen chloride.
2. In the condensed phase, dehydrochlorination leads to formation of intermolecular cross-links and carbonization disturbing heat and mass exchange between the flame zone and the material.
3. Injection of antimony trioxide leads to antimony trichloride formation, which inhibits combustion on the polymer surface.

## 2. Organic Compounds Increasing Fire Shield Properties of Materials

Organic fillers by their nature and element composition are mostly combustible materials. However, their application is able to change favorably the process of polymeric matrix degradation and combustion as a whole. Let us turn back to the elementary model of polymeric material combustion and decide which processes the organic filler injected can affect. Firstly, the organic compound degrading releases a great amount of gaseous products, among which there are combustible and incombustible ones. Degradation is accompanied by various heat effects with heat absorption and release, which can show up essentially on the heat balance of thermal degradation of the entire material. Moreover, additives or fillers injected can affect structuring processes in the polymeric matrix during both processing and preparation of materials, and at the stage of operation under thermal influence, aiming the process at cross-linking macromolecules and coke formation in the condensed phase. This reduces formation of combustible gaseous products. Moreover, the fillers can affect the heat exchange by changing conditions of heat and mass transfer in both condensed and the gas phase due to properties of the fillers themselves and their degradation products.

Hence, it is clear that injection of fillers changes polymeric material combustion conditions in three main ways.

Compounds complexly affecting thermal transformations in the material are the most profitable.

Among previously discussed inorganic compounds, aluminum hydrochloride possesses the most full complex of these properties. Analysis of data from literature suggests that several inorganic fillers are similar to aluminum hydrochloride by the decomposition type and the presence of thermal effects.

Melamine cyanuride, azoisobutyronitrile, azodicarbonamide, pentaerythritol, and benzguanidine, the thermal degradation of which is accompanied by essential endoeffects, are known as organic fillers degrading with release of a great amount of gaseous products. In the case of the first three fillers, investigation of fire and heat shield properties under thermal influence displays material degradation proceeding during the first five minutes. This is associated with active gas formation in the material mass, which leads to occurrence of internal stresses degrading the material. The use of pentaerythritol and benzguanidine combined with silicon dioxide as the fillers produces efficient fire and heat shield materials.

Study of pentaerythritol and benzguanidine by the method of differential thermal analysis, the results of which are shown in Figure 3, showed the presence of significant endoeffects during thermal degradation of these products. Tests of composite materials containing these fillers displayed a dependence of fire and heat shield properties on the filler amount in the system passing through the optimum. The optimal concentration of pentaerythritol is 20 – 30 wt. parts.

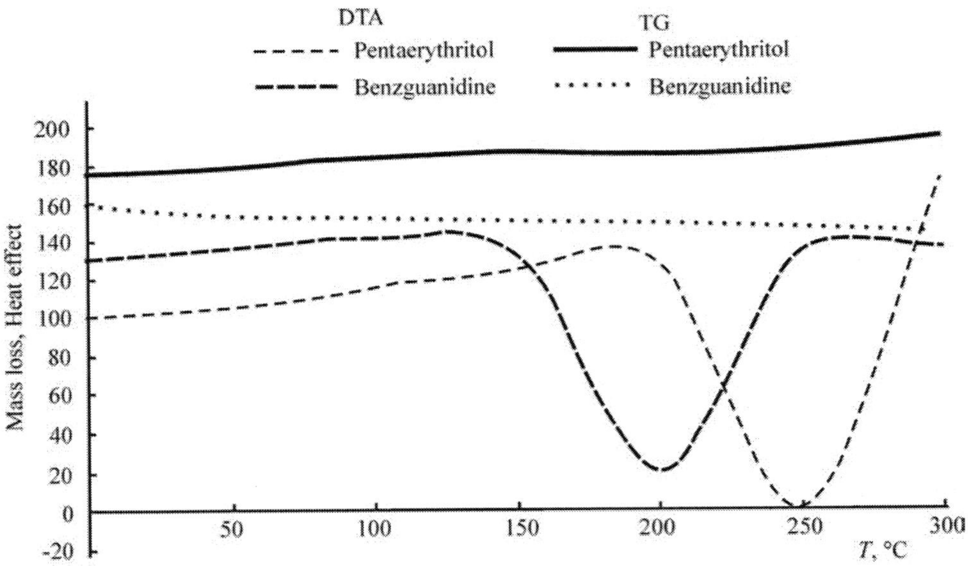


**Figure 3.** Benzguanidine and pentaerythritol DTA results

Figure 4 shows the dependence on benzguanidine concentration of fire and heat shield properties of the material based on sulfochlorinated polyethylene.

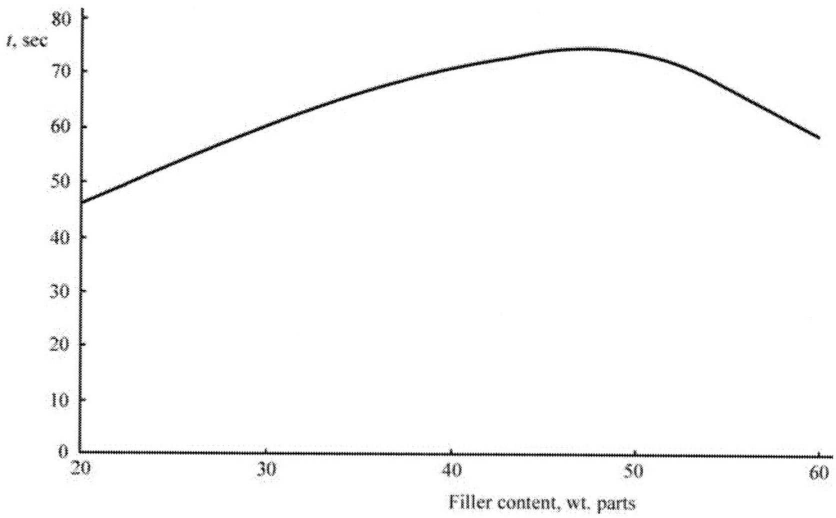


**Figure 4.** Influence of benzguanidine concentration on time taken to reach 400°C on the protected surface

It can be concluded from the test results that the optimal concentration of this compound equals 50 wt. parts per 100 wt. parts of sulfochlorinated polyethylene.

However, insufficiently good fire and heat shield properties of the materials obtained necessitated injection of a complex of mineral and organic fillers. Results of these tests are shown in Table 7.

**Table 7. Temperature dependence of the protected surface on the filler type**

| Filler | Temperature during tests by minutes, °C | | | Time taken to reach 400°C, min |
|---|---|---|---|---|
| | 5 | 10 | 15 | |
| Antimony oxide | 170 | 350 | > 400 | 11 |
| Antimony oxide + chloroparaffin | 160 | 280 | > 400 | 13.5 |
| Benzguanidine | 111 | 400 | - | 10 |
| Benzguanidine + antimony oxide | 192 | 400 | - | 10 |
| Benzguanidine + silicon dioxide | 138 | 213 | 400 | 15 |
| Benzguanidine + silicon dioxide + antimony oxide | 99 | 213 | > 400 | 12 |

Analysis of Table data shows that the complex combining benzguanidine and silicon dioxide is the most effective by the operation duration. In the initial stage of operation (5 min), the lowest temperatures of the protected surface can be obtained with benzguanidine application.

# 3. Combustion Decelerators. Operation Mechanism and Injection Methods

Fire resistance of sulfochlorinated polyethylene vulcanizates is quite high owing to chlorine content in the polymer. Injection of mineral and organic fillers increases fire resistance and, thence, fire shield properties of the materials.

Fire resistance of the materials can be improved by using efficient combustion decelerators.

Analysis of the data from literature indicates that fire shield properties are possessed by certain specific elements. Among them are halogens, phosphorus, nitrogen, several metals, platinum, in particular. However, injection of pure elements into polymeric materials is a quite difficult operation sometimes because of their physical state or poor miscibility with the polymer. So, a great number of different compounds, containing the above-mentioned elements, have been developed. The main general requirements imposed upon polymer combustion decelerators need to be considered in choosing these compounds. They are miscibility with polymer, stability at temperature of material or composite processing into articles, preservation of technological effectiveness, quite good physicomechanical and other operation properties.

Mineral filler (antimony trioxide), for which chlorine-containing compounds are synergists, is the effective antipyren for chlorinated polymers.

Compounds with phosphorus, nitrogen and halogens were taken up as effective combustion decelerators. However, compounds containing these elements altogether are the most effective, but in this case, a definite ratio between the elements must be preserved. The ratio is unique for every particular polymer. Bis-(oxymethyl)-phosphinic acid derivatives can be used as combustion decelerators. Using this example, the influence of the element composition and structure of combustion decelerators on heat shield properties was estimated. Phosphorus-organic compounds with or without halides were studied and the effect of the following structural features was noted:

1. Nature of functional groups bonded with phosphorus by P-S bonds. The structural formula is the following:

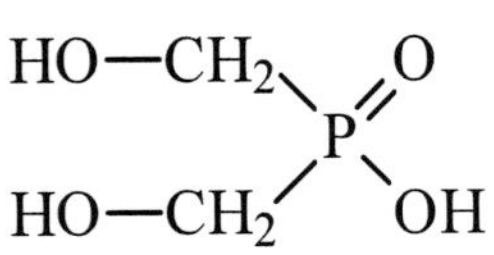
(0)

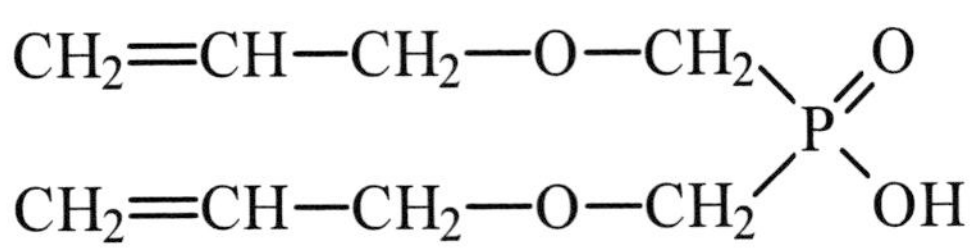
(1)

2. Halogenation degree:

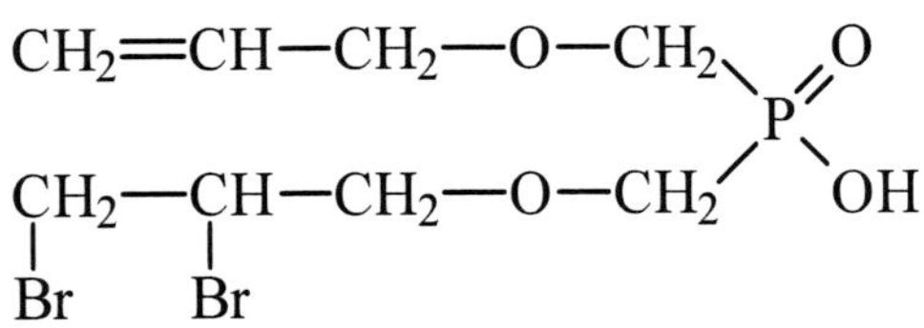
(2)

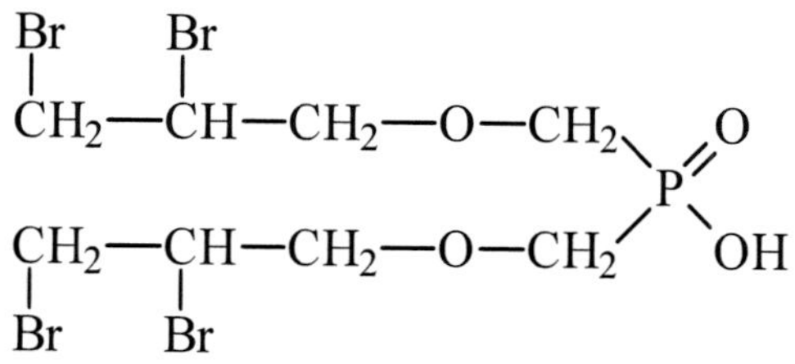

(3)

3. Various functional groups at phosphorus atom:

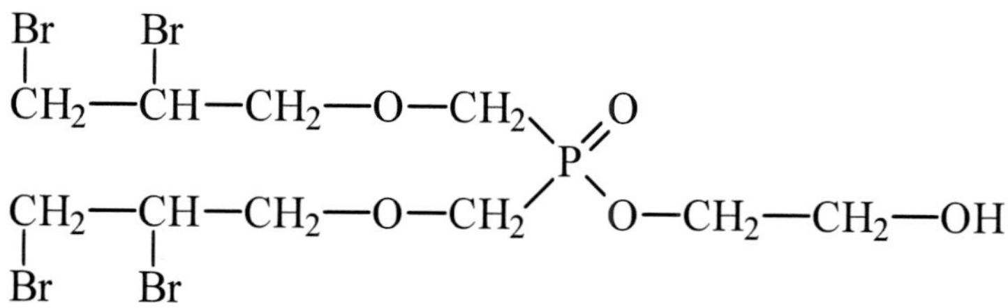

(4)

Estimation of fire and heat shield properties of materials based on composites containing the above-mentioned compounds shows that phosphorus-organic compounds containing no halogens do not improve fire and heat shield properties. Comparing properties of composites containing compounds (3) and (4) with equal amount of bromine, the advantage of compound (3) differing by the crystalline structure of the solid phase (the rest compounds represented viscous fluids) was observed. Tests results of fire and heat shield properties of the composites containing these combustion decelerators are shown in Table 8.

**Table 8. Dynamics of protected surface temperature obtained by the use of different combustion decelerators**

| Compound No. | Temperature during tests by minutes, °C | | | Time taken to reach 400°C |
|---|---|---|---|---|
| | 5 | 10 | 15 | |
| - | 145 | 250 | 370 | 17 |
| 0 | 180 | 294 | 400 | 15 |
| 1 | 140 | 315 | 400 | 15 |
| 2 | 153 | 237 | 400 | 15 |
| 3 | 135 | 220 | 300 | 19 |
| 4 | 150 | 240 | 400 | 15 |

Table data show the advantage of the compound No. 3. This may be due to the following factors: higher bromination degree combined with the presence of an acid group, crystalline state of the compound allowing its distribution in the polymeric matrix shaped as microcrystals or aggregates, which form a somewhat crystalline phase in the polymeric matrix, the melting of which consumes an additional heat amount.

One of the prospective directions for increasing fire resistance of sulfochorinated polyethylenes is the application of acid-base complexes of the salt type, formed by alkoxymethylphosphinic acids and organic bases, as combustion decelerators.

It is advisable to use an organic base containing bromine atoms as the main component of complexes of this type. Bromine content produces a favorable effect on fire shield properties. The compound 2,4,6-tribromoaniline capable of forming the following complexes was studied as the above-mentioned additive:

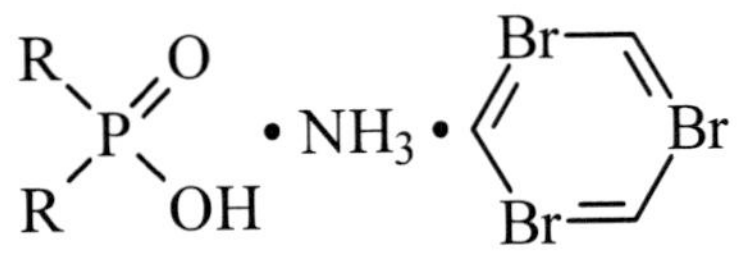


(5)

Compound (5) represents the complex of compound (0) with brominated organic base, and compound (6) is based on compound (1) and the same brominated base. Compound (7) is based on compound (3), and compound (8) on compound (4) combined with the same brominated base. Test results of fire and heat shield properties of materials based on sulfochlorinated polyethylene possessing the above-mentioned compounds in the composition are show in Table 9.

**Table 9. Fire and heat shield properties of materials with phosphinic complexes**

| Complex | Temperature during tests by minutes, °C | | | Time taken to reach 400°C, min |
|---|---|---|---|---|
| | 5 | 10 | 15 | |
| - | 145 | 250 | 370 | 17 |
| 5 | 156 | 260 | 400 | 15 |
| 6 | 140 | 300 | 400 | 15 |
| 7 | 129 | 219 | 264 | 22 |
| 8 | 168 | 312 | 400 | 15 |

Analysis of the study held shows that compound (3) displays the highest indices, which exceed the properties of the initial material. Improvement of fire and heat shield properties by injecting this complex compound into the composite material is probably due to formation of an ionic bond between complex formers.

To clarify the role of complex formation and relative contribution, composites with different ratios of complex forming components (mol %) were taken. Test results of fire-and heat shield properties are shown in Table 10.

Analysis of test results shown in the Table show that phosphinic acid mixed with tetrabromoaniline deteriorates fire shield properties compared with the initial phosphinic acid in all the cases, except one of 50:50 ratio. The sequence of compositions studied displays the clear dependence of fire and heat shield properties on the part of components included into the complex.

Investigations of phosphinic acid salts of nickel, cobalt, iron, ammonium, cadmium and aluminum as combustion decelerators show that the initial phosphinic acid is the most effective. Application of ammonium salt improves physicomechanical properties.

**Table 10. Fire and heat shield properties of materials with the complex of phosphinic acid and tetrabromoaniline**

| Test duration, min | Temperature at phosphinic acid and tetrabromoaniline ratio per minute of test | | | | | |
|---|---|---|---|---|---|---|
| | **100:0** | **80:20** | **60:40** | **50:50** | **40:60** | **20:80** |
| 5 | 135 | 178 | 166 | 129 | 158 | 180 |
| 10 | 220 | 374 | 348 | 219 | 328 | 268 |
| 15 | 300 | - | - | 264 | - | - |
| Time taken to reach 400°C, min | 19 | 12 | 13 | 22 | 13 | 14 |
| Amount of components included into the complex, wt.% | 0 | 20 | 80 | 100 | 80 | 20 |

Another class of combustion decelerators is represented by flamals, which include compounds containing halogens, phosphorus, nitrogen, and aromatic rings in different combinations and positions. These substances can be subdivided into three types on the basis of chemical structure:

- Phosphinic acid ethers;
- A single or several aromatic rings linked by carbon or sulfur atom, with or without substituents;
- Compounds including both previous types of compounds.

Application of these compounds combined with mineral fillers allowed an essential improvement of fire and heat shield properties of materials based on sulfochlorinated polyethylene. Therewith, the rate of temperature increase of the protected surface was decelerated and time taken to reach 400°C on it was increased.

Analysis of the element composition effect on fire and heat shield properties shows that the best properties are possessed by halogen-containing compounds. In this case, bromine is preferable. The presence of aromatic rings in the structure also improves protective properties. Compounds containing sulfur, carbon or phosphorus bridges between rings are also effective. In the case of rings bonding by oxygen or nitrogen, the efficiency of compounds is low. Being the substituent at the bridge element between two rings, chlorine does also reduce efficiency of compounds.

## 4. Methods of Antipyren Injection into Materials

The method of combustion decelerator injection may exert an essential influence on properties of the entire material.

The technological process of rubber and rubber-like material production is usually composed of the following operations:

1. Preparation of components: drying, cutting, weighing.
2. Rubber plasticization.

3. Mixing of components on rolls or in rubber mixture.
4. Calibration of raw rubber sheet.
5. Preparation pattern cutting.
6. Preparation and loading of a press mold.
7. Article molding and vulcanization.
8. Prepared article removal.

Combustion decelerators can be injected into polymeric matrix starting from the moment of polymer synthesis. However, this is rather difficult, because this requires meddling in the synthesis, which is very sensitive to any additives. In this case, the method of polyolefin modification by chlorine injection is applied in industry. The next stage allowing injection of a combustion decelerator into the polymeric matrix is mechanical mixing of components on rolls or in a mixer. Therewith, a series of requirements are imposed on modifying additives. The main ones among them are the following – the components injected must:

- preserve their properties during technological processing;
- not decompose at vulcanization temperature;
- not complicate the technological process;
- not interact with other components of the rubber mixture;
- not cause subvulcanization or premature vulcanization during processing;
- not deteriorate physicomechanical and operation properties of prepared material;
- not sweat out and decompose during storage and operation;
- distribute regularly in the polymeric matrix.

The order of loading components into the mixing device can also show up during mixing. Working off the laboratory technology of preparing fire and heat shield materials shows that injection of combustion decelerators is desirable after injection fillers, but before vulcanization agents. However, because of comparatively low amounts of injected substances, their uniform distribution is difficult. Some combustion decelerators, such as phosphoric acid and red phosphorus, are difficult to inject into the rubber mixer because of increased corrosion activity, explosion hazard, and a series of other reasons.

Modifying additives can be injected into the polymeric matrix applying them preliminarily to the surface of mineral filler. Being the carrier of combustion decelerators, silicon dioxide, which also itself reduces combustibility of materials is the most suitable compound for fire and heat shield materials.

Problems concerning interaction of combustion decelerators with fillers during combustion of filled composite are quite thoroughly discussed in literature. However, it is common knowledge that silicon dioxide is capable of chemical interaction with phosphoric acid at increased temperatures [13], so polymeric composite systems containing phosphorus compounds and silicon dioxide simultaneously have been studied in detail.

The filler influence on properties of the polymeric composite material is mostly defined by processes proceeding on the polymer-filler phase boundary. Of great practical and theoretical interest is the study and application of the methods of changing the chemical nature of the surface of the widespread highly dispersed mineral fillers, i.e. their modification.

Analysis of scientific and technical literature allows us to divide methods of silicon dioxide filler processing into three stages: gas-phase, liquid-phase and mechanochemical. The gas-phase method of powder material surface modification represents processing by various gasified chemical compounds. This process includes preparation of the powder surface for modification, chemical reaction proceeding and elimination of side reaction products in a wide range of temperatures. The liquid-phase method involves a single stage of the processing performed by an individual chemical compound in liquid or diluted in the excess of solvent state. The mechanochemical method differs from the previous ones in that chemical bonding of the reagent with the mineral powder surface is reached by active surface formation with mechanical rupture of bonds between atoms of the solid. This is achieved by performing the chemical process at continuous intensive mechanical action, for example, in the ball or vibratory mill. Modifying reagent can exist in the solid, liquid or the gas state.

The method of processing powder-like materials from silicon dioxide in the gas phase is determined by physicochemical properties of the modifier. Described in the literature is obtaining of surface chemical compounds on silicon dioxide in the form of silica gel and aerosyl by treating by phosphorus trichloride and oxychloride, which represent acid chlorides of phosphorous and phosphoric acids, respectively.

Ref. [14] studied sorption of phosphorus trichloride on the aerosyl surface from the vapor phase at temperature for 25 to 350°C. The authors have observed complete elimination of hydroxyls at room temperature during 35 hours and suggested interaction by the sorption mechanism, which proves reduction of phosphorus amount in aerosyl at high rates of the reaction mixture heating. Modified aerosyl interacts with alcohols and primary and secondary aliphatic amines at room temperature. Ref. [15] is devoted to the study by the methods of infrared spectroscopy and electron microscopy of aerosyl surface, modified by a series of halogens, including $PCl_3$ with future hydrolysis by water vapors and dehydration at 600°C. After dehydration performance in compounds, besides the absorption band of valence oscillations of silanol surface groups (3,750 $cm^{-1}$), a new band at 3,669 $cm^{-1}$ was observed, belonged to free P-OH groups. Besides the band of silanol groups, the band of P-OH groups was also observed at consecutive modification after hydrolysis and hydration. In the authors' view, a discontinuous layer of new oxides involving separate clusters with diameter from several hundreds to thousands of angströms is formed on the aerosyl surface during modification. Hydration-dehydration processes and consecutive modification lead to increase of cluster diameter.

Data obtained by kinetic, IR-spectroscopic and chemical investigation methods [16] show that the reaction of dehydrated silicon dioxide with phosphorus trichloride proceeds rapidly at temperatures above 200°C. Hydrolysis of processed samples leads to secondary occurrence of some silanol groups, as well as small bands devoted to POH groups. The number phosphorus atoms on the surface is 20 – 25% of the total amount of OH groups. The ratio between chloride and phosphorus on the surface exceeds 3:1.

The study of phosphorus trichloride interaction with silica gel gave somewhat different results. The authors of ref. [17] have found that on completely hydroxylated surface of silica gel phosphorus trichloride molecules interact with hydroxyl groups at silicon atoms. Calcination of samples at 600 – 700°C promotes this reaction forming a six membered ring including atoms of oxygen, silicon and phosphorus.

The detailed study of phosphorus trichloride reaction with hydrated aerosyl surface in a wide range of temperature by gravimetry methods combined with chemical analysis [18] suggested progress of two parallel reactions – nucleophilic substitution of hydroxyl group at silicon atom on the surface and electrophilic hydrogen substitution in the hydroxyl group with formation of ≡SiCl and ≡Si-O-$PCl_2$ surface compounds, respectively. Therewith, the degree of phosphorylation of the surface reduces with temperature, and the degree of substitution of silanol groups by chlorine increases.

Preparation of an applied catalyst for olefin polymerization using phosphorus trichloride and a solid oxide carrier is described in the literature [19, 20]. In the case of silica gel application, the authors describe formation of functional ≡Si-O-$PCl_2$ groups on the carrier. Obtaining a chromium and phosphorus oxides layer of the silica dioxide surface by consecutive adsorption of chromium chloride and phosphorus trichloride in the inert gas flow at temperature below or equal to 200°C with further treatment by water vapor and calcination has been described [20].

Processing of silica gel surface by phosphorus oxychloride is thoroughly discussed in ref. [21]. The authors have found that chemosorption on hydrated silicon dioxide proceeds with $(\equiv SiO)_3P(O)Cl$ formation. Therewith the substitution degree of hydroxyls by these groups equals 0.37. The increase of phosphorus oxychloride concentration induces a rise in the amount of phosphorus in the sample with simultaneous hydrolysis of every molecule at the sacrifice of two hydroxyls causing $(\equiv SiO\text{-})_2P(O)Cl$ group formation. The amount of the latter increases as the surface is filled up, and concentration of $(\equiv SiO)_3PO$ groups remains constant. The application of silica gels of this type as water vapor adsorbents is suggested [22]. The method of obtaining adsorbent in water vapors including silica gel treatment by a phosphorus compound differs in that silica gel is dried before treatment, and the treatment is performed by phosphorus oxychloride vapors followed by processing in water vapors and elimination of reaction gas products. Silica gels are dried at 200 – 300°C. Treatment by phosphorus oxychloride vapors is performed at 200 – 300°C up to chlorine concentration of 1 – 1.6 mg-at/g in the final product.

Studies of phosphorus oxychloride interaction with aerosyl surface by the method of infrared spectroscopy show that at room temperature silanol groups participating in the adsorption complex composition are formed on the surface. Hydrolysis of the surface chemical compound by water vapor with subsequent evolution of heat leads to occurrence of absorption bands, corresponding to valence oscillations of P-OH groups. Occurrence of an absorption band corresponding to valence oscillations of phosphonilic P=O groups, is observed simultaneously [23]. Analogous absorption bands are observed at modification of silicon dioxide surface in the presence of various forms of adsorbed water. One can suggest that surface chemical compounds of the silicophosphate type, possessing the structure shown below, are formed by oxychloride reaction with silicon dioxide surface and consecutive hydrolysis, as well as in the presence of water:

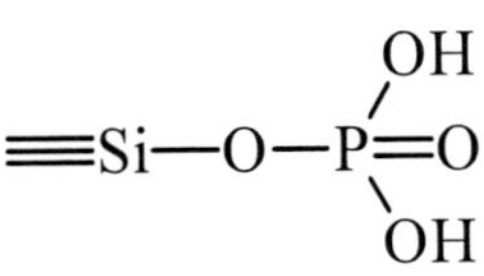


Besides phosphorus oxychlorides, the surface of dispersed silicon dioxide in the gas phase can be treated by phosphates. This is of special interest, because these compounds are described as antipyrens [24].

Triethyl phosphonate reacts quite slowly with hydroxyl groups of the aerosyl surface substituting one-third of the hydroxyl groups at temperature of 200°C. Such phosphates as triethyl phosphite and diethyl phosphite react much more actively [25]. Infrared spectroscopic study of the products obtained revealed decomposition of ethoxy groups on the silicon dioxide surface accompanied by formation of bulky silanol groups [26,27]. Simultaneously, Si-O-P bond is thermally stable, which is indicated by invariability of silicophosphate composition at heating up to 1,000°C [28]. Investigations show that surface ethoxylation of silanol groups with formation of poly(silicic acid) dominates in the chemical reaction of triethyl phosphite with dehydrated aerosyl. Probably, the interaction proceeds according to the scheme of trialkylphosphite dealkylation by acids [29].

It is interesting to note that a P-H bond is formed on hydrated aerosyl at triethyl phosphate adsorption. The absorption band typical of valence oscillations of silamol groups in the formed adsorbed complex, disappearing at the end of the reaction, when all hydroxyl is reacted, is also observed [30].

The presence of various adsorbed water forms on the silicon dioxide surface, significantly increasing the amount of acidic proton centers, changes the direction of the reaction with triethyl phosphite. In this case, surface phosphorylation becomes predominant, whereas diethyl phosphite serves as the phosphorylating agent independently of the presence of adsorbed water forming hydrophosphorylic groups on the silicon dioxide surface. Temperature rise of triethyl phosphite reaction with dehydrated aerosyl up to 340°C also leads to preferable phosphorylation of the surface.

Hence, in reactions with fine silicon dioxide surface depending on the reaction conditions, triethyl phosphite may serve as ethoxylating or phosphorylating reagent, whereas diethyl phosphite is a phosphorylating reagent in any case.

The chemical nature of silicon dioxide surface is changed by modification, at which the inorganic matrix is completely or partly covered by chemosorbed compounds. In the case of applying such filler to polymeric systems, one can suggest that properties of silicon dioxide will vary not only due to the screening effect of the modifying cover, but also by electron density redistribution on the silicon dioxide surface already at low degree of filler particle surface coating by the modifier.

Application of phosphorus trichloride as the modifier of dehydrated silicon dioxide gives incomplete phosphorylation of the surface. Future processing of the surface by methanol vapor causes formation of $\equiv Si\text{-}O\text{-}CH_3$ groups on the surface, and formation of $\equiv Si\text{-}O\text{-}P(OCH_3)_2$ groups may occur.

The above-mentioned information about the influence of phosphorus atoms present in a silicon dioxide surface layer on the chemical structure and, respectively, on thermal properties of phosphorus-containing silicon dioxides illustrates the expediency of applying fillers with antipyretic properties.

## 5. Application of Phosphorus-Containing Compounds on the Silicon Dioxide Surface from the Liquid Phase

The method of modifying the surface of dispersed materials by phosphorus compounds by solution treatment is associated with properties of the modifier. They are mostly complex oxygen-containing phosphorus compounds, nonvolatile under technological conditions and nondegraded by heating. Practically all methods of powder treatment by phosphorus-containing compounds, described in the literature, involve their solutions in appropriate solvents. This process is most often associated with production of catalysts.

Olefin hydration catalyst [31] consists of a carrier impregnated with an acid or from silicic acid. The catalyst is obtained by impregnating a solid porous gel of silicic acid containing 99% or more of silicon dioxide by diluted phosphoric acid and drying it. The gel used is resistant to effects of water solutions and water vapor. The method of forming the modified surface by consecutive treatment of it by various compounds, including phosphorus-containing ones, is also known [32]. This method provides for consecutive treatment of silicon dioxide surface by solutions of metal compounds and phosphorus-containing ones with further thermal treatment. Simultaneous treatment of glassy articles by a mixture of metal compounds and phosphoric oxyacid or a compound capable of forming such oxyacid in solution is also possible [33]. Processing temperature is also varied in a wide range after carrier impregnation.

Mixed dyes are known [34] that represent fine-grained silicate with cover from hydrated amorphous phosphate, phosphite or borophosphite applied on its particles. To apply the cover, the amount of which in the dye varies from 4 to 65%, metal compounds are injected into water suspension of the silicate, a cover precipitating reagent is added (for example, phosphoric or phosphorous acid), and then the suspension is rubbed up.

Various dispersed fillers for polymeric materials treated by phosphorus compounds are described in scientific and technical literature [35].

Interaction products of metal oxides or salts and organic phosphite-phosphonate compounds are applied in natural polymeric materials as antipyrenic smoke-suppressing additive [36].

Of special interest are the data on phosphoric acid effect on the silica gel structure, discussed in ref. [37]. Silica gel samples were impregnated with water solutions of phosphoric acid under atmospheric pressure or in vacuum. When washed off, the samples were dried and calcinated in a muffle furnace. All changes therewith proceeding in the silica gel structure are associated with chemical interaction between phosphoric acid and silicon dioxide. Condensation of phosphoric acid with formation of phosphorous acid polymers also interacting with silicon dioxide proceeds in parallel.

As the mineral filler for fire shield materials based on sulfochlorinated polyethylene, of special interest is silicon dioxide, applied in the silica white form. To increase efficiency of this filler, it was treated by phosphoric acid solution. Hence, filler samples containing 3, 5 and 10% of phosphorus were obtained. Rise of phosphorus concentration in the sample is accompanied by pH decrease of the water suspension at equal time of hydrolysis. However, the increase of modified product treatment temperature reduces its ability to hydrolysis. One may suppose that condensed compounds of silicophosphates and condensed phosphoric acid polymers with P-O-P and Si-O-P bonds are formed on the silica white surface during thermal treatment.

Acidity is reduced due to a decrease of the total number of hydroxyl groups as the amount of oxygen bridges between phosphorus atoms increases. This confirms pH increase. Concentration of hydroxyl groups capable of detaching a proton increases during hydrolysis, and pH value is reduced. At increased temperature, the samples obtained display mass loss, most noticeable at higher phosphorus concentration. The mass loss rate increases at 360°C, which can be induced by the following processes:

- elimination of water formed during condensation of chemosorbed oxycompounds;
- dry distillation of formed phosphorous anhydride modifications.

Differential thermal studies show that the highest rates of the mass change are observed at 140 and 280°C. The end of intensive mass losses is observed at 340°C. Further heating up induces insignificant monotonic mass decrease. DTA curve displays two endothermal peaks at temperatures of 140 and 300°C and one exothermal effect with the maximum at 600°C.

Comparing the above-mentioned data with the results of thermal analysis of silicon-phosphorus-organic compounds [38] (with results of the differential thermal study of crystalline phosphorus pentaoxide with endothermal effect at 341°C and model mixtures based on it), it is permissible to suggest the type of processes proceeding on the silica white surface modified by phosphoric acid during thermal treatment. The effect at 200°C is probably associated with physical removal of sorbed water. Polyphosphinic structures are formed in the temperature range of 200 – 400°C, which is accompanied by degradation and water removal from the system. At heating above 400°C, anhydride polyphosphoric compounds begin interacting with the silica matrix. Therewith, a silicophosphate structure is formed. Occurrence of the exothermal effect with the maximum at 600°C on the DTA curve may be related to transformations with silicophosphate formation. Analogous effects are described by thermal analysis of silicon-phosphorus-organic compounds performed in the above-mentioned work.

The presence of free phosphoric acid on the surface of modified silica white can unfavorably show up on both technological and operation properties of composite materials based on sulfochlorinated polyethylene. In this connection, nitrogen-containing compounds (ammonia gas and urea) were used for neutralizing excessive amounts of acidic groups, formed during silicon dioxide treatment by phosphoric acid water solution. Differential thermal analysis shows an essential change of the curve type. In the case of ammonia gas use, three peaks are observed on DTG corresponding to temperatures at which mass losses are the most intensive. Products of such reactions are eliminated at 130, 230 and 450°C. A broad peak on the DTA curve with the maximum at 550°C may reflect formation of silicophosphate structures.

At neutralization of phosphorus-containing samples by urea, results of differential thermal analysis are of somewhat different form. Three endothermal effects on the DTA curve coincide with the mass loss areas observed in the DTG one. Investigations show the absence of physically sorbed water and free forms of urea. Chemical surface compounds are present on the silicon dioxide surface. They are removed with decomposition in the temperature area of 200, 400 and 540°C. It should be noted that the exothermal effect corresponding to formation of silicophosphate groups is absent on the DTA curve.

The described methods of silicon dioxide surface modification produces fillers possessing different element composition and injection of additives even incompatible with polymer into the polymeric matrix.

## 6. Obtaining of Phosphorus-Containing Silica Filler by Mechanochemical Method

Acceleration and initiation of chemical reactions in gases, liquids and solids resulted in elastic energy absorption related to mechanochemical phenomena. Elastic energy transformation into chemical energy implies mechanical rupture or activation of chemical bonds [39].

Modified products manufacture with the help of mechanical activation is suitable, because it requires the simplest equipment, and material and energy consumption is reduced to the minimum.

The method of surface modification of powder-like materials is based on the surface activation at mechanical destruction of a solid particle. Therewith, active centers formed possess high reactivity, which promotes proceeding of chemical reactions between fine-grained mixture components. In the simplest case, this is reduced to surface oxidation, if the grind is performed in air. It is common knowledge [40] that the possibility of proceeding, direction, and rate of heterogeneous mechanochemical reactions are mostly determined by properties of newly formed surface of solids. The rate of gas or reaction mixture interaction with the solid surface is high at the moment of mechanical treatment and decreases abruptly after its termination. In this case, modification should be performed at combined grinding of mineral filler with modified reagent.

Aerosyl and silica white are synthetic silicon dioxides. They differ by the fact that aerosyl is produced in the gas phase, and silica white – in the liquid phase. Dispersion degree of particles is low, for example, aerosyl particles are sized 4 – 40 nm. At such size of particles, further grinding is hardly possible under mechanical influence. This supposition is confirmed in ref. [39] showing studies of mechanically active silicon dioxide influence on polyethylene resistance to thermooxidation. These studies indicate that new active surface is not formed on aerosyl.

Mechanochemical modification method can be used for obtaining products based on silica white and phosphorus compounds, present in the solid state under normal conditions. In this method, phosphates of alkaline and alkaline-earth metals and red phosphorus can be used as modifiers. Large particles of phosphorus compounds are grained during combined grinding, the solid phase surface increases, and particles of dispersed silicon dioxide prevent their conglomeration locating on the activated surface of the modifier. The filler obtained in this case produces more homogeneous distribution of components in the polymer mass.

Of special interest are fillers containing red phosphorus obtained by mechanochemical mixing. Application of red phosphorus as combustion decelerator of polymeric materials is described in various works. In some of them special attention is paid to the necessity to protect red phosphorus particles. For example, to obtain incombustible composites based on thermoplasts, red phosphorus is suggested containing aluminum hydroxide precipitated on the surface [41]. To increase fire resistance of polymeric materials, red phosphorus was also suggested, shaped as particles of size 200 μm, encapsulated into the cover from polyester resin [42]. At combined grinding with dispersed silicon dioxide, red phosphorus disperses uniformly over the filler surface coloring it pink, which changes to white during storage.

Mechanochemical treatment is simple to perform in industry.

Basing on the above-described methods of obtaining modified mineral fillers, a series of phosphorus-containing mineral fillers based on silicon dioxide was developed.

In the gas phase, phosphorus trichloride, phosphates, triethyl phosphate, triethyl phosphite, and diethyl phosphite were applied to the filler surface.

Silicon dioxide surface was modified in the liquid phase by phosphoric acid solution with further neutralization by ammonia and urea, as well as by the compound with the formula 3-iso-$C_{10}H_{21}OP(O)(OH)_2$.

A mechanochemical method was used to apply phosphates of alkaline and alkaline-earth metals and red phosphorus.

As a result, samples of fillers were developed, characteristics of which are shown in Table 11.

**Table 11. Physicochemical data on samples of modified fillers**

| Sample name | Phosphorus content, % of $SiO_2$ content | Bulk density, g/l |
|---|---|---|
| Treatment by phosphoric acid solution | | |
| BSF-3 | 3 | 311.3 |
| BSF-6 | 6 | 392.7 |
| BSF-10 | 10 | 394.9 |
| Treatment by phosphoric acid and ammonia mixture solution | | |
| BSFA-3 | 3 | 324.8 |
| BSFA-6 | 6 | 360.2 |
| BSFA-10 | 10 | 426.2 |
| Treatment by phosphoric acid and urea mixture solution | | |
| BSFM-3 | 3 | 552.9 |
| BSFM-6 | 6 | 511.0 |
| BSFM-10 | 10 | 575.1 |

Silicon dioxide surface was also treated by organic compounds used as combustion decelerators of polymeric materials. These compounds were applied by solution treatment in butanol. Characteristics of the “flamal” series of products obtained (named after the modifier) are shown in Table 12.

**Table 12. Physicochemical data on samples of the "flamal" series**

| Sample name | Modifier amount, wt.% $SiO_2$ | Mass losses at calcination, % | Bulk density, g/l | Water suspension pH |
|---|---|---|---|---|
| BS-flamal-449 | 15 | 26.33 | 365.7 | 8.25 |
| BS-flamal-315A | 30 | 25.45 | 323.4 | 6.47 |
| BS-flamal-408 | 15 | 27.97 | 319.6 | 8.16 |
| BS-flamal-351 | 15 | 22.61 | 350.1 | 8.32 |
| BS-flamal-369 | 30 | 24.53 | 346.2 | 2.92 |

Filler samples with red phosphorus or phosphates were prepared by the mechanochemical method. Physicochemical data of samples modified by phosphates are shown in Table 13.

**Table 13. Physicochemical data on samples obtained by mechanochemical method**

| Sample name | Modifier amount, % | Mass loss at calcination, % | Bulk density, g/l | Water suspension pH |
|---|---|---|---|---|
| BS/$Mg(HPO_4)_2$ | 50 | 11.45 | 363.17 | 7.85 |
| BS/$Na_3PO_4$ | 50 | 15.95 | 411.99 | 10.2 |
| BS/$KH_2PO_4$ | 50 | 6.36 | 487.62 | 5.34 |

# 7. Study of Biological Activity of Phosphorus-Containing Silicon Dioxides[1]

Biological activity of dispersed materials insoluble in liquids is determined by processes proceeding on the contact border of the surface and these materials with biological objects.

The effect of dispersed silicon dioxide and various silicate powders on living organism tissues is studied in numerous works in chemical, biochemical and medical and biological investigations. The direction of chemical modification of the dispersed filler surface, intensively developed in recent years, and its application for laboratory, test industrial and industrial purposes requires estimation of biological activity of the products obtained. The elementary composition is changed in the range of 0.5 – 10% during chemical modification. However, it includes the surface of particles directly in contact with a biological object. This must determine a more significant change of the total biological activity or specific biological effect on living organisms.

Among various functional active groups, phosphates are of special importance: they participate directly in the energy supply of cells and ATP (adenosine triphosphoric acid) formations, which determines activity of growth and development of the individual. ATP is a component of human and animal tissues. It is formed in the oxidation reactions and during glycolytic decomposition of hydrocarbons. Dephosphorylation releases free energy for

[1] The investigation was carried out in L.V. Pisarzhevski Institute of Physical Chemistry, Academy of Sciences of the UkrSSR, in the department of surface chemistry by V.M. Bogatyirev et al.

muscle activity and takes part in formation of intermediate metabolic products [43]. Taking into consideration that the number of modifications of dispersed silicon dioxides increases, including phosphorus-containing ones, they should be compatible with biological material.

Aerosyl (the most suitable dispersed silicon dioxide for model studies), modified by phosphorus oxychloride, was used in investigations. Chemical transformations proceed on the silicon dioxide surface during synthesis. Test determinations were executed on two types of cells: mast cells of rat abdominal cavity exudate and reproductive cells of horned cattle.

Mast cells were chosen as an alternative of cells possessing high sensitivity to effects of foreign substances and fast response. They belong to a specific functional group of cells from connective tissue, characterized by the presence of granules in cytoplasm, which when colored with aniline dyes indicates the state of the cells. Mast cells are carriers of biologically active substances, released from a cell into the surrounding medium in different immune and allergic situations.

Investigations with mast cells were carried out with the help of microscopic studying cells, colored in neutral red. In this case, the total number of mast cells is counted at the beginning and at the activation maximum, as well as the number of dead cells. For activation, the cell state was taken, characterized by membrane deformation and ejection of granules through the most overloaded areas of it. Aerosyl A-300 was used as the control sample.

As would be expected, mast cells were found active in relation to different types of aerosyl. Different cell responses and selective cell activity were observed. Hence, the test reaction with mast cells indicates a change of silicon dioxide biological activity due to chemical modification. However, the type of changes observed could not be fully estimated without consideration of physiological functions of mast cells in the living organism.

Reproductive cells of horned cattle (bull sperm) were used in the second test-reaction. These cells represent the system of carbohydrate-containing saturated polymers both in plasma and on the cell surface. Mobility of these cells is provided by ATP-synthesizing reactions. Considering participation of phosphate groups in these reactions, the effect of phosphorus-containing aerosyls on moving cells should be tested.

Studies performed on reproductive cells show that the control aerosyl A-300 stabilizes cell surface structures, but does not destroy them. On the contrary, a phosphorus-containing aerosyl destroys cells. Cell tail loss, observed in this case, indicates the inhibiting effect on the energy supply system. The same reactions performed in plasma show a lesser effect.

The results obtained in the reaction with reproductive cells explain also the mechanism of interaction with mast cells. Probably, the ATP-phase present in membranes is locked, so transmissivity of the membrane is changed, and cell content escape is observed.

Hence, phosphorus-containing aerosyl directly contacting a living cell disturbs energetic processes proceeding in it. In the presence of natural medium, the effect is reduced significantly which indicates the protective functions of the medium components.

# 8. Study of Fire and Heat Shield Properties of Materials Filled with Modified Silicon Dioxide

Rubber-like materials, filled with modified silica whites, were prepared on the base of sulfochlorinated polyethylene. The materials were produced according to traditional rubber technology. To estimate efficiency of fillers, studying of fire and heat shield properties was performed on the one-side thermal impact unit.

Silicon dioxides treated by chromium, aluminum, titanium and phosphoric acids were used as the fillers. The results obtained are shown in Table 14.

Analysis of the table data shows the highest efficiency of phosphorus-containing compounds for silicon dioxide treatment. The consecutive work has studied fillers modified in the gas and the liquid phase, and by the mechanochemical method. Studies performed show that the filler efficiency is first affected by the amount of injected phosphorus, but not the nature of modifying agent or modification method. This conclusion is illustrated in Table 15 showing test results of fire and heat shield properties of materials filled with silicon dioxide surface treated by different phosphorus compounds.

**Table 14. Change of protected surface temperature with time for samples of shielding materials filled with modified silicon dioxide**

| Treating substance | Protected surface temperature by minutes of test, °C | | | Time taken to reach 400°C, min |
|---|---|---|---|---|
| | 5 | 10 | 15 | |
| - | 145 | 250 | 370 | 17 |
| Chromium | 175 | 310 | 385 | 16 |
| Aluminum | 175 | 308 | 420 | 14 |
| Titanium | 145 | 270 | 390 | 15 |
| Phosphoric acid | 135 | 240 | 340 | 19 |

**Table 15. Temperature change on protected surface with time for shielding materials filled with silicon dioxide treated by phosphorus compounds**

| Phosphorus compound | Phosphorus content in filler, % | Protected surface temperature by minutes of test, °C | | | | Time taken to reach 400°C, min |
|---|---|---|---|---|---|---|
| | | 5 | 10 | 15 | 20 | |
| - | 0 | 145 | 250 | 370 | > 400 | 17 |
| Trisdibromo-propyl phosphate | 0.7 | 144 | 228 | 345 | > 400 | 18 |
| Phosphoric acid | 1.3 | 138 | 222 | 264 | > 400 | 19 |
| Phosphoric acid | 2.2 | 150 | 246 | 378 | > 400 | 17 |
| Red phosphorus | 3.0 | 165 | 228 | 228 | 325 | 23 |
| Phosphorus oxychloride | 3.4 | 147 | 195 | 246 | > 400 | 19 |
| Red phosphorus | 6.0 | 186 | 234 | 400 | > 400 | 15 |

Table 15 shows that heat shield properties of materials are improved with phosphorus concentration independently of the type of compound injected. For materials based on sulfochlorinated polyethylene, the optimal phosphorus concentration is about 3% of silicon dioxide content, which gives 1.5% converted to polymer, because the amount of the mineral filler injected is 50 wt. parts per 100 wt. parts of the polymer.

Even compounds poorly compatible with polymer and deteriorating physicomechanical properties of composite materials can be introduced into rubber mixtures by applying on the mineral filler surface. Table 16 shows test results of physicomechanical properties of materials filled with silicon dioxide, treated by different mineral and organic compounds.

**Table 16. Physicomechanical properties of materials filled with treated silicon dioxide**

| Treating substance | Phosphorus content in oxide, % | Tensile strength limit, kg/cm$^2$ | Relative elongation, % |
|---|---|---|---|
| - | - | 120 | 300 |
| Phosphoric acid | 3 | 102 | 200 |
| | 6 | 110 | 280 |
| | 10 | 116 | 230 |
| Phosphoric acid + ammonia | 3 | 158 | 160 |
| | 6 | 158 | 170 |
| | 10 | 127 | 170 |
| Phosphoric acid + urea | 3 | 160 | 127 |
| | 6 | 165 | 130 |
| | 10 | 165 | 150 |
| Magnesium biphosphate | 3 | 107 | 280 |
| Sodium phosphate | 3 | 140 | 200 |
| Potassium biphosphate | 3 | 140 | 200 |
| Boric acid: 20 wt. parts | - | 60 | 290 |
| 40 wt. parts | - | 156 | 335 |
| Aluminum chloride + urea | - | 50 | 270 |

Tests of material samples performed on a one-side heating unit show the highest efficiency of materials filled with silicon oxide, treated by phosphoric acid combined with urea. As estimation of the heat amount transmitted through the sample, the surface of which is affected by heat flux at 1,000°C (Figure 5) show that the efficiency of silicon dioxide treated by phosphoric acid is increased by additional treatment by boric acid or ammonia.

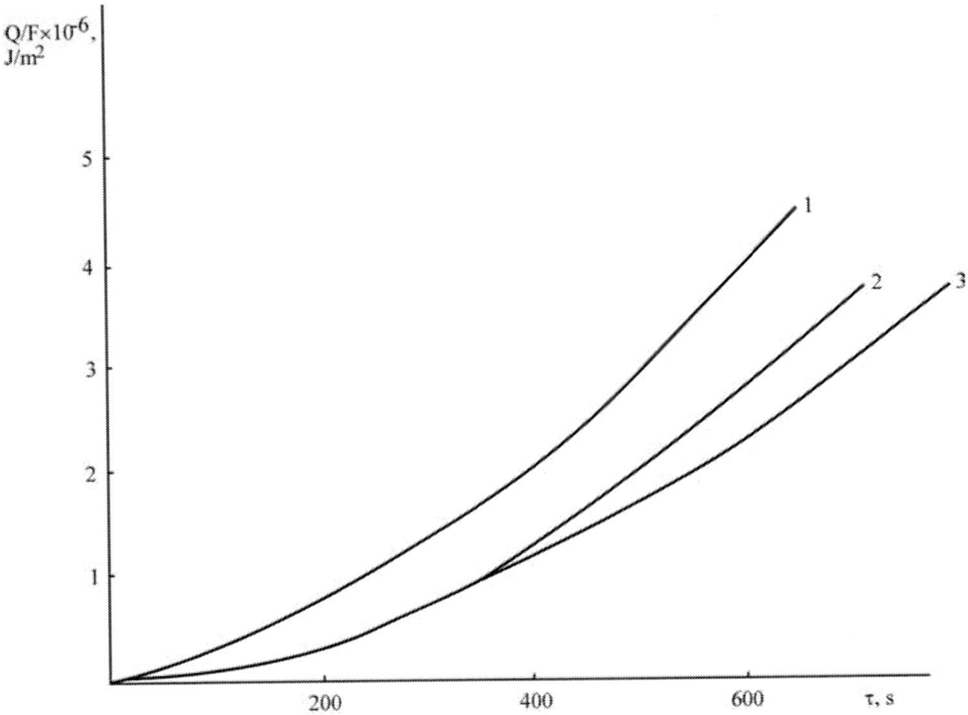


**Figure 5.** The amount of heat transmitted through the sample

## REFERENCES


[1] Hartey P., *Des. Eng.*, 1972, Oct., p. 81.

[2] *USA Patent* No. 3,878,167.

[3] *USA Patent* No. 3,874,889.

[4] *Japan Patent* No. 366,246.

[5] *USA Patent* No. 3,816,367.

[6] Dontsov A.A., Lozovik G.A., and Novitskaya S.P., *Chlorinated Polymers*, Moscow, Khimia, 1979, 232 p. (Rus)

[7] Ronkin G.M., Khromenkov L.G., and Iliin B.A., Thermal Analysis of Carbochain Polymers, *Thes. Proc. 18th All-Union Conference on High-molecular Compounds*, Kazan, 1973, p. 125. (Rus)

[8] Dontsov A.A., *Structuring Processes in Elastomers*, Moscow, Khimia, 1978, 288 p. (Rus)

[9] Dolgov B.N., *Catalysis in Organic Chemistry*, Moscow-Leningrad, Goskhimizdat, 1949, 560 p. (Rus)

[10] Comido G., Costa L., and Luda M.P., *Fourth Annual BBC Conference on Flame Retardancy: Recent Advances in Flame Retardancy of Polymeric Materials*, Stanford, Connecticut, 1992, May 18 – 20.

[11] Perelman V.I., *Concise Chemist's Reference Book*, Moscow, Khimia, 1964, 620 p. (Rus)

[12] Papa Anthony L., Flame-retarding polyurethanes, *J. Flame Retard. Polym. Mater.*, 1975, New-York, vol. **3**, pp. 1 – 133.

[13] Ailer R.J., *The Chemistry of Silica: Solubility, Polymerization, Colloid and SurfaceProperties and Biochemistry of Silica*, Wiley, New York, 1979, 712 p. (Rus)

[14] Neshevich L.A., Vinogradova A.S., and Kachan A.A., Interaction between Phosphorus Trichloride and Aerosil, *Ukr. Khim. Zh.*, 1976, vol. **42**(3), pp. 1109 – 1113. (Rus)

[15] Fink P., Camara B., Welz E., Pham. Dink TV, Adsorption Properties and Structure of Chemically Modified Silicon Dioxide and Silicate Surfaces; Hydroxyl Groups and Structure of Silicon Dioxide Surface, Modified by $PCl_3$, $BCl_3$, $GeCl_4$ and $SnCl_4$, *Z. Chem.*, 1971, vol. **11**(12), pp. 473 – 474.

[16] Hair M.Z. and Whertl ?., Silica Surface Chlorination, *J. Phys. Chem.*, 1973, vol. **77**(17), pp. 2070 – 2075.

[17] Koltsov S.I., Volkova A.N., and Aleckobski V.B., Influence of Silicagel Dehydration Degree on Phosphorus Trichloride Chemosorption Mechanism, Collection, *Inst. Fiz. Khim.*, 1979, vol. **44**, pp. 2246 – 2250. (Rus)

[18] Bogatyrev V.M. and Chuikov A.A., Phosphorus Trichloride Interaction with Dehydrated Aerosil on Its Surface, *Ukr. Khim. Zh.*, 1984, pp. 831- 835. (Rus)

[19] *USSR Patent* No. 468,503, C08F 10/00, B01j 21/00. (Rus)

[20] *USSR Patent* No. 422,446, B01j 11/46. (Rus)

[21] Volkova A.N., Malygin A.A., Koltsov S.I., and Aleskovski V.B., Study of Phosphorus Oxychloride Reaction with Hydrated Silica, *Inst. Obshch. Khimii*, 1973, vol. **43**(4), pp. 724 – 728. (Rus)

[22] *USSR Patent* No. 997,795, B01j 20/10, 1983. (Rus)

[23] Bellami L.J., *Infrared Spectra of Complex Molecules*, Wiley, New York, 1960, 558 p.

[24] *FRG Application* No. 2,013,547, C08h 1/60, 39b 1/60.

[25] Tertyikh V.A., Pavlov V.V., Mashchenko V.M., and Chuiko A.A., Forms of Absorbed and Structural Water on Surface of Dispersed Silica, *Doklady AN SSSR*, 1971, vol. **201**(4), pp. 913 – 916. (Rus)

[26] Kiselev A.V. and Lyigin V.I., Studies of Silica De-ethoxylation by IR- and ESR-spectroscopy Methods, *Kolloidny Zh.*, 1975, vol. **38**(2), pp. 347 – 351. (Rus)

[27] Chuiko A.A., Pavlov V.V., and Sherstyuk A.I., Study of Methoxyl Chemical Compounds Degradation on Silica Surface, *Kolloidny Zhurnal*, 1974, vol. **36**(5), pp. 1012 – 1015. (Rus)

[28] Murashkevich A.N., Pechkovski V.V., Chubarov A.V., and Ivkovich N.A., Studies of Amorphous Silica Interactions with Phosphoric Acid, *Zh. Neorg. Khim.*, 1983, vol. **28**(9), pp. 2190 – 2195. (Rus)

[29] Arbuzov A.E., *Selected Works on Chemistry of Phosphorus-organic Compounds*, Moscow, Nauka, 1976, 560 p. (Rus)

[30] Pavlov V.V., Tertyikh V.A., Chuiko A.A., and Bogatyrev V.M., *Dopovidi AN UkrSSR*, Ser. B, No. 8, 1979, pp. 639 – 641. (Ukr)

[31] *FRG Application* No. 2,015,536, B01j 11/48, 12g 11/48.

[32] *France Patent* No. 1,289,128/23-4, B01j 11/46.

[33] *Great Britain Application* No. 1,406,410, B05D 7/00, C01B 25/37, B2E, C1A.

[34] *Great Britain Application* No. 13,773,047, C01B 33/24, 33/26, C09C 1/28, C08G 39/08, C09D 3/26.

[35] *Japan Application* No. 52-56,145, 25(1), A210.1, C08K 9/04.

[36] *USA Patent* No. 4,313,761, C08K 5/59, C08K 5/53.

[37] Kostrich L.P., Mangasaryan N.A. et al., Influence of Phosphoric Acid on Silicagel Structure, *Kolloidny Zhurnal*, 1983, vol. **45**, pp. 337 – 341. (Rus)

[38] Kirichenko E.A. and Kamaev V.V., DTA of Several Silicon-Phosphorus-organic Compounds, *Trudy MKhTI im. D.I. Mendeleeva*, 1974, Iss. **79**, pp. 153 – 154. (Rus)

[39] Skorik Yu.N., Khoinike G., Kugaeva S.K., and Khenning Kh.P., Influence of Mechanically Active Silica on Polyethylene Resistance to Thermooxidation, AS USSR, Central Institute of Physical Chemistry AS GDR, *Zh. Prikl. Khim.*, 1975, vol. 48(5), pp. 1168 - 1170. (Rus)

[40] Agaev Yu.N. and Butyagin P.Yu., About Short-living Active Centers in Heterogeneous Mechanochemical Reactions, *Doklady AN SSSR*, 1972, vol. **207**(4), pp. 892 – 895. (Rus)

[41] *FRG Application* No. 2,623,112 C08K 0/02.

[42] *France Application* No. 2,373,575, C08K 3/32.

[43] Mashkovski M.D., *Medicinal Means*, Moscow, Meditsina, 1967, vol. 1, pp. 636 – 637. (Rus)

In: Chemical Physics of Pyrolysis, Combustion and Oxidation ISBN 1-59454-163-9
Editors: A. A. Berlin et al., pp.175-187 

***Chapter 17***

# FORMATION PRINCIPLES OF FIRE PROOF MATERIALS CONTAINING NANOSTRUCTURES

***V. I. Kodolov, A. M. Lipanov and S. G. Shuklin***
Basic research-high educational center of Chemical Physics and Mesoscopy,
Udmurt Scientific Center, Ural Division, Russian Academy of Sciences.
222 Gorky St., Izhevsk 426001, Russia, e-mail: *kodol@istu.udm.ru*

## ABSTRACT

Based on the proposed criterion parameters for forming coke surface layers comparative appraisal of possible variants of polymeric materials fire protection was carried out. The efficiency of nanostructure application in polymeric materials to decrease their ignition and combustibility levels was substantiated. The possible ways for developing the modeling of carbonization processes and foamcoke formation during the action of high-temperature flows and flame upon fireproof materials were explained.

**Key words:** fire resistant materials, intumescent fireproof coatings, carbonization, pyrolysis, carbon metal nanostructures, nanophases, nanoreactors, modeling, quantum chemical calculations.

## INTRODUCTION

Previously [1] a number of criteria for providing the formation of fireproof surface layers were proposed, including the carbon ones, the formation of which is connected with several reactions proceeding during the action of heat flows upon the materials. At the same time the approaches for assessing the possibility of such layers formation and methodology for selecting components of fireproof polymeric materials and coatings were developed. The assessments and methodology proposed are quite applicable and can be adapted to fireproof nanocomposites. It is of interest to consider generalized results based on experimental data and theoretical models of fireproof materials and coatings formation taking into account the formation of nanostructures of certain shape and composition in them.

# 1. Comparative Evaluation to Protect Materials from Fire

Based on the experience gained on protecting combustible materials from fire the main variations are in decreasing the possibility of exothermic redox reaction, this can be expressed in oxidizer and/or reducer concentration decrease below a certain limit, temperature drop below critical. This implies considerable changes during ignition and combustion. The ways of protecting polymeric materials from high-temperature flows or flame when using fireproof coatings or introducing fire-retardants into materials are known [1]. However this and other ways for reducing combustibility have certain disadvantages. First, this can cause coating peeling during exploitation or combustion source action. Second, the changes in material basic characteristics, sometimes considerable, can sharply reduce its application fields. Therefore, when fireproof materials are applied, the changes in thermal-physic and adhesive characteristics during fire action are very important, but when fire-retardants are introduced into polymeric materials it is necessary to decrease the content of additives introduced into the material to an effective minimum, increasing their activity to reduce combustibility. In both cases nanophases of certain extension are formed, thus stimulating the formation of organic and inorganic cokes preventing the destruction of basic mass of polymeric materials. During the combustion of fireproof polymeric materials the mixed combustion takes place [2], i.e. in coking process the regularities of heterogeneous combustion appear when unstable combustion takes place [3], if a coked surface occupies over 80% of the surface subject to combustion. When introducing active additives representing nanostructures, that can have the properties of nanocatalysts, into polymeric matrixes, nanophases stimulating self-organization of matrix macromolecules into the "blank" of future coked surface ($S_c$) can be formed. In this case $S_c$ can be written down as a formula resembling Silberberg equation [4]:

$$S_c = l \cdot E(\%) \cdot k_s \cdot k_e,$$

where l – coordination number of fire retarder active element participating in the interaction or maximum possible number of active groups in fire retarder or fire-retardant system; E(%)·- active element optimum content in fire-retardant; $k_s$ – number of macromolecules of first layer adsorbed on fire-retardant active surface; $k_e$ – number of macromolecules of second layer included into the nanophase being formed.

Product $k_s \cdot k_e$ depends on adsorption variants and interactions between fire-retardants and macromolecules of matrix polymers. The decrease of fire-retardant particles and their uniform distribution in a material contributes to the increase of nanophase and "blank" of coked surface, and, consequently, to the decrease of active element content. The combustion values of phosphorus-containing polyesters and semi-empirical calculations that arise from the following aspects are given as an example:

1. Increase in active group or active element content increases their inter-molecular interaction thus leading to the association and formation of particle aggregates, the surface of which is depleted by the centers actively interacting with macromolecules.
2. Decrease in active element content during the dispersion degree growth contrary to the aforesaid intensifies the chemisorption and structuring of polymer macromolecules by analogy with structure-forming agents [3].

3. Since the coked surface cannot exceed 100%, the number of active phosphorous-active groups is less than 4, and experimental effective phosphorous content is 5,9% [1], the product of $k_s \cdot k_e$ approaches 4. At the same time with the decrease of adsorbed molecules number on active centers, the amount of changes in the structure of macromolecules of second layer increases, i.e. coefficient $k_s$ characterizing the chemisorption of macromolecules on active surfaces of fire-retardant decreases together with the growth of active element content, and coefficient $k_e$, characterizing the aggregation degree increases.

From the above theses it follows that the dependences of combustibility and other characteristics, for instance, physic-mechanical ones, density properties, upon fire-retardant content will be described with curves having extremes (maximums and minimums). The combustibility assessment during weight losses can be inversely proportional to coked surface percentage in combustion area (Fig. 1).

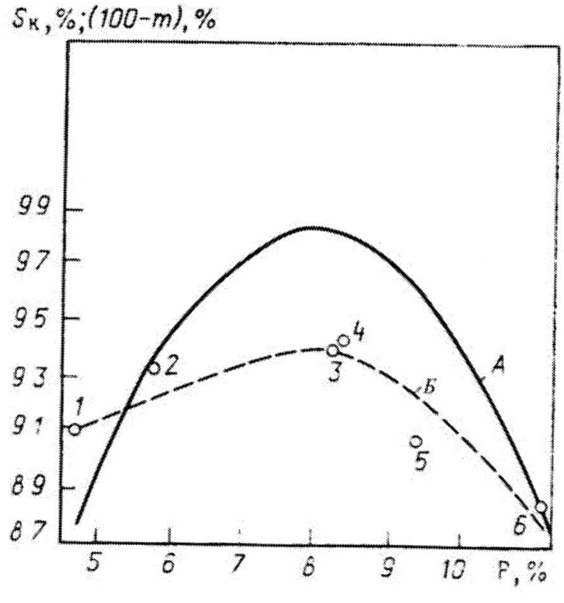


**Fig. 1.** Dependence of $S_c$ upon phosphorus content.

Block curve is calculated on the basis of the aforesaid theses, and dashed-line curve represents experimental data for phosphorus-containing polyesters of similar structure (Table), when Sc ~ (100–m)%.

**Table 1. Changes in weight losses during the combustion with phosphorus content changes.**

| Phosphorus-containing polyesters | Phosphorus content, % | Weight losses during combustion, % | 100–m % |
|---|---|---|---|
| PPE 1 | 4,7 | 10 | 90 |
| PPE 2 | 5,7 | 6,5 | 93,5 |
| PPE 3 | 8,0 | 5,5 | 94,5 |
| PPE 4 | 8,2 | 5,4 | 93 |
| PPE 5 | 9,0 | 7,0 | 93 |
| PPE 6 | 11,2 | 11,5 | 88,5 |

Surely there is no complete coincidence of curves, but their character and location of maximums conforms to assessments made.

Analogous dependences were obtained experimentally for phosphorus-vanadium-containing polyamides (Fig. 2). In this case the combustibility is determined by the duration of independent burning, and physic-mechanical properties are represented by bending breaking stress.

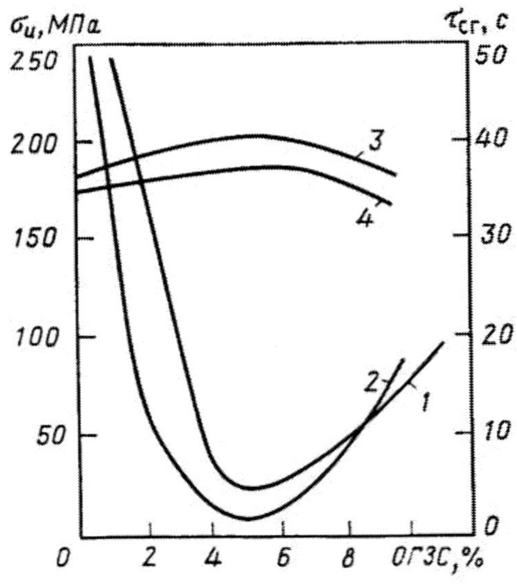


**Fig. 2.** Time changes in independent combustion $\tau_{cr}$ (1,2) and bending breaking stress (3,4) depending upon the content in polyamides (kapron 1,3 and nylon 2,4). Fire-retardant systems (FRS) [P-Ba$(VO_3)_2$]

The fire-retardant content and the proportion of components in complex fire-retarding systems are suggested to be assessed taking space-energy parameters into consideration. If protective carbon layer is formed on material or coating surface, the fire-retardants applied must be transformed into a complex structure with interface distance close or slightly over the interface distances of graphite-like substances

Energy characteristics must correspond to the transition of chemical bonds of one type into another. If the half-sum of destructing and forming bonds is less or equal to the adsorption potential, the process is possible. Such an approach is based on concepts of multiple catalysis theory. For polyesters containing phosphorus-oxide groups the schemes of bond transformation in polyester with couples P–O (Fig. 3) can be recorded.

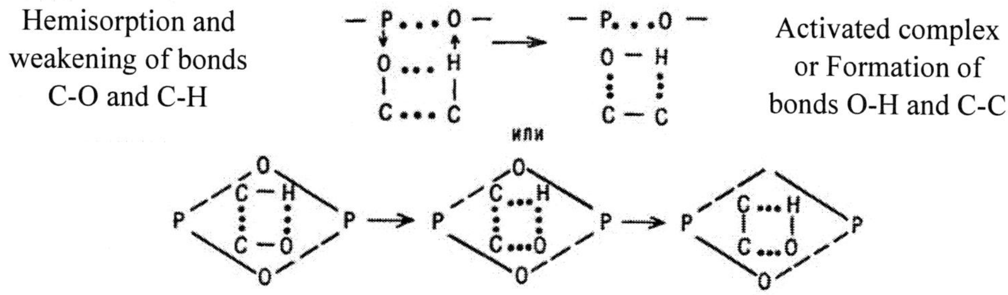


**Fig. 3.** Scheme of the formation of multiplets in phosphorus-containing polyester

Interface layers d (polyphosphates or polyphosphoric acid) are taken as equal to the sum of Van der Waals radii $d = R_p + R_0 = 3{,}3$ pm. Adsorption potential q is expressed through half-sums of electronegativities of atoms participating in the interaction of active centers of fire-retardants and chemisorbed atoms of macromolecule fragments and equals to $q = 2{,}72$. Half-sum of energies of destructing and forming bonds is calculated by the analogy through the half-sums of electronegativities of atoms participating in the bonds. For polyesters – $\sum E/2 \approx 2{,}65$. Since $q \geq \sum E/2$, the reaction is directed at the primary water and carbon substances formation. The application of phosphorus-vanadium-oxide fire-retarding systems, when assessing carbonization products, is more favorable in comparison with phosphorus-oxide systems or vanadium-oxide systems, since the percentage of fire-retarding phosphorus-vanadium-oxide system appears to be less due to the increase of coordination number 1. Depending upon vanadium coordination saturation, the content of coked surface changes: the thickness of boundary layers and "embryo" of coked surface increase together with coordination number growth. Thus, the registration of coordination processes during the formation of compositions and fire-retardant systems gives the possibility to predict and assess the efficiency of fireproofed material or coating. At the same time the assessment is made on the level of molecular and atomic interactions with nanophase formation, the appearance of which is conditioned by the formation of boundary layers.

## 2. Efficiency of Nanostructure Application in Polymeric Materials

The formation of nanophases and nanostructures of different composition and structure is possible when fire-retardant systems in the form of powders or liquids are introduced into the composition.

In case of liquid components mixing their thermodynamic interaction with the formation of extensive phases is possible. However, it is known [5, 6] that even polymerhomologs restrictedly dissolve in each other. Therefore in multi-component composites isotropy and density of materials are ensured by the distribution of components and phases from the surface to the volume of samples. At the same time boundary layers are formed between phases, which have properties slightly different from the properties of contacting phases and which can be identified with nanophases. The density of nanophases can be intermediate between the planes of interacting phases (Fig. 4a), or can be below the density of both phases if there is no interaction between the phases (4b).

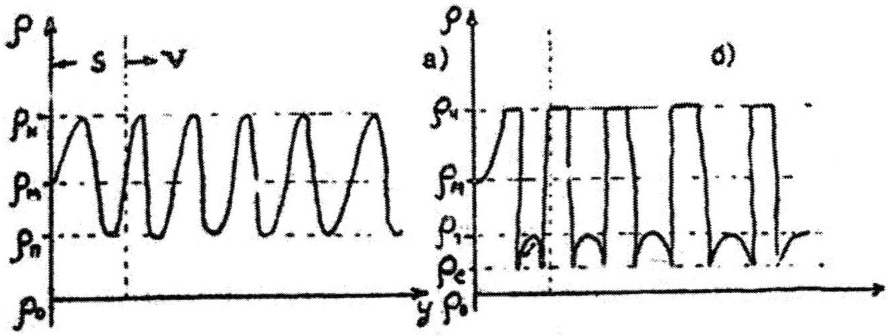


**Fig. 4.** Diagram of density with the formation of nanophase with high density (a) and nanophase with low density (b).

When the sizes of phases decrease, their activity and interaction between them increase, thus leading to the growth of nanophase density. If the sizes of component phases are in the range of nanophases, then the composite density as a whole increases together with heat capacity due to the appearance of dense boundary layers (nanophases).

Therefore the formation of nanostructures in composites can results in the decrease of temperature conductivity of the material. This is confirmed by experimental data based on thermalphysic characteristics of elastic material, in which, when introducing fire-retarding phosphorus-vanadium-containing system, nanophases are formed and when the temperature grows from 20 to 200 °C the heat capacity changes to a greater extent than thermal conductivity:

| | non-modified material | modified material |
|---|---|---|
| $C_p$, kJ/kg·degree | 1,47 → 1,89<br>30 → 200 °C | 1,66 → 2,54<br>20 → 200 °C |
| λ, Wt/m$^2$·degree | 0,26 → 0,39<br>30 → 200 °C | 0,27 → 0,33<br>30 → 200 °C |

In case fire-retardant systems of lamellar structure or lamellar-forming structure ones are introduced, when distributing in polymeric matrix the conditions for the formation of extended nanophases can arise, in which one of the dimensions exceeds the limits of nanometer formations. These can be fibrillar, lamellar or tubular structures with the length over 1 mcm. In such formations as nanocontainers the transformation of chemical particles with the formation of new nanophases increasing the total density and stability of materials to the action of heat and flame sources are possible.

At the same time as follows from literature, for instance [6], there are interactions, and on the primary structure the secondary one – "directed" is formed. After fire-retardants are introduced into the compositions, the formation of nanophases contributing to heterogeneous coke-formation catalysis leads, in some cases, to their "memorizing" and repetition of corresponding reflexes in difractograms of pyrolysis or coke residues (Fig.5). As shown in the Figure, the nanophase appeared during the introduction of phosphorus-vanadium-containing fire-retarding system is preserved on broad-angle difractograms of the residues of fire-protected polycarbonate pyrolysis.

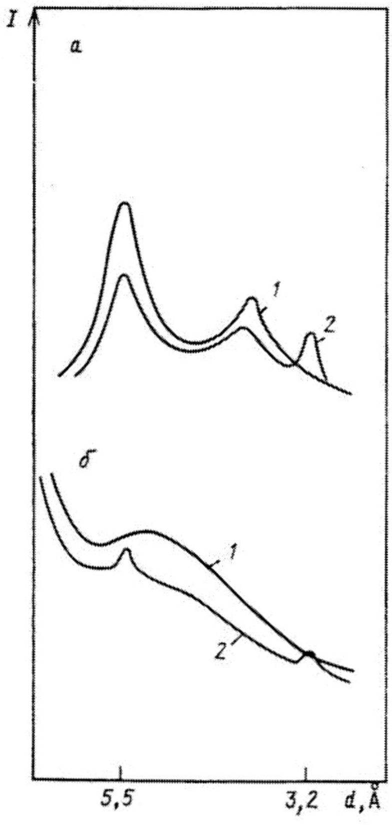


**Fig. 5.** Broad-angle difractograms of polycarbonates not containing (1) and containing (2) phosphorus-vanadium-containing system (a), and corresponding pyrolysis residues (b).

Thus, nanophases appearing in polycarbonate are stable to high thermal loads. Therefore it can be expected that the formation of nanocomposites containing a lot of nanophases can lead to the increase of their stability to mechanical and thermal loads.

At the same time, the presence of directed processes along the layers formed in the material can lead to material reinforcement and anisotropy appearance. For fire-retardant coatings this property contributes to heat flow along the surface in a certain direction. The appearance of nanostructures of certain directivity in the material contributes to the formation of double electric layers and catalytic processes of fire-retardant coating formation. Perspective are fire-retardant intumescent coatings the introduction of nanostructures into which allows increasing the coating adhesion to polymeric material being protected and the strength of foamcoke being formed. What is the action mechanism of nanostructures being formed or introduced? When fire-protected polymeric materials are formed by adding combustion fire-retardants to polymeric compositions, which are structure-formers and nanophase formation stimulators in materials, the increase of "embryos" of coked surface is observed when the temperature goes up. In turn, the quick formation of coked surface layer having good adhesion to the bulk of material prevents the development of combustion process under the action of one and the same fire source. More often the heat flow along the surface increases, and in some cases this leads to the significant decrease of mass combustion rate at comparatively high flame spreading velocity. In these cases it is necessary to carry out additional surface treatment of polymeric materials with thermal shocks. At the same time, additional nanostructures contributing to the decrease of amount of volatile and highly inflammable products under the action of fire sources onto the materials appear in interface spaces, nanocontainers, in a way. The density of surface layers and thermal capacity considerably increase, and thermal conductivity decreases along the normal to the surface, therefore sharp decrease of temperature conductivity is observed.

The introduction or formation of nanostructures in fireproof intumescent coatings leads to the formation of foamcokes with regularly located closed pores and sufficiently strong bubble walls. Moreover, depending upon the location and shape of nanostructures the gas bubbles from gas-forming agent are distributed in foamcoke bulk in a certain order. If fine powders of ammonium polyphosphate (APP) or its derivatives are used as gas-forming agents and dispersed with ultrasonic field in the composition, it is possible to approach the model system, in which the distribution of fire-retardant and gas-forming agent particles, as APP, is practically uniform (Fig. 6). The Figure shows a simplified scheme of particle location. The particles have a ball shape, their size is about 10 mcm, the distance between them is about 1mm. The foamcoke formed represents carbonaceous material with bubbles filled with the mixture of ammonium and water. At the same time a carbonaceous barrier containing the layer of polyphosphoric acid on the inner surface is formed between the bubbles. In contrast to the previous coatings, in which, according to X-ray photoelectron spectroscopy data the surface is carbonized, in intumescent coatings first the carbonization of inner wall of bubbles being formed takes place [7]. The introduction of carbon-metal-containing nanostructures obtained from polyvinyl alcohol and metal salts into intumescent fireproof compositions [8, 9] contributes to the increase of location regularity of bubbles being formed from PVA in foamcokes. At the same time the strength of foamcoke formed increases [10]. Such a result is conditioned by the decreased velocity of ammonium bubble formation due to the coordination of ammonium molecules on metal clusters, for instance, copper, nickel or cobalt, which are placed in carbon shells open from the ends. Such carbon-metal-containing tubules are filled with metal phases up to 60-70%. They mainly represent twisted branched structures 40-60 nm in diameter and 300-1000 nm long. The bubbles formed from APP partly sorb on the surface of these structures. At the same time foamcokes during their formation become more stable to mechanical loads due to the formation of new carbon nanophases on the "frame" of nanostructures introduced. Thus, when the intumescence degree slightly goes down, the foamcoke strength and stability under various loads increase.

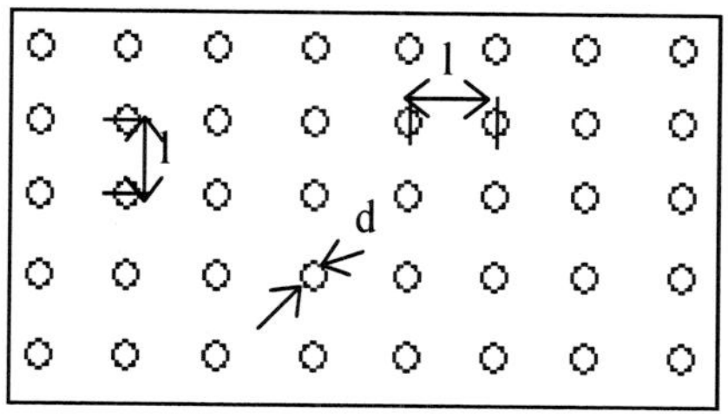


**Fig. 6.** APP distribution in epoxy composition.

## 3. Possible Models of Carbonization Processes and Foamcokes Formation

Modeling of carbonization process at the temperatures of polymeric materials decomposition can proceed with and without formation of liquid phase [1]. The first variant can be represented in one or two stages. In two-stage process intermediate formation of liquid stage is possible, which speedily "structures" into solid amorphous and crystalline products or transforms into gaseous products.

When putting a task to model processes in two-stage process, which in practice is used more often, it is necessary to take into consideration possible reactions responsible for structuring and reactions resulting in coking. In general form reactions proceeding during structuring can be given as:

$$k_{is} = (1 - \beta)^{n_{is}} k_{is}(T),$$

where β – share of decomposed active component, $n_{is}$ – order of 1$^{st}$ structuring reaction. The main difficulties when building up the models, given in detail in monogram [1], arise when determining the proceeding possibilities of this or another reaction, its way, velocity and activation energy constant. When building up the models the application of computation schemes of quantum chemistry, molecular mechanics and thermal-dynamic concepts based on activated complex theory is possible. It is advisable to use adapted programs Hyperchem, Gamess and Gaussian. Fragments of molecular structures are picked out, based on which interactions can proceed and which can be identified with reaction centers. In reaction centers the transformation processes of one bond into another take place. The calculations of energies of molecules and molecular fragments participating in the interaction are carried out with the help of Hartry-Fock method in minimum basis 3-21G. In this process program complex GAMESS is used [11]. In the frameworks of energy model mentioned, the balanced atomic geometries of fragments and interaction energy of reaction centers are determined. When kinetic parameters such as activation energy and pre-exponential factor $k_0$ are determined, *ab initio* calculations using activated complex theory are carried out [12]. In case of reactions connected with intermolecular rearrangement of chemical bonds, the system transition state was looked for. Afterwards the system movement along the reaction coordinate via transition state obtained (activated complex) was investigated. The reaction activation energy is calculated as the difference between the energies of activated complex and initial state taking into account the energies of zero oscillations. This value is a criterion of reaction proceeding possibility. The pre-exponential factor is calculated based on the analysis of statistical sums of transition state and initial molecular fragments or reaction centers in accordance with the following equation:

$$k_0 = \frac{k_в T}{h}\frac{z_{act}}{z_0},$$

where $k_в$ – Boltzmann constant, $h$ – Plank constant, $z_{act}$ – statistical sum of transition state, $z_0$ – statistical sum of initial molecular fragments.

If reactions were connected with molecular fragment breaking-off, the activation energy was assessed as the difference between the total energy of isolated molecular fragments formed during the reaction and initial state energy. At the same time the transition state energy is usually higher than the total energy of fragments formed. Therefore both approximations partly compensate each other. To assess errors calculations were made in the frameworks of semi-empirical method PM3 taking configuration interaction C1 into account [11]. The surroundings influence upon the process was estimated with the help of molecular-dynamic modeling in the frameworks of the model of molecular mechanics MM+ [13]. The proposed approach was used in all the cases when the directivity of fast-proceeding reactions including combustion, carbonization and heterogeneous catalysis processes was predicted. The process can proceed with liquid phase formation, in which convection processes can occur. The process gets complicated in case of radiation from soot particles to the surface. Therefore it is necessary to take these circumstances into account when carrying out a computation experiment for reactions proceeding in corresponding areas and phases considering nanophases and interactions between them. Computing experiment of pore-formation acquires extreme importance when modeling carbonization or mineralization processes. If volume shares of crystalline and amorphous cokes increase in the process of liquid decomposition with the formation of gaseous phase, the porosity of material condensed layers being formed decreases. Bubbles, appearing as a result of the gas-formation and evaporation processes exceeding over condensation, form a summation distributed in accordance with sizes and satisfying continuous Zeldovich equation [14]. The bubbles are supposed to grow further until their volume share reaches the porosity value. Afterwards the bubbles can merge into one big bubble and gaseous substances will flow out from it as a result of bubble integrity breakage due to the decrease of bubble wall adhesion strength with basic partly pyrolyzed material. The problem of obtaining a stable strong foamcoke is solved in the models of fireproof intumescent coatings [15], on the example of epoxy-polymer hardened with polyethylene-polyamine with polyammonium-phosphate (PAP) added. At the same time minimum eight gross-reactions are recorded that proceed in temperature-time and spatial zones. These zones are limited by gas-forming agent surface, inner surface of bubble being formed (polymer-gas boundary) and bubble cavity, in which the mixture ammonium-water is being formed. At high destruction temperatures and oscillations by bond with low frequency, statistic oscillating sums can be used – $z_k = \frac{kT}{h\nu}$, where ν– oscillation frequency of breaking chemical bond. If the bond oscillations correspond to the transfer along the reaction coordinate, it can be said that $k_0 = \frac{kT}{h}\frac{z^{\dagger}}{z_A} \square \partial$, and since the oscillations of chemical bonds are characterized by wave numbers, $k_0 \square \omega c$, where $\omega$ – wave number, $c$ – light speed. The calculations of pre-exponential factor for monomolecular reactions carried out with the help of quantum-chemical methods are correlated with the values obtained experimentally, and without experimental data with numeric values of frequencies of breaking bonds. For bimolecular reactions the calculations of $k_0$ were carried out using semi-empirical method by statistic rotational sums –

$$k_0 = \frac{h^2 Z_{RT}}{(2\pi)^{3/2} k_B^{1/2} (m^*)^{3/2} T^{1/2} (Z_A)_{RT} (Z_B)_{RT}}$$

for reaction A + B → C + D, where $m^* = \frac{m_A \cdot m_B}{m_A + m_B}$ reduced mass of fragments A and B, $(Z_A)_{RT}$ and $(Z_B)_{RT}$ – statistic rotational sums A and B, $Z_{RT}$ – statistic rotational sum of transition state. Computing of statistic rotational sums is carried out based on the following formula:

$$Z_{RT} = \frac{8\pi^2 (k_B T)^{3/2} (8\pi^3 I_A I_B I_C)^{1/2}}{\sigma^* h^3}$$

where σ* corresponds to the section of impacts and is found in accordance with the following formula:

$$\sigma^* = \frac{2^{1/3}}{8N^{2/3}}\left[\left(\frac{M_A}{\rho_A}\right)^{1/3} + \left(\frac{M_B}{\rho_B}\right)^{1/3}\right]^2$$

$M_A$ and $M_B$ – molar masses of interacting fragments, $\rho_A$ and $\rho_B$ – densities of corresponding reagents, $I_A$, $I_B$, $I_C$ – inertia moments of fragments. To find inertia moments it is necessary to pick up the coordinate system and rule to select distances in activated complex. It is decided that reacting atoms in activated complex are located at the distance equal to the sum of Van der Waals radii, and changes in valence angles in activated complex are insufficient. Inertia moments are found in relation to the selected center in coordinate system based on the following scheme:

$$I_x^{(A)} = \sum M_i \left(z_i^2 + y_i^2\right) \quad I_{xy} = \sum M_i x_i y_i$$
$$I_y^{(A)} = \sum M_i \left(x_i^2 + z_i^2\right) \quad I_{xz} = \sum M_i x_i z_i \quad I_A I_B I_C = \begin{vmatrix} I_x & -I_{xy} & -I_{xz} \\ -I_{xy} & I_y & -I_{yz} \\ -I_{xz} & -I_{yz} & I_z \end{vmatrix}$$
$$I_z^{(A)} = \sum M_i \left(x_i^2 + y_i^2\right) \quad I_{yz} = \sum M_i y_i z_i$$

It should be noted that $I_X$, $I_Y$, $I_Z$ – axial inertia moments in relation to any three mutually perpendicular axes going through the gravity center of interacting functional groups, and $I_{XY}$, $I_{XZ}$, $I_{YZ}$ – products of inertia.

At the same time the coordinates of system gravity center are found from the following relations:

$$x^* = \frac{\sum M_i x_i}{\sum M_i}; \quad y^* = \frac{\sum M_i y_i}{\sum M_i}; \quad z^* = \frac{\sum M_i z_i}{\sum M_i};$$

Calculated value of $k_0$ by the equation

$$k_0 = \frac{h^2 \sigma^* \left(I_A^{\neq} I_B^{\neq} I_C^{\neq}\right)}{(2\pi)^{3/2} k_B^{3/2} \left(m^*\right)^{3/2} T^{1/2} I_A^A I_B^A I_C^A}$$

for the reaction

- P(O)OH + HO(O)P -  →  - P(O) – O – (O)P -  +  $H_2O$
  with each P bearing an $ONH_4$ group: $ONH_4$ $ONH_4$ → $ONH_4$ $ONH_4$

equals to $10^{-14}$ $cm^3/s$ that approximately corresponds to experimental values and values obtained with the help of quantum-chemical calculations.

Spatial-energy parameters can be applied to determine nanophase distribution in fire-protected composites and coatings, as well as to determine the interactions between [16, 17].

## REFERENCES


[1] V.K. Bulgakov, V.I. Kodolov, A.M. Lipanov. Modeling of polymeric materials combustion. M.: *Khimiya*, 1990, 238pp.

[2] L.E. Maharinsky et al. *Physics of combustion and explosion*, No 5, p. 83, 1983.

[3] G.G. Sheluhin, V.D. Buldakov, V.P. Belov. *Physics of combustion and explosion*, No 1, p. 42, 1969.

[4] Silberberg A. *Pure and Appl. Chem*.,Vol. 20, №3, p. 585-591,1971.

[5] Yu. S. Lipatov. *Colloid chemistry of polymers*, Naukova dumka, Kiev, 1984, 343pp.

[6] Yu. S. Lipatov. *Interface phenomena in polymers*, Naukova dumka, Kiev, 1980, 259pp.

[7] S.G. Shuklin, V.I. Kodolov, A.P. Kuznetsov, *Chemical physics and mesoscopy*, Vol. 2, No 1, p. 5-11, 2000.
[8] A.A. Didik, V.I. Kodolov, A.Yu. Volkov, E.G. Volkova, K.H. Halmayer, *Inorganic materials*, Vol. 39, No 5, p. 693-697, 2003.
[9] A.A. Didik, V.I. Kodolov, A.Yu. Volkov, E.G. Volkova, *Chemical physics and mesoscopy*, , Vol. 4, No 2, p. 214-223, 2002.
[10] S.G. Shuklin, A.A. Didik, S.G. Bystrov, A.P. Kuznetsov, V.I. Kodolov, in this book.
[11] Schmidt M.W. et al *J. Comput. Chem.*, Vol. 14, p. 1347-1363, 1993.
[12] N.M. Emmanuel, L.G. Knorre. *Course in chemical kinetics*, Vysshaya shkola, Moscow, 1969, 431 pp.
[13] J.J.P. Stewart *J. Comput. Chem.*, Vol. 12, p. 320-333, 1991.
[14] V.P. Skripov. *Metastable liquid*, Nauka, Moscow, 1972, 312 pp.
[15] S.G. Shuklin, E.N. Klimenko, V.I. Kodolov. *Chemical physics and mesoscopy*, Vol. 3, No 2, p. 113-125, 2001.
[16] G.A. Korablev. *Application of spatial-energy concepts in prognostic assessment of phase-formation of solid solutions of refractory and relative systems*. IzhSAA, Izhevsk, 1999, 290 pp.
[17] G.A. Korablev, V.I. Kodolov, *Chemical physics and mesoscopy*, Vol. 3, No 2, p. 243-254, 2001.

In: Chemical Physics of Pyrolysis, Combustion and Oxidation ISBN 1-59454-163-9
Editors: A. A. Berlin et al., pp.189-197 

*Chapter 18*

# FOAM AND COKE STRUCTURE REGULATION WHEN INTRODUCING CARBON METAL-CONTAINING NANOSTRUCTURES INTO FIRE PROOF INTUMESCENT COATINGS

***S. G. Shuklin*, A. A. Didik*, S. G. Bystrov**, A. P. Kuznetsov* and V. I. Kodolov****

*Basic Research-High Educational Center of Chemical Physics and Mesoscopy, Udmurt Scientific Center, Ural Division, RAS
**Physical-Technical Institute, Ural Division, RAS
426069 Izhevsk, 7 Studencheskaya St., *shuklin@istu.udm.ru*

## ABSTRACT

The paper is dedicated to foamcokes structure regulation when introducing carbon metal-containing nanostructures into fireproof intumescent coatings.

Results are reported of research on the influence of fireproof intumescent coatings based on epoxy resin hardened with polyethylene-polyamine and containing, as gas-forming agent and carbonization stimulator, ammonium polyphosphate and insignificant additives (up to 5% weight) of carbon metal-containing nanostructures with copper, nickel and cobalt included, on intumescence process of modified epoxy compositions (EC), as well as on the structure, physical-mechanical and thermal-physical properties of carbon foamcokes being formed.

It is found out that carbon metal-containing nanostructures influence the structure regulation, strength and thermal-physical properties of foamcokes due to high surface energy and small sizes.

**Key words:** fireproof intumescent coatings, carbon metal nanostructures, foamcoke, thermal-physical properties, density, physical-mechanical properties, atomic force microscopy.

## INTRODUCTION

One of the effective methods of polymer material combustibility decrease is the application of intumescent coatings onto the surface to be protected [1]. The efficiency of fireproof intumescence coatings is determined by the intumescence repetition factor, thermal-physical properties, structure regulation and strength of carbon foamcoke formed, as well as by chemical processes accompanying the intumescence of fireproof materials. Carbon metal-containing nanostructures introduced into fireproof intumescent coatings improve foamcoke structure. Besides, intumescent coatings have to possess good adhesion to the material being protected [2].

## EXPERIMENTAL: OBJECTS AND METHODS

Epoxy compositions contained epoxy resin based on diphenylolpropane cross-linked with polyethylene polyamine and modified by ammonium polyphosphate (APP), calcium borate, manganese dioxide and also by metal-carbon nanostructures. In its turn, nanostructures basically represented tubules containing metals. Tubules containing chromium and nickel were obtained from anthracene in eutectic alloys following the methods described in paper [3]. These tubules and additives were introduced into the first series of compositions (Table 1). After the efficiency of tubules in the compositions of the first series was determined, in the second series the extended investigation of the influence of tubules containing copper, nickel and cobalt on the combustibility was carried out (Table 2). In the secone series the tubules represented multiwall nanotubes filled with metallic and metal-oxide phases. The synthesis was carried out by polyvinyl alcohol (PVA) in the presence of copper (II), nickel (II) and cobalt (II) chlorides [4]. Molar proportions of polyvinyl alcohol and metal chloride varied in the range (PVA/salt) (20-1)/1. The structure of the product obtained in the presence of copper chloride is shown in Fig. 1. In Fig.2 there is electron microdiffraction picture of the chosen area: changes indicating the product cylindrical symmetry are well-seen. With the help of X-ray-phase analysis metal compounds and metals are found. At the same time with the help of X-ray photoelectron spectroscopy $sp^2$ and $sp^3$-hybridized carbon atoms are found. Metals and their compounds are not found by this method, thus indicating their incapsulation into carbon shells.

**Table 1. Component contents of the composites investigated (Series A).**

| Components | Composition (fraction of total mass) | | | | | | | | | | | |
|---|---|---|---|---|---|---|---|---|---|---|---|---|
| | 1A | 1Ab | 2A | 3A | 3Ab | 4A | 4Ab | 5A | 5Ab | 6A | 7A | 8A |
| ED-20 | 10.0 | 10.0 | 10.0 | 10.0 | 10.0 | 10.0 | 10.0 | 10.0 | 10.0 | 10.0 | 10.0 | 10.0 |
| APP | 2.0 | 3.0 | 2.0 | 2.0 | 3.0 | 2.0 | 3.0 | 2.0 | 3.0 | 3.0 | 3.0 | 3.0 |
| Tubules (Ni) | | | | | | | | | | | | 0.8 |
| Tubules (Cr) | 0.4 | 0.4 | 0.4 | | | | | | | 0.4 | 0.8 | |
| Calcium borate | 0.4 | 0.4 | | 0.4 | 0.4 | 0.4 | 0.4 | | | | | |
| $MnO_2$ | | | | 0.4 | 0.4 | | | | | 0.4 | 0.8 | 0.8 |
| PEPA | 1.5 | 1.5 | 1.5 | 1.5 | 1.5 | 1.5 | 1.5 | 1.5 | 1.5 | 1.5 | 1.5 | 1.5 |

**Table 2. Components content in compositions of B Series.**

| Component | № composition | | | | | | | | | | | |
|---|---|---|---|---|---|---|---|---|---|---|---|---|
| | **1B** | **2B** | **3B** | **4B** | **5B** | **6B** | **7B** | **8B** | **9B** | **10B** | **11B** | **12B** |
| ED-20 | 10,0 | 10,0 | 10,0 | 10,0 | 10,0 | 10,0 | 10,0 | 10,0 | 10,0 | 10,0 | 10,0 | 10,0 |
| PVA | 3,0 | 3,0 | 3,0 | 3,0 | 3,0 | 3,0 | 3,0 | 3,0 | 3,0 | 3,0 | 3,0 | 3,0 |
| Cu 1:1 | 0,5 | | | | | | | | | | | |
| Cu 5:1 | | 0,5 | | | | | | | | | | |
| Cu 20:1 | | | 0,5 | | | | | | | | | |
| Ni 1:1 | | | | 0,5 | | | | | | | | |
| Ni 5:1 | | | | | 0,5 | | | | | | | |
| Ni20:1: | | | | | | 0,5 | | | | | | |
| Co 1:1 | | | | | | | 0,5 | | | | | |
| Co 5:1 | | | | | | | | 0,5 | | | | |
| Co 20:1 | | | | | | | | | 0,5 | | | |
| graphite | | | | | | | | | | 0,5 | | |
| $C_{act}^{*}$ | | | | | | | | | | | 0,5 | |
| PEPA** | 1,5 | 1,5 | 1,5 | 1,5 | 1,5 | 1,5 | 1,5 | 1,5 | 1,5 | 1,5 | 1,5 | 1,5 |

* activated carbon, ** polyethylene-polyamine

The preparation of the intumescent compositions consisted in the preparation of mixture from the components in the proportions shown in Tables 1 and 2. The components were mixed under normal conditions as follows. The weighed amounts of powdered components were successively put into the vessel containing the measured quantity of epoxy resin without a hardener and carefully mixed. Then the weighed amount of hardener was put into the vessel, and the components were mixed afterwards for another 2 minutes. The ready mixture whose life was 1,5 – 2 hours was protected from moisture and kept for less than 30 minutes before use.

Intumescence repetition factor, thermal-physical parameters, density of foamcokes and also their strength were determined by conventional methods. Atomic force microscopy (AFM), local force microscopy were applied during the investigations. Both inner and outer layers of foamcokes were investigated. The investigations were carried out using the instrument P4-Solver produced by NT-MTD and silicon probes CS12 from the same manufacturer. The coating strength was assessed based on the minimum needle pressing force to the sample, when there are stripes left on the coating after repeated scanning of the area chosen. The pressing force was adjusted by changing the instrument set point. During scanning the probe moved with the velocity of 6 mcm/sec, the area scanned was 1 $mcm^2$. The stepped load was applied (Fu) 1.Fu = 1,35 mcN, 2.Fu= 2,7 mcN, 3.Fu = 4,05 mcN. The coating polarity change depending upon the foamcoke recipe, outer and inner surfaces was assessed by the change in interacting force between the sample and the probe [5, 6], when taking the curves: force – distance (F-d mode).

## RESULTS AND DISCUSSION

The preparation of intumescent composition consists in the preparation of the components mixture in the ratios shown in Table 1.

The introduction of such structure-formers allows changing or influencing not only the foamcoke structure, but also the foamcoke structure influences the shape and location of bubbles, In the picture of Sample 1A the pores are located along the surface, therefore the surface and volume height ranges are similar. In Sample 1 the breaking load is over 3,2nN in volume and surface. The volume and surface structures are similar. When FA (Ni) is introduced into Sample 2A, the same regularities are present, but the structure is more ordered in comparison with Sample 1A. The volume and surface structures are similar.

In Sample 3A the difference in orientating the volume foamcoke structure as compared with the surface one is connected with nickel quantity increase in tubule or nanotube. The use of different metal (manganese) as a structure-former allows to obtain a less orientated (ordered) foamcoke structure in comparison with Sample 2A where nickel was used, but more similar in surface and volume orientation, therefore breaking loads are similar (for surface – 1,8 and for volume – 3,2). But foamcoke is more ordered in volume.

The ordered foamcoke structure in the volume is shown in picture (Fig.4) Sample 2A. The bubble shapes are different in all the samples. The foamcoke has a lamellar structure, it is possible that bubbles are formed due with the help of these faces. The bubbles are extended upwards (Samples 2, 3, 4 of A series) and along the surface (Samples 1, 5, 6 of A series).

Since the product is of nanosized structure and contains a large number of interface boundaries, it was supposed that it could act as a structure-forming agent stimulating foamcoke formation at the aforesaid boundaries when being introduced into intumescent fireproof compositions.

Heat conductivity of natural graphite along the axis parallel to the layers is ~400 Wt/(m·K) at 25 °C, and along the axis perpendicular to the layers ~80 Wt/(m·K), thus compared with copper heat conductivity (388 Wt/(m·K)) [7]. Heat conductivity of polycrystalline graphite usually does not exceed ~200 Wt/(m·K), but can be as high as ~500 Wt/(m·K). The heat conductivities of FC obtained are practically the same and are less than polycrystalline graphite heat conductivity at the average in ~550 times (Table 4). The less is foamcoke heat conductivity, the higher is its heat-shielding. In this respect the foamcokes of systems 4-6, 9 of B series having minimum values of the aforesaid magnitude are the most effective. It should be pointed out that heat conductivity of isolated SWNT symmetry (10,10) theoretically calculated is very high and amounts to ~6600 Wt/(m·K), however the value obtained for compact samples in practice is considerably less due to interlayer interaction and does not exceed ~0,7 Wt/(m·K) [8], therefore the contribution of small nanostructural additives for increasing the heat conductivity is insufficient. Specific heat capacity of foamcokes (Table 3) exceeds by an order the heat conductivity of graphite at room temperature (~720 J/(kg·K) [7], this is connected with foamcoke porous structure, nanophase formation and gas considerable contribution in pores.

**Table 3. Coke investigation results using AFM method.**

| Sample | Properties | | | |
|---|---|---|---|---|
| | Adhesion force,Fa, nN | Breaking load, Fd, nN | Statistics | |
| | | | Heights range, Sa, nm | Mean-square roughness, Sg, nm |
| №1 Surface | 680 | >3,2 | 7 | 8 |
| №1 Volume | 20 | >3,2 | 9 | 10 |
| №2 Surface | 8 | >3,2 | 18 | 21 |
| №2 Volume | 35 | >3,2 | 59 | 69 |
| №3 Surface | 340 | 0,5 | 9 | 11 |
| №3 Volume | 450 | >3,2 | 76 | 95 |
| №4 Surface | 90 | 1,8 | 20 | 24 |
| №4 Volume | 935 | 3,2 | 10 | 12 |
| №5 Surface | 755 | 0,5 | 16 | 19 |
| №5 Volume | 214 | 3,2 | 41 | 49 |
| №6 Surface | 135 | 0,5 | 34 | 39 |
| №6 Volume | 90 | >3.2 | 98 | 57 |

Fa – probe adhesion force to the sample, Fd – normal load under which the sample surface destructs during scanning, nN

**Table 4. Heat conductivity (λ) and specific heat capacity ($C_m$) of foamcokes obtained from compositions 1-12 of B series at 800 °C**

| № | $C_m$, J/(kg·K) | λ, Wt/(m·K) | № | $C_m$, J/(kg·K) | λ, Wt/(m·K) |
|---|---|---|---|---|---|
| 1 | 6146 | 0,384 | 7 | 5763 | 0,385 |
| 2 | 6216 | 0,382 | 8 | 5773 | 0,363 |
| 3 | 4916 | 0,412 | 9 | 7470 | 0,350 |
| 4 | 5481 | 0,332 | 10 | 5975 | 0,384 |
| 5 | 5772 | 0,362 | 11 | 5784 | 0,388 |
| 6 | 5881 | 0,329 | 12 | 5176 | 0,377 |

Actual densities of foamcokes with nanostructures do not exceed ~1,4 g/cm$^3$ and practically do not depend on the type of metal contained in nanostructures (Table 5).

Graphite maximum possible density is 2,265 g/cm$^3$, however the density of even natural graphite rarely exceeds 2 g/cm$^3$, and usually amounts to 1,6 g/cm$^3$ and lower; the densities of carbon transition forms also do not exceed ~1,5-2 g/cm$^3$ [8]. Densities of FC are lower than densities of different carbons in 1,2-1,5 times (Table 5). This can be connected with the closed pore system, the hollows of which are not accessible for liquid in picknometer, or imperfect packing of carbon layers and crystallites in foamcoke structure.

**Table 5. Densities of foamcokes (d) obtained from compositions 1-12 of B series at 800 °C**

| № FC | d, g/cm$^3$ | № FC | d, g/cm$^3$ |
|---|---|---|---|
| 1 | 1,280 | 7 | 1,304 |
| 2 | 1,249 | 8 | 1,151 |
| 3 | 1,335 | 9 | 1,401 |
| 4 | 1,276 | 10 | 1,497 |
| 5 | 1,131 | 11 | 1,445 |
| 6 | 1,183 | 12 | 1,451 |

Such index as intumescence repetition factor (Table 6) showing the relative volume increase of carbon foam in comparison with the initial composition was studied for foamcokes

$$K = \frac{V_{FC}}{V_{EC}}, \tag{1}$$

where K- intumescence repetition factor; $V_{FC}$- effective (external) volume of foamcoke; $V_{EC}$- volume of initial epoxy composition.

**Table 6. Intumescence repetition factors of foamcokes obtained from compositions 1-12 of B series at 800 °C**

| № FC | K | № FC | K |
|---|---|---|---|
| 1 | 23,59 | 7 | 31,02 |
| 2 | 30,21 | 8 | 38,28 |
| 3 | 18,43 | 9 | 33,54 |
| 4 | 38,66 | 10 | 9,59 |
| 5 | 33,80 | 11 | 20,20 |
| 6 | 30,12 | 12 | 12,62 |

Intumescence repetition factor for epoxy compositions of B series with the addition of nanostructures is approximately 2-3 times greater than for initial epoxy resin with ammonium polyphosphate (Table 6). Maximum value of intumescence repetition factor is observed in additives containing nickel and cobalt – ~3 times greater than for compositions without additives. The addition of activated carbon also increases intumescence repetition factor in nearly 2 times, addition of graphite powder, on the contrary, decreases it – in 1,3 times.

To assess the heat-shielding properties, the temperature conductivity coefficient of foamcokes intumescent at 800 °C was calculated based on expression in [9].

$$a = \frac{\lambda}{C_m d^*_{\mathrm{FC}}}, \qquad (2)$$

where λ- specific heat conductivity, Wt/(m·K); $C_m$- specific heat capacity, J/(kg·K). $d_{FC}^*$- apparent foamcoke density taking intumescence repetition factor into consideration.

$$d^*_{\mathrm{FC}} = \frac{m_{\mathrm{FC}}}{\mathrm{V}^*_{\mathrm{FC}}}, \qquad (3)$$

where $m_{FC}$ и $V_{FC}^*$- weight and volume of FC intumescent sample, respectively. Taking into account that $m_{FC}=m_0 \cdot \Delta m/100\%$ (where $m_0$ and Δm(%)- respectively the weight of initial EC and EC weight loss after heat treatment at 800°C), $V_{FC}^*=V_0 \cdot K$ (where $V_0$ и K- respectively EC initial volume and intumescence coefficient) and EC initial density $d_0=m_0/V_0$, we get

$$d^*_{\mathrm{FC}} = d_0 \frac{\Delta m}{K \cdot 100}, \qquad (4)$$

where $d_{FC}^*$, $d_0$ in g/cm$^3$, Δm in %.

Actual and apparent densities of foamcokes differ by 1-2 orders(Table 7).

**Table 7. Actual densities of EC ($d_0$) and apparent densities ($d^*_{FC}$) of foamcokes obtained from them at 800°C**

| № FC | Δm,% | $d_0$, g/cm$^3$ | $d_{FC}^*$, g/cm$^3$ | № FC | Δm,% | $d_0$, g/cm$^3$ | $d_{FC}^*$, g/cm$^3$ |
|---|---|---|---|---|---|---|---|
| 1 | 88,2 | 1,254 | 0,046 | 7 | 87,7 | 1,248 | 0,035 |
| 2 | 88,7 | 1,250 | 0,037 | 8 | 87,6 | 1,245 | 0,029 |
| 3 | 89,1 | 1,257 | 0,061 | 9 | 88,1 | 1,249 | 0,032 |
| 4 | 87,9 | 1,242 | 0,028 | 10 | 96,0 | 1,224 | 0,122 |
| 5 | 87,9 | 1,240 | 0,032 | 11 | 96,4 | 1,211 | 0,058 |
| 6 | 88,3 | 1,242 | 0,036 | 12 | 97,3 | 1,219 | 0,094 |

Values of foamcokes temperature conductivity for (a) epoxy compositions with nanostructures added (№1-9) are 1,5-2 times greater than for foamcokes obtained from epoxy compositions №10-12 (Table 8).

**Table 8. Heat conductivity coefficients for foamcokes obtained from compositions 1-12 of B series at 800°C**

| № FC | $a \cdot 10^{-6}$, $m^2/s$ | № EC | $a \cdot 10^{-6}$, $m^2/s$ |
|---|---|---|---|
| 1 | 1,33 | 7 | 1,91 |
| 2 | 1,66 | 8 | 2,25 |
| 3 | 1,37 | 9 | 1,42 |
| 4 | 2,16 | 10 | 0,53 |
| 5 | 1,96 | 11 | 1,16 |
| 6 | 1,55 | 12 | 0,78 |

Compression resistance of foamcokes (Table 9) was determined following the aforesaid methods.

Approximate heat conductivity value calculated for polycrystalline graphite based on average indexes for polycrystalline graphite at 25°C (density ≈1,6 g/cm$^3$, $\lambda \approx 200$ Wt/(m·K), $C_m \approx 720$ J(kg·K) [9]) amounts to ~1,74·10$^{-4}$ m$^2$/s, i.e. approximately by 2 orders greater than the values obtained for foamcokes. Values obtained for foamcokes from epoxy compositions with nanostructures added are a little greater (in ~1,5 times) than for foamcokes №10-12, this can be explained by the presence of small amounts of metals with great temperature conductivity values. Among the foamcokes obtained from epoxy compositions with nanostructures added foamcokes №1, 6, 9 have the least temperature conductivity (Table 8).

Strength of foamcokes after the additives are introduced (№1-9) increases considerably (in ~2,5-3 times), but the introduction of graphite powder does not practically result in EC strength increase, and activated carbon – in insignificant increase (in ~1,25 times) (Table 9). Thus the introduction of gel heat treatment products sufficiently increases the mechanical strength of epoxy composition foamcokes. Previously, based on the example with fullerenes $C_{60}$, it was shown [10] that carbon nanostructures are better bound by polymers containing aromatic rings (polystyrene) than by not containing them (polymethylmethacrylate). Similarly SWNT contained in epoxy resin lead to some decrease in its bending strength [11]. In this respect the strength increase of foamcokes with nanostructures can be connected with this fact since fragments of aromatic macromolecules are apparently present in foamcoke structure. Foamcoke strength is a very important factor since a fireproof effect, unlike a heat-shielding one, also depends on foamcoke destruction stability under flame influence and mechanical stresses appearing in intumescent coating.

**Table 9. Compression resistance of foamcokes obtained from compositions 1-12 of B series at 800°C**

| № composition | F, g/cm$^2$ | № composition | F, g/cm$^2$ |
|---|---|---|---|
| 1 | 12,10 | 7 | 12,12 |
| 2 | 12,39 | 8 | 11,87 |
| 3 | 12,85 | 9 | 13,21 |
| 4 | 14,38 | 10 | 4,95 |
| 5 | 14,03 | 11 | 3,99 |
| 6 | 13,02 | 12 | 4,07 |

Analyzing foamcoke compression resistance values and repetition factors of composition intumescence it can be asserted that the best additives are gel heat treatment products with the following compositions: PVA:$NiCl_2$ 1:1, 5:1 and PVA:$CoCl_2$ 20:1 – for containing them epoxy compositions with nanostrucutres added foamcoke maximum strength value and intumescence repetition factor are observed.

## REFERENCES


[1] S.G. Shuklin, V.I. Kodolov, K.I. Larionov, S.A. Tyurin, *Physics of combustion and explosion*. №2, p. 73, 1995.

[2] V.I. Kodolov, K.I. Larionov, S.G. Shuklin et al. *Method of surface preparation of items made of organic- and fiberglass plastics before applying fireproof coatings*. Russian Patent No 2010620 dated 25.10.91.

[3] V.I. Kodolov, A.P. Kuznetsov., Didik A.A., Makarova L.G., Shabanova I.N., Volkov A.Yu., Volkova E.G., *Chemical Physics and Mesoscopy*,.Vol. 3, N 1, p. 7, 2001.

[4] A.A. Didik, V.I. Kodolov, A.Yu. Volkov, E.G. Volkova, *Chemical Physics and Mesoscopy*, Vol.4, N 2, p.188, 2002.

[5] S.K. Sinniah, A.B. Steel, C.J. Miller, J.E. Reutt-Robey, *J.Am. Chem. Soc*., Vol.118, p. 8925, 1996.

[6] S.N. Maganov, M-H Whangbo. *Surface Analysis with STM and AFM, VCH*, Weinheim, New York, Basel, Cambridge, Tokyo, 1996, 50pp.

[7] A.R. Ubbelode, F.A. Lewis. *Graphite and its crystalline compounds* / Translated from English, Mir, Moscow, 1965, 256 pp.

[8] S. Berber, Y.–K. Kwon, D. Tománek *Physical review letters*, Vol. 84, № 20, p. 4613, 2000.

[9] A.V. Lykov. *Heat conductivity theory*. Vysshaya shkola, Moscow, 1967, 599 pp.

[10] E.V. Anufrieva, M.G. Krakovyak, T.D. Ananjeva, T.N. Nekrasova, R.Yu. Smyslov. *Fizika Tverdogo Tela*, Vol. 44, No 3, p. 443, 2002.

[11] K.-T Lau, D. Hui *Carbon*, Vol. 40, № 9, p. 1605, 2002

In: Chemical Physics of Pyrolysis, Combustion and Oxidation ISBN 1-59454-163-9
Editors: A. A. Berlin et al., pp.199-217 

*Chapter 19*

# Description of Impacts on the Polymer Structure in the Framework of the Cluster Model

***G. V. Kozlov and G. E. Zaikov****
Chemical Faculty, Kabardino-Balkar State University,
173 Chemyshevski Street, Nalchik 360004, Russia
*N.M. Emanuel Institute of Biochemical Physics, RAS,
4 Kosygin Street, Moscow 119991, Russia

## 1. Crystallization

Actually, there are certain reasons why crystallization of polymers cannot be described in the framework of the cluster model. Nevertheless, there is a definite connection between short-range (local) order and long-range order formation [1]. Heat motions in a liquid induce continuous formation and termination of heterophase fluctuations, which represent potential focuses of future crystalline structures [2]. As temperature falls below the glass transition temperature, chemical potential of molecules in crystal is lower than in liquid. That is why in heterophase fluctuation the nucleation center can become stable and start propagating spontaneously [2].

It is assumed that configuration, morphology and mechanism of crystalline phase nucleation and propagation will be highly determined by the origin of heterophase fluctuations in the amorphous phase. For example, the assumption of heterophase fluctuation from folded chains (the Yech model [3]) leads to formation of crystallites with folded chains (CFC). An alternative model of heterophase fluctuation was suggested [2], where the presence of parallel chain segments of different macromolecules in it is assumed. The use of these models means the possibility of formation of CFC, crystallites with extended chains (CEC) or some intermediate crystalline morphology [4]. In its physical essence, the type of heterophase fluctuation suggested [2] is analogous to that of the local order zone in the cluster model [5]. Application of the mentioned model allows quantitative description of such heterophase fluctuations and, as a consequence, analysis of variations in the nucleation mechanism and crystallization morphology [6]. Such analysis has been performed on the example of oriented crystallization of two cross-linked polymers [1]: low density polyethylene (LDPE) and polychloroprene (PCP).

It has been shown [7] that the increase of stretching, $\lambda$, during oriented crystallization induces elongation of the statistical segment, $l_{st}$. Thus as the critical condition $l_{st} = l^*$ (where $l^*$ is the crystallization nucleus length) is reached, the cluster becomes such nucleus and the crystallite morphology depends upon the cluster morphology. Dependence $l_{st}(\lambda)$ can be calculated as follows. If one assumes homogeneous deformation of the cross-linked carcass, the distance between neighbor chemical cross-link points, $R_c$, will change proportionally to $\lambda$. Then the convolution degree of macromolecules, $\beta$, is determined as follows [8]:

$$\beta = \frac{\lambda R_c}{L_{seg}}, \quad (1)$$

where $L_{seg}$ is the chain segment length between chemical cross-link points, determined in accordance with [7].

In its turn, $\beta$ is associated with the number of statistical segments, $n_{st}$, in the chain segment $L_{st}$ long as follows [8]:

$$\beta = \left(\frac{3}{2} n_{st}\right)^{-1/2}, \quad (2)$$

and parameters $n_{st}$ and $l_{st}$ are bound to one another by correlation (5.57) that allows calculation of $l_{st}$ as the function of $\lambda$.

Figure 1 shows $l^*(\lambda)$ dependencies for LDPE, composed by data from the literature [4] (curves 1 – 4), and dependencies $l_{st}(\lambda)$, composed in accordance with equation (1) –(2) (curves 5 and 6 for LDPE and PCP, respectively). Vertical lines (7 and 8) indicate $\lambda^*_{cr}$ values (critical stretching at transition from mono- to multimolecular nucleation) for LDPE and PCP, respectively. Obviously, comparison of these data show that $\lambda^*_{cr}$ values estimated from the condition $l_{st} = l^*$ conform well to the experimental values of this parameter ($\lambda^*_{cr} \approx 1.6$ for LDPE [9] and 2.3 for PCP [10]). For HDPE, crystallization in the CFC form is assumed (experimental $\lambda^*_{cr}$ value corresponds to the crossing point of curves 4 and 5); for PCP formation of mixed crystalline morphology is supposed, where the part of crystallites with extended (fibrillar) chains gives 30%, approximately (experimental $\lambda^*_{cr}$ value corresponds to the crossing point of curves 1 and 6). It should be noted that for CPC such assessment is rather approximate, because curves 1 – 4 are plotted in accordance with the data on polyethylene [1]. Nevertheless, the data in Figure 1 demonstrate correctness of the cluster identification as heterophase fluctuations, which subsequently become crystalline phase growth nuclei. From these data it also follows that the increase of experimental $\lambda^*_{cr}$ values means decrease of the part of fibrillar crystallites in mixed crystalline morphology due to increase of the CFC part. The data obtained prove conclusions [10] assuming formation of mixed-type crystallites at PCP oriented crystallization.

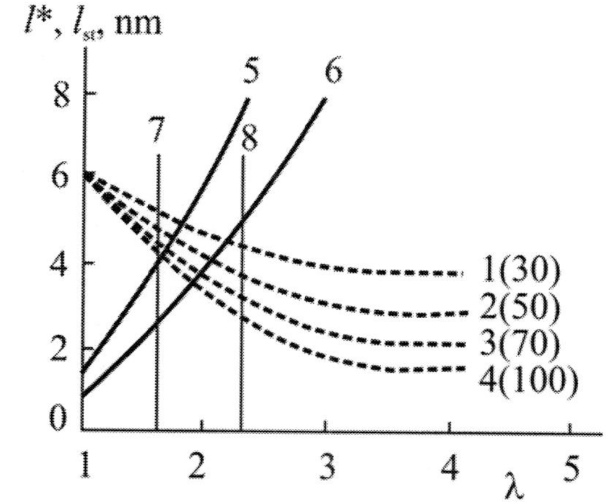


**Figure 1.** Dependencies of crystalline nucleus size, $l^*$ (1 – 4), and length of statistical segment, $l_{st}$ (5, 6), on the stretch degree, $\lambda$, for LDPE (1 – 5) and PCP (6). Straight lines (7) and (8) indicate critical values, $\lambda^*_{cr}$, for LDPE (7) [9] and PCP (8) [10]. Digits at $l^*(\lambda)$ curves give percentage of CEC content [1]

Besides longitudinal size of critical heterogeneous fluctuation, $l^*$, there is one more critical size of this fluctuation, the transversal one [4, 6]. The above-mentioned size is determined as follows:

$$(S\nu^*)^{1/2},$$

where $S$ is the macromolecule cross-section; $\nu^*$ is the functionality of critical nucleus. For polyethylene, dependence of $(S\nu^*)^{1/2}$ on $\lambda$ is shown in the work [6]. For the cluster, analogous parameter can be determined as follows:

$$(SF^*)^{1/2},$$

where $F$ is the cluster functionality. For LDPE, Figure 2 shows dependencies of $(S\nu^*)^{1/2}$ in accordance with the data from [6] and $(SF^*)^{1/2}$ on the stretching, $\lambda$. Clearly in accordance with the condition of equality of transverse dimensions, the cluster becomes the critical nucleus of crystallization at $\lambda^*_{cr} \approx 1.75$, which is close to experimental value of this parameter ($\approx 1.60$) [9].

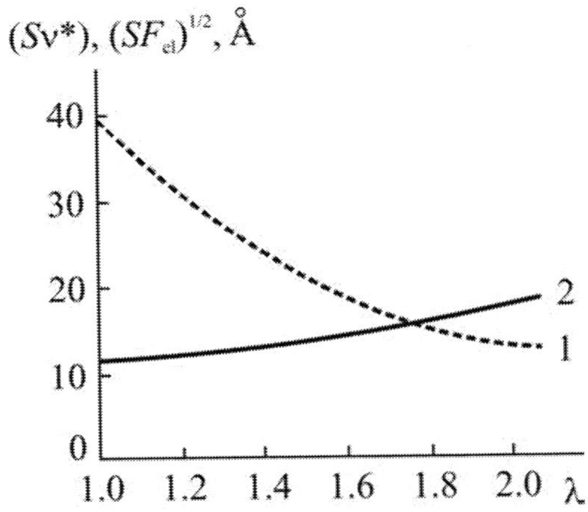


**Figure 2.** Dependencies of critical crystallization nucleus $(S\nu^*)^{1/2}$ (1) and cluster $(SF^*)^{1/2}$ (2) transverse dimensions on the stretch rating, $\lambda$, for LDPE [1]

Thus clusters by both physical meaning (remind that they are of thermofluctuational origin [5]) and their critical dimensions correspond to heterophase fluctuations, which may become crystallite nuclei. It should be noted that in contrast with the glassy state, dynamic local order zones in the devitrified state may partly consist of folded chains. Results obtained for PCP indicate its mixed crystalline morphology including both folded and fibrillar crystallites. The increase of part of the latter leads to decrease of the critical stretch rating, $\lambda^*_{cr}$, at which transition from mono- to multimolecular nucleation is observed.

## 2. Plasticization

Plasticization is introduction of organic compounds into a polymer for the purpose of attaching elasticity and frost resistance to them, and decreasing processing temperature [11]. In recent 50 years, due to high practical importance of the plasticization effect, great attention was paid to the study of different aspects of it. The so-called anti-plasticization effect, which is the increase of polymer rigidity and strength with plasticizer concentration in them, has been discovered [11]. At present, the approach based on the assumption of occurrence (or modification) of ordering at the introduction of low amounts of plasticizer that induces polymer rigidity increase, is the most reasonable one [11 – 14]. Further increase of plasticizer concentrations does not cause ordering of the polymer structure: polymer rigidity and strength are decreased. However, so far these structural changes have been described qualitatively only. That is why structural changes proceeding in polymers at plasticizer injection and determining the well-known plasticization and anti-plasticization effects were described quantitatively [15]. The above-mentioned changes were described quantitatively in the framework of the cluster model on the example of polycarbonate (PC) plasticized by dibutylphthalate (DBP) [5].

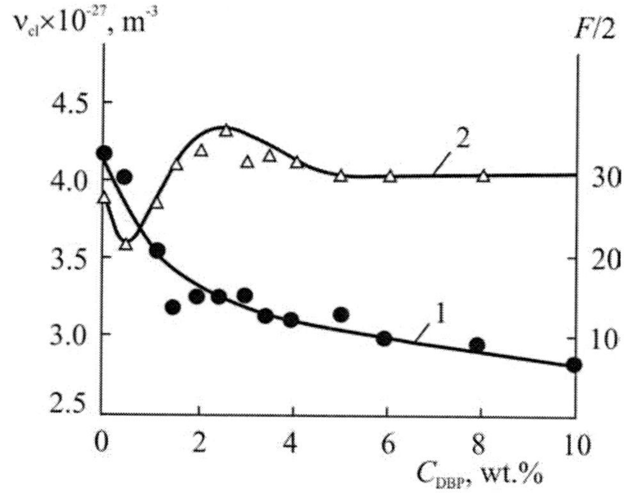


**Figure 3.** Dependencies of macromolecular entanglement cluster network frequency, $\nu_{cl}$ (1), and the number of segments in the cluster, $F/2$ (2), on dibutylphthalate concentration in PC, $C_{DBP}$ [15]

Figure 3 shows dependencies of $\nu_{cl}$ and $F/2$ on DBP plasticizer concentration in PC, $C_{DBP}$. Of special interest is different behavior of $\nu_{cl}$ and $F/2$ parameters with the concentration. The former decreases monotonously; to put it differently, local ordering in PC is monotonously decreased by the plasticizer. However, the extreme increase of the number of statistical segments in the cluster (equal $F/2$) indicates cluster restructuring. Simultaneous consideration of $\nu_{cl}(C_{DBP})$ and $F/2(C_{DBP})$ dependencies suggests that plasticizer injection leads to degradation of lower stable (with lower $F/2$) clusters and simultaneous formation of more stable (with higher $F/2$) local order zones [15].

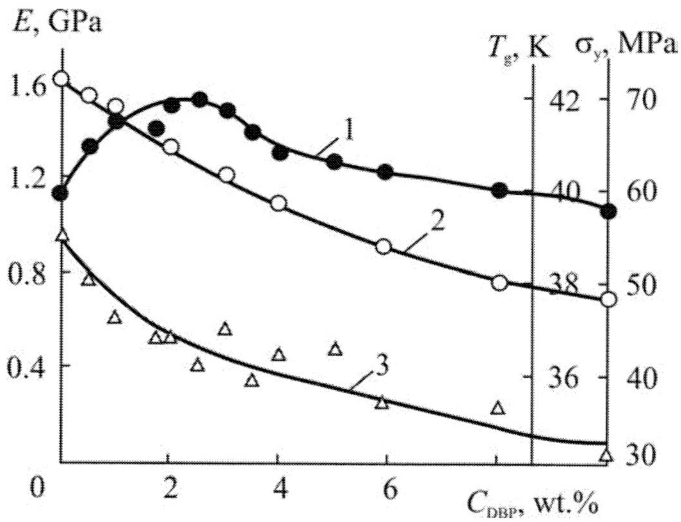


**Figure 4.** Dependencies of elasticity modulus $E$ (1), glass transition temperature $T_g$ (2) and yield stress $\sigma_y$ (3) on dibutylphthalate concentration in PC, $C_{DBP}$ [15]

Figure 4 shows dependencies of elasticity modulus $E$, glass transition temperature $T_g$ and yield stress $\sigma_y$ on the plasticizer concentration in PC, $C_{DBP}$. Two latter parameters decrease monotonously with $C_{DBP}$ increase, and the former one ($E$) passes through a maximum at $C_{DBP}$ = 2.5 wt.% that conforms to $F/2$ maximum in Figure 3.

Comparison of Figures 3 and 4 indicates complete determination of $T_g$ and $\sigma_y$ by $\nu_{cl}$ parameter (e.g. the local order degree), and more complicated dependence between $E$ and the structure. The first assumption is proved by plots in Figure 6.5 presenting dependencies of $T_g$ and $\sigma_y$ on the entanglement cluster network frequency $\nu_{cl}$ (or the number of closely packed segments per specific volume). Clearly both plots are linear. Moreover, extrapolation of $T_g(\nu_{cl})$ dependence to $\nu_{cl}$ = 0 demonstrates that the glass transition temperature of PC will equal the test temperature, i.e. under such conditions PC becomes the rubber-like composite. To put it differently, the polymer devitrification condition is disappearance of the local order. Naturally, under these conditions $\sigma_y = 0$, which is displayed by $\sigma_y(\nu_{cl})$ curve extrapolation to $\nu_{cl} = 0$ (Figure 5).

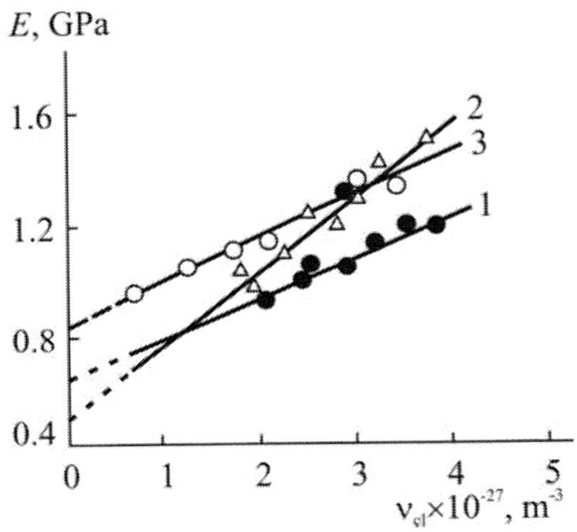


**Figure 5.** Dependencies of elasticity modulus $E$ on entanglement cluster network frequency $\nu_{cl}$ for primary PC (1) and PC with DBP contents as follows: 2.5 (2) and 10 (3) wt.% [15]

Figure 6 shows dependencies of elasticity modulus ($E$) on entanglement cluster network frequency ($\nu_{cl}$) for primary PC (line 1), PC containing 2.5 wt.% of DBP (anti-plasticized one, line 2), and the one containing 10 wt.% of DBP (plasticized one, line 3). In contrast with $\sigma_y(\nu_{cl})$ plot (Figure 5, these lines at $\nu_{cl}$ = 0 are extrapolated to non-zero $E_0$ values. This indicates that $E$ is determined by both $\nu_{cl}$ value and packless matrix rigidity [16]. Comparison of three plots shown in Figure 6.6 allows the following conclusions. Firstly, anti-plasticized PC possesses the minimum of $E_0$ at $\nu_{cl}$ = 0. This assumes that anti-plasticization effect is induced by formation of more stable clusters and, the more so, by more loosening of the packless matrix that results in decrease of its rigidity. Secondly, anti-plasticized PC displays the greatest slope of $E(\nu_{cl})$ line. To put it differently, more stable clusters are the most rigid ones. Thirdly, at equal slopes of $E(\nu_{cl})$ dependencies $E_0$ at $\nu_{cl}$ = 0 is greater for plasticized PC rather than for the primary one. It may be assumed that increased rigidity of plasticized PC packless matrix is stipulated by free volume microcavity filling in by plasticizer, which is indicated by monotonous decrease of gas permeability with $C_{DBP}$ increase.

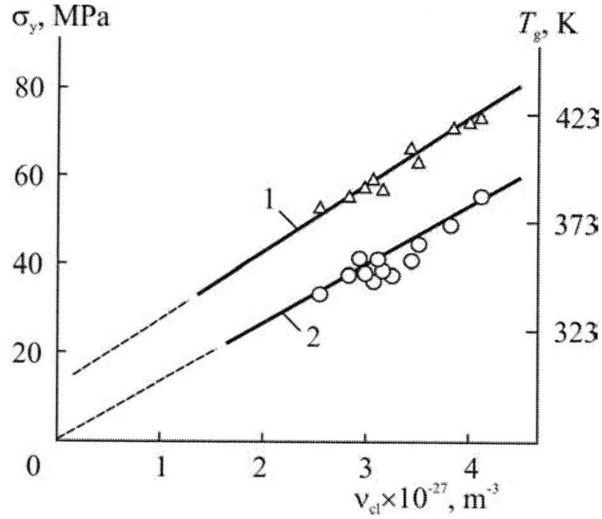


**Figure 6.** Dependencies of glass transition temperature $T_g$ (1) and yield stress $\sigma_y$ (2) on entanglement cluster network frequency $\nu_{cl}$ for PC, plasticized by DBP [15]

The mentioned curves (Figures 3 – 6) give demonstrative quantitative description of structural changes in PC under plasticizer injection into it. Note also that interpretation of plasticization and anti-plasticization effects of amorphous glassy-like PC in the framework of the cluster model correlates with the results obtained by different authors [12 – 14].

## 3. Hydrostatic Pressure

Usually, changes in the structure of amorphous polymers under hydrostatic pressure are interpreted in the framework of free volume models [17 – 19]. As shown before [20] for glassy-like epoxy polymer, dependence of the relative part of the fluctuation free volume, $f_f$, on external pressure, $P$, possesses three typical areas displaying different mechanisms of polymer structure consolidation. For example, in the pressure range $P$ = 0 - 200 MPa fluctuation microcavities "implode", and in the range $P$ = 200 – 600 MPa consolidation is stipulated by changes in macromolecule conformations (entropy decrease). Finally, at $P \geq 800$ MPa entire macromolecules are compressed that requires extremely high pressures. In this range, $f_f$ reaches its maximum [20].

The same data on $f_f(P)$ dependence are considered [21] in the framework of the fractal analysis [22] and the cluster model of the polymer amorphous state [5]. For epoxy polymer derived from bisphenol A diglycidyl ester, experimental data obtained by the positron annihilation spectroscopy method were used [18].

Parameter $f_f$ and fractal dimensionality of the structure $d_f$ are determined from equations (2) and (3), respectively. Combination of these equations gives the following correlation between $f_f$ and $d_f$ [21]:

$$f_f \approx 8.5\times10^3\left(\frac{d_f}{d-d_f}\right). \qquad (3)$$

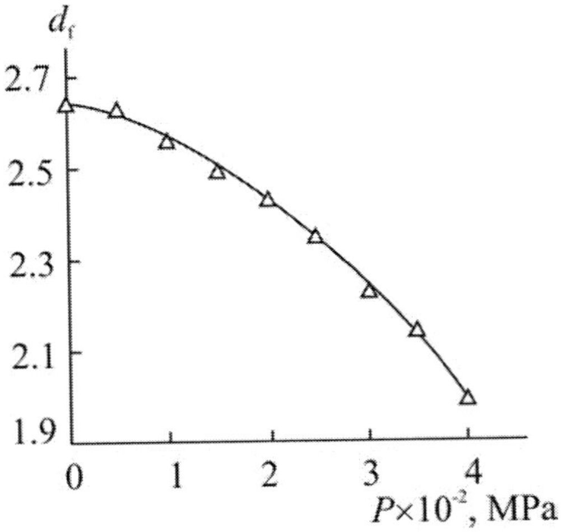


**Figure 7.** Dependence of epoxy polymer fractal dimensionality, $d_f$, on hydrostatic pressure, $P$ [21]

Figure 7 shows dependence $d_f(P)$, calculated by the equation (3) using parameter $f_f$ as the function of $P$ with regard to previously obtained data [20]. Clearly increase of $P$ induces monotonous decrease of $d_f$. From these values of $d_f$ using equations (1) - (3) the relative part of clusters $\varphi_{cl}$ was calculated as a function of $P$ (Figure 8) under the conditions $S$ = const and $C_\infty$ = const. As indicated in the Figure, $\varphi_{cl}$ is rapidly increased in the range of $P = 0 - 200$ MPa. At $P \approx 200$ MPa, the relative part of clusters reaches its maximum, $\varphi_{cl} = 1.0$, and is kept at this level during further pressure increase: $\varphi_{cl}$ = const = 1.0 [21].

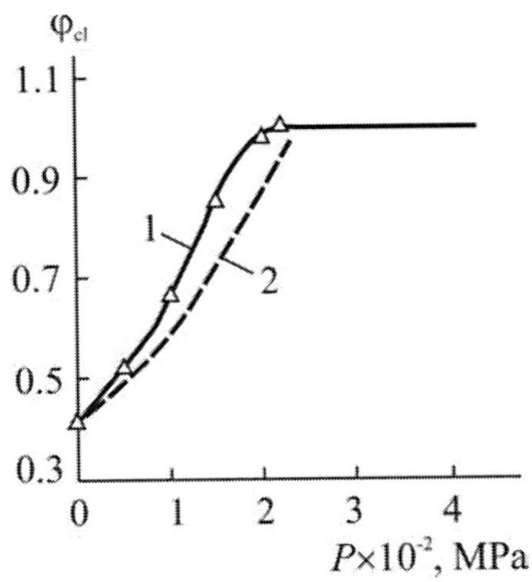


**Figure 8.** Dependence of the relative part of epoxy polymer volume, incapable of relaxation, $\varphi_{cl}$, on hydrostatic pressure, $P$: 1 – calculated by equation (1) at $C_\infty$ = const; 2 – calculated by equation (3) with regard to $C_\infty$ assessment [21]

As accepted [5] from the cluster model, the entire fluctuation free volume, $f_f$, is concentrated in the packless matrix. Under the impact of hydrostatic pressure applied to the polymer, fluctuation free volume microcavities "implode" and $\varphi_{cl}$ is increased. Obviously there are no reasons for the supposition that increase of epoxy polymer structure ordering is induced by pressure. In this case, more general definition of parameter $\varphi_{cl}$ can be given: it is the part of the polymer volume incapable of relaxation. Primarily, this part of the volume includes local order zones. In its turn, incapability of relaxation is stipulated by an abrupt decrease of $f_f$ or real absence of the fluctuation free volume. To put it differently, "porosity" degree of the packless matrix is determined by $f_f$ value; as $f_f \approx 0$, the matrix obtains a new quality: it behaves itself similar to closely packed local order zones [21].

Revert to the condition $\varphi_{cl}$ = const = 1.0 at $P > 200$ MPa. Obviously at $P > 200$ MPa $S$ remains constant, too ($S$ = const), because compression of macromolecules themselves requires much higher pressures. That is why $C_\infty$ is the only variable parameter, remained in the equations (1) – (3). Figure 9 shows dependence $C_\infty(P)$ under the supposition of its constancy at $P \leq 200$ MPa and $\varphi_{cl}$ = const = 1.0, and $S$ = const at $P > 200$ MPa. In the latter case ($P > 200$ MPa), an abrupt decrease of $C_\infty$ is clearly observed, and at the limit $C_\infty$ approaches 1. It is common knowledge [23] that in accordance with the equation (1) the characteristic relation $C_\infty$ is the function of internal rotation angle of chemical valence bonds, $\theta$. Obviously for $C_\infty \to 1$, $\cos\theta \to 0$ and $\theta \to 90°$. Thus as assumed before [20], pressure increase in the range $P > 200$ MPa will induce conformational changes in macromolecules, characterized by $C_\infty$ decrease and $\theta$ increase.

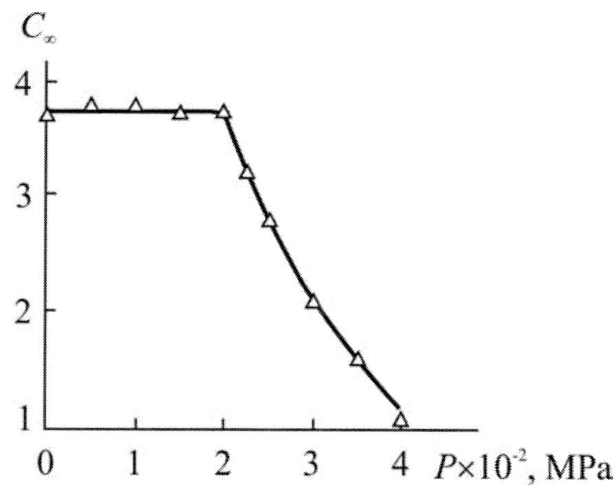


**Figure 9.** Dependence of characteristic relation, $C_\infty$, of epoxy polymer on hydrostatic pressure, $P$, calculated by the equations (1) – (3) under the assumption $C_\infty$ = const at $P < 200$ MPa [21]

Assuming that a change of $d_f$ induces variation of $C_\infty$ equal that for the case without application of hydrostatic pressure, the interconnection between $C_\infty$ and $d_f$ is set by the equations (1) – (3). In Figure 8 dashed line presents dependence $\varphi_{cl}(P)$, calculated by the equations (1) – (2) under the condition of variable $C_\infty$. Clearly qualitatively analogous result is obtained: the conditions $\varphi_{cl} = 1.0$ is reached at $P$ =250 MPa.

Thus the fractal analysis methods combined with the cluster model of the polymer amorphous state structure allow quantitative description of structural changes in epoxy polymer impacted by external hydrostatic pressure. The results obtained correlate well with the data followed from kinetic theory of the fluctuation free volume [20].

# 4. HEAT (PHYSICAL) AGING

Polymer heat aging processes are studied in many works. It is found that the effect of aging on structure and properties of these polymers is caused by thermodynamic non-equilibrium of the glassy state, and aging itself can be considered as slow approaching of the equilibrium state of a system displaying a broad distribution of relaxation times [24]. It has been assumed [25 – 27] that aging of amorphous polymers at temperatures below their glass transition points, $T_g$, induces consolidation of their liquid-like packing, which, in its turn, approaches their structure to the equilibrium glassy state. There are two points of view on particular ways of the mentioned approaching. The first is [28] that aging causes changes of the free volume in polymers. The second states [29] that the effects of aging are caused by thermally reversible molecular restructurings. These processes were specified by other authors [20]. It has been supposed that aging causes rather fine changes in conformations, smaller than, for example, transition from *cis*- to *trans*-conformation. The authors have also suggested that the free volume decrease is the demonstration of aging, but not the main reason for changes in physical properties of the polymer. The results of heat aging for PC were also described in the framework of the cluster model [31], which indicated that this interpretation generalizes both points of view.

Figure 10 shows dependencies of the entanglement cluster network frequency, $\nu_{cl}$, on test temperature, $T$, for PC, primary and annealed at 393 K during 6 hours. Clearly $\nu_{cl}$ is higher for aged PC, i.e. it displays higher local order degree. This tendency conforms completely to the amorphous polymer structure approaching to the equilibrium state, because the latter is the ordered state. Moreover, increase of the number of segments in clusters correlates with the conclusions made in ref. [30] in the point that structural changes are stipulated by relatively fine conformational changes.

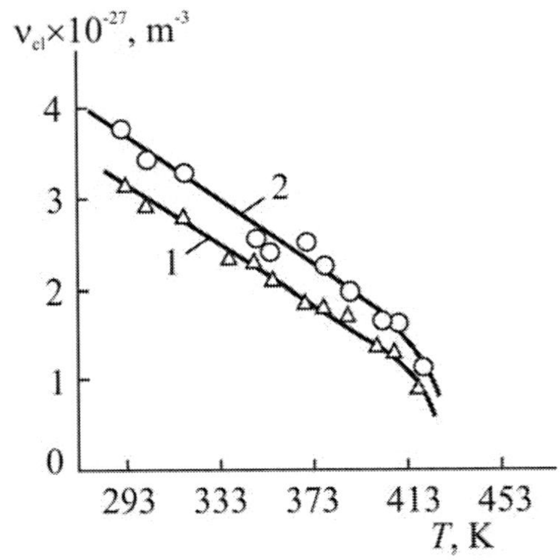


**Figure 10.** Dependencies of entanglement cluster network frequency, $\nu_{cl}$, on test temperature, $T$, for PC: primary (1) and aged (2) [31]

Discussed below are changes of the relative fluctuation free volume, $f_f$, during heat aging of PC. Dependencies of $f_f$ on test temperature $T$ are shown in Figure 11. It is shown that in the whole temperature range $f_f$ is higher for primary PC rather than for the aged one. This conforms to the conclusions made in ref. [29]. Much more interesting conclusions can be made on the basis of $f_f(\nu_{cl})$ dependencies, shown in the insertion to Figure 11. For primary and aged PCs, these dependencies do not coincide giving lower $f_f$ for aged polymer. Since the entire fluctuation free volume is concentrated in the packless matrix, at heat aging the latter is consolidated. This conclusion conforms to positron annihilation spectroscopy data on polyarylate annealed at temperature below $T_g$ [32].

Of special interest is also extrapolation of $f_f(\nu_{cl})$ dependencies to $f_f = 0$ and $\nu_{cl} = 0$. Obviously at $f_f = 0$ a closely packed polymer structure - the unique super-cluster, in principle, must be obtained. For this structure, $\nu_{cl}$ can be assessed as $L/l_{st}$. In the case of PC (either primary or aged), this gives $\nu_{cl} = 7.04\times10^{27}$ m$^{-3}$. In the insertion to Figure 11, extrapolation of $f_f(\nu_{cl})$ dependence to $f_f = 0$ gives quite close $\nu_{cl}$ value, $6.8\times10^{27}$ m$^{-3}$, approximately.

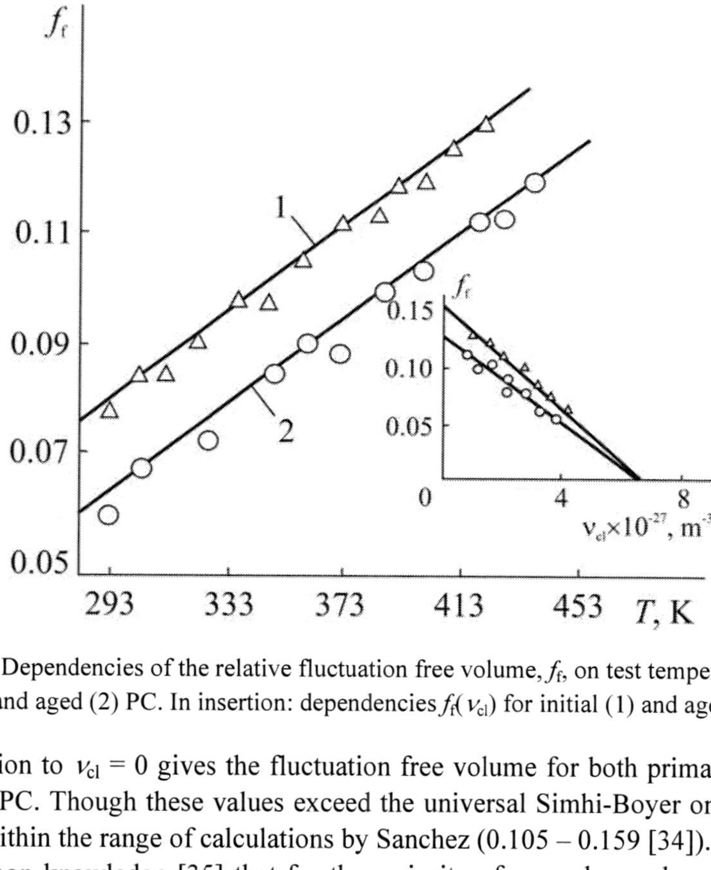


**Figure 11.** Dependencies of the relative fluctuation free volume, $f_f$, on test temperature, $T$, for primary (1) and aged (2) PC. In insertion: dependencies $f_f(\nu_{cl})$ for initial (1) and aged (2) PC [31]

Extrapolation to $\nu_{cl} = 0$ gives the fluctuation free volume for both primary (~0.150) and aged (~0.133) PC. Though these values exceed the universal Simhi-Boyer one for $f_f$ (~0.122 [33]), it falls within the range of calculations by Sanchez (0.105 – 0.159 [34]).

It is common knowledge [35] that for the majority of amorphous glassy-like polymers, the Prigozhin-Defye criterion is not met. This means that for characterization of the structure of these polymers, the single order parameter is not enough. On the example of PC, the reasonableness of this argument is demonstrated by $f_f(\nu_{cl})$ dependence. Obviously different values of $f_f$ at the same $\nu_{cl}$ indicate that besides $\nu_{cl}$, the order parameter is required that characterizes structural state of the packless matrix. Thus in the framework of the cluster model, interpretation of physical aging of amorphous glassy-like PC allows quantitative description of structural changes happening.

As shown in Section 2, fractal dimensionality, $D$, of the chain segment between entanglement or cross-link points characterizes mobility of the chain. The interconnection of mobility and local order was studied on the example of two series of epoxy polymers (primary (EP-1) and aged during 3 years under natural conditions (EP-2)) [36]. Dependence $D\left(\nu_{cl}^{-1/3}\right)$ in Figure 12 indicates that increase of $\nu_{cl}$ or local order degree induces decrease of $D$ and, consequently, suppresses chain mobility. Any decrease of the chain mobility, independently of its reasons, induces polymer friability increase [37], which is always observed for physical aging of these materials. Comparison of dependencies $D\left(\nu_{cl}^{-1/3}\right)$ and $\nu_{cl}(K_{st})$, shown in Figure 13, allows a conclusion that in the present case, this process is stipulated by polymer aging [36].

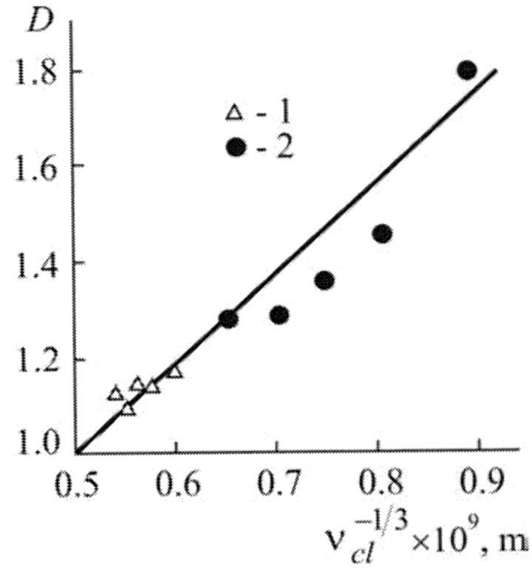


**Figure 12.** Dependence of the fractal dimensionality, $D$, of chain intersect between cross-link points on parameter $\nu_{cl}^{-1/3}$ for EP-1 (1) and EP-2 (2) [36]

As mentioned in Chapter 2, the initial temperature of the glass transition interval corresponds to the glass transition point of the packless matrix, and the final temperature corresponds to the glass transition point of the polymer [38]. Therefore, the glass transition interval width, $\Delta T_g$, is determined by the number of segments in the cluster, $n_{cl}$. Based on these data and Figure 14, one can state that the aging of polymers promotes increase of the cluster structure homogeneity [36].

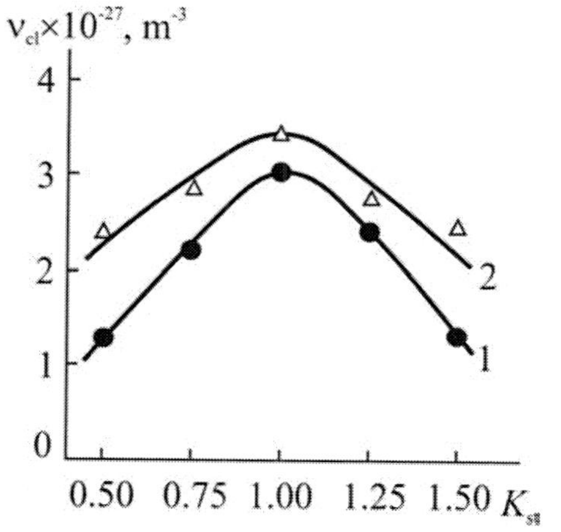


**Figure 13.** Dependence of the cluster network frequency, $\nu_{cl}$, on the curing agent:oligomer ratio, $K_{st}$, for EP-1 (1) and EP-2 (2) [36]

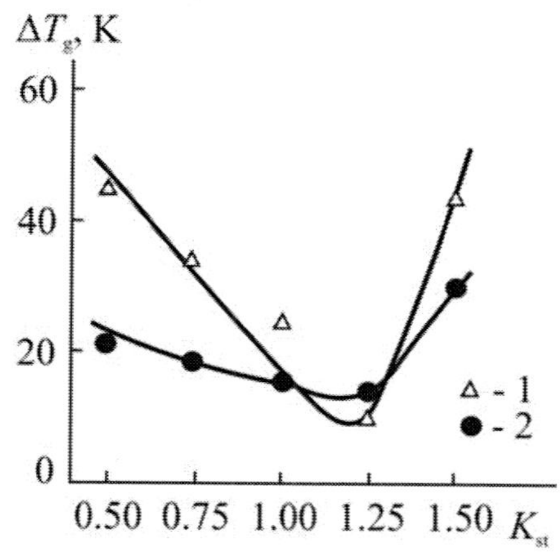


**Figure 14.** Dependence of the glass transition interval width, $\Delta T_g$, on the curing agent:oligomer ratio, $K_{st}$, for EP-1 (1) and (EP-2) (2) [36]

Thus physical aging of amorphous glassy polymers represents changes in structure and properties with time and is the consequence of thermodynamically non-equilibrium origin of these polymers. At quite low (about room level) temperatures, this process proceeds slowly, and significant changes in properties are realized during several years [36, 39]. That is why such events must be forecasted with regard to both spatial and time disorder of the systems subject to aging [40]. Spatial structural inhomogeneity of amorphous glassy-like polymers makes the range of microscopic rates of relaxation transitions rather wide, i.e. spatial disorder induces the time one [40].

Relating to description of the physical aging in the framework of the fractal analysis, the following facts should be noted. The study of reaction dynamics in amorphous systems indicate [40] that both types of disorder must be considered. At long time periods for the majority of reactions the subordination expressed by the relation below [40] takes place:

$$\Phi(t) \sim f(t^{\alpha\gamma}), \tag{4}$$

where $\Phi(t)$ is an arbitrary function, independent of time $t$; $\alpha$ and $\gamma$ are exponents taking into account the presence of spatial and time disorder, respectively.

Relating to the physical aging, structural changes with time can be described based on the general correlation (4) as follows [41]:

$$\nu_{cl} = c\nu_{cl}^{0} t^{\alpha\gamma}, \tag{5}$$

where $\nu_{cl}$ and $\nu_{cl}^{0}$ are entanglement cluster network frequencies for aged and primary samples, respectively; $c$ is the proportion coefficient considering replacement of the proportion sign in the correlation (4) by the sign of equality in equation (5).

In the equation (5) the physical meaning of exponents $\alpha$ and $\gamma$ shall be determined. In the framework of the fractal analysis, exponent $\alpha$ characterizing spatial disorder can be determined as $(d_f - 2)$, where $d_f$ is the fractal dimensionality of the primary epoxy polymer structure (EP-1), due to the following reasons. Variation of $d_f$ in the range of 2 -3 characterizes the disorder degree of the polymer structure, and the basic information about this degree is carried by fractional part of $d_f$: the greater $d_f$ is, the lower the local order degree is and, consequently, higher spatial disorder of the system is [42]. Essentially, the integer part of $d_f$ carries the information about dimensionality of the Euclidean space, to which the fractal object is buried. Moreover, it should be noted that the exponent $\alpha = d_f - 2$ coincides with the fractional index in the theory of fractional integration-differentiation [43], in which this index characterizes the part of system states, not varied during evolution [44]. To put it differently, in the framework of the fractional integration-differentiation it is assumed that the ordered part of the polymer structure will be changed during physical aging, which is clearly displayed by the equation (5).

As shown in Chapter 3, there is a quasi-equilibrium state of polymers, approached by the polymer structure with increase of its thermodynamic equilibrium degree, characterized by the condition $d_f = 2.5$. This condition relates to completely relaxed polymer with a narrow distribution of free volume cavity sizes [45] and, consequently, to the shortest time disorder characterized by the relaxation time spectrum width. That is why, henceforth, the following approximation will be used [41]:

$$\gamma = d_f - 2.5. \tag{6}$$

Thus the equation (5) can be reduced to the form as follows [41]:

$$\nu_{cl} = c\nu_{cl}^{0} t^{(d_f-2)(d_f-2.5)}, \tag{7}$$

where empirically determined constant $c$ equals ~0.14, and the aging time $t$ is given in seconds.

Figure 15 shows dependencies of $\varphi_{cl}$, calculated by equations (7) and (4), for primary (EP-1) and aged (EP-2) epoxy polymers on the curing agent:oligomer ratio, $K_{st}$, as well as experimental ones. The data shown indicate good correspondence between theory and experiment. Even at $K_{st} = 0.50$ and 1.50, at which deviation is maximal, it does not exceed

~25%. It has been concluded [41] that the mentioned deviation is associated with the calculation method for the Poisson parameter, $\nu$, and, consequently, $D_f$ (equation (4)), namely, with application of equation (7)and strong dependence of $\nu_{cl}$ on $d_f$ in equation (7) for this purpose. It is known [46] that for epoxy polymers, inaccuracy of $\nu$ assessment by the equation (3) increases with test temperature, $T$. At high temperatures this method can give $\nu$ by ~7% higher than those, determined by other methods. Since at $K_{st}$ = 0.50 and 1.50 the highest values of $\nu$ are observed (0.364 and 0.370, respectively), the highest inaccuracy shall be expected for them. Decrease of both these values of $\nu$ to 0.360 (i.e. by 1 – 3% only) and recalculation of $\varphi_{cl}$ by equations (7) and (3) give $\varphi_{cl}$ values much closer to experimental ones (the dashed line in Figure 15).

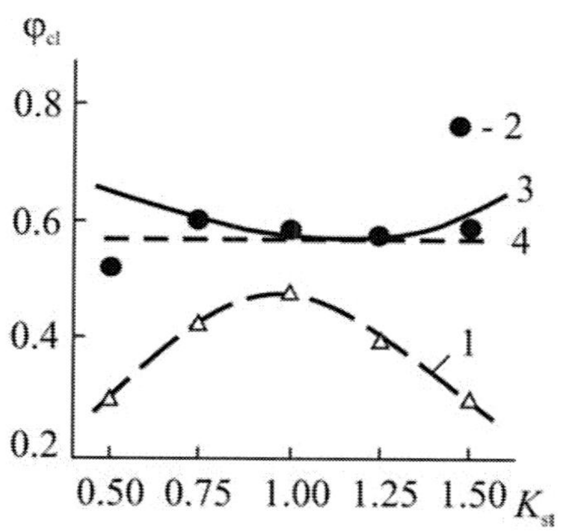


**Figure 15.** Dependencies of the relative part of clusters, $\varphi_{cl}$, on the curing agent:oligomer ratio, $K_{st}$, for EP-1 (1) and EP-2 (2 – 4); 1, 2 – experimental data; 3 – calculation by equations (7) and (4); 4 – calculation by the same equations with adjustment of $\nu$ values for EP-2 with $K_{st}$ = 0.50 and 1.50 (refer to the text) [41]

The second important aspect of physical aging also follows from data of the Figure 15. For example, for different $K_{st}$ (and, consequently, for different frequencies of chemical cross-link points $\nu_c$ [47]) intensities of aging, expressed by difference of $\varphi_{cl}$ values for EP-2 and EP-1, are significantly different. Expectedly [42], minimal intensity of aging is observed for the system with stoichiometric value $K_{st}$ = 1.0, and $K_{st}$ deviation from this value leads to the increase of physical aging intensity [36]. As follows from equation (7), the effect observed is induced by increase of both $\alpha$ and $\gamma$ for systems with highest deviations from $K_{st}$ = 1.0 or, to put it differently, with increase of fractal dimensionality of the structure, $D_f$ [41].

As shown in Chapter 4, correlations (5), (6) and (7) are true for polymers, for which changes of supermolecular structure parameters ($d_f$ and $D_f$) and molecular index (statistical chain rigidity, characterized by $C_\infty$) during physical aging are assumed. Finally, physical aging levels the structure of epoxy polymers independently of $\nu_c$, which varies by, approximately, an order of magnitude for them [47]. Naturally, such structure leveling also determines a significant approach of properties for EP-2. For example, the elasticity modulus $E$ in squeeze tests varies for EP-1 in the range of 2.50 – 3.60 GPa and for EP-2 in the range of 1.60 – 1.90 GPa only [36].

Thus the technique suggested in [41] allows calculation of changes in the structure of epoxy polymers during their physical aging using ideas of the cluster model and the fractal analysis. The coefficient $c$ in equation (7) may possibly be the function of aging temperature and, in principle, an analytical correlation for its determination can be deduced. Of much higher probability is that the increase of $\nu$ (and, consequently, $d_f$) with temperature will take into account the aging intensification. These assumptions require to be proved experimentally. Of importance is that structural restructuring during physical aging involve both molecular and supermolecular levels [42].

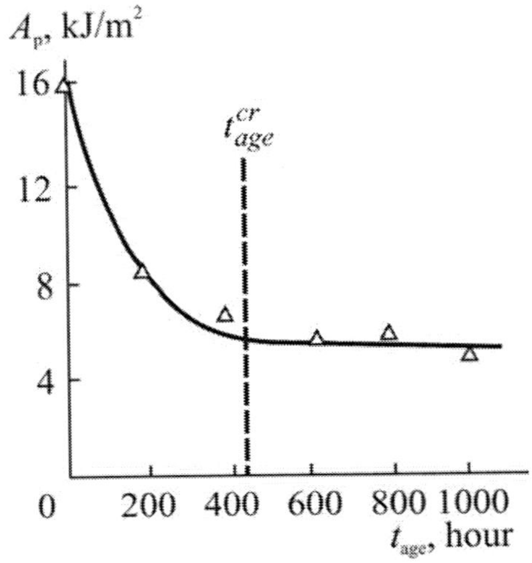


**Figure 16.** Dependence of blow viscosity, $A_p$, on aging time, $t_{age}$, for PC. Vertical dashed line shows $t_{age}^{cr}$ value, calculated by equation (7) [48]

The above-considered concepts are used for description of PC blow viscosity, $A_p$, dependence on physical aging duration. Expectedly, due to general ideas [37], PC plasticity is decreased during physical aging, which is reflected by $A_p$ decrease with aging time, $t_{age}$ (Figure 16). Clearly the data in the Figure indicate $A_p$ decrease during initial ~500 hours of aging, then it fits an asymptotic curve with low absolute values of $A_p$ (of about 4.5 kJ/m$^2$). This run of $A_p(t_{age})$ dependence can be explained by reaching some border state of the structure, which is defined as the quasi-equilibrium state (refer to Chapter 4). For this state, $D_f$ value $\left(D_f^{cr}\right)$ is determined by the equation [22]:

$$D_f^{cr} = \frac{4\pi T}{\ln\left(1/f_f\right)T_g}. \tag{8}$$

As shown in Chapter 5, polymer structure can be simulated as a thermal cluster, i.e. the cluster for which formation both geometrical and thermal interactions are important. The order parameter of such cluster, $\varphi_{cl}$, is determined from the comparison (7).

Figure 17 shows dependence $d_f(t_{age})$, calculated in accordance with equations (3) and (4), illustrating evolution of PC structure during physical aging. Clearly at $t_{age}$ > 500 hours $d_f$

reaches the asymptotic value equal ~2.46. Further on, $D_f^{cr}$ was calculated [48] by equation (8), and then the appropriate $d_f$ value $\left(d_f^{cr}\right)$ was estimated using the equation (7). In accordance with equations (8) and (7) the following values of fractal dimensionalities were obtained: $D_f^{cr} \approx 2.68$ and $d_f^{cr} \approx 2.41$. The latter value correlates well with the above-shown $d_f$ for asymptotic branch of dependence $d_f(t_{age})$ (Figure 17). Thus $A_p$ fitting the asymptotic branch, shown in Figure 16, can be explained by reaching quasi-equilibrium state by PC structure during physical aging.

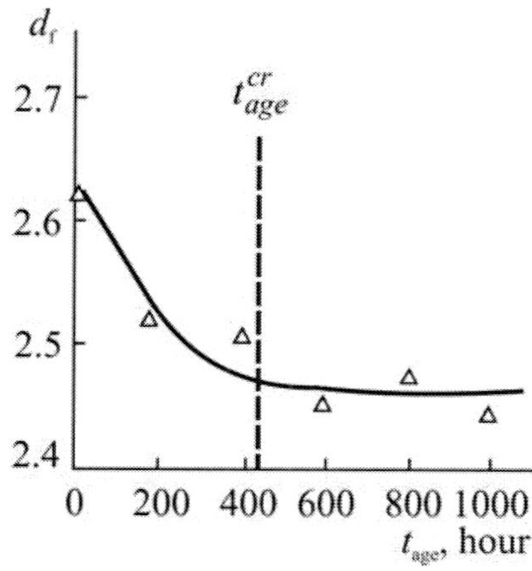


**Figure 17.** Dependence of the fractal dimensionality, $d_f$, of PC structure on aging time, $t_{age}$. Vertical dashed line shows $t_{age}^{cr}$ value, calculated by equation (7) [48]

Then in the quasi-equilibrium state $\varphi_{cl}$ can be estimated by the equation (4) via determination $\beta_T = \beta$ in accordance with the equation (5), i.e. under the assumption that indices of thermal and percolation clusters are equal. This calculation gives $\varphi_{cl}^{cr} = 0.619$. Subsequently, one can estimate $d_f^{cr}$ value from the general correlation (5), which gives $d_f^{cr} = 2.45$. This value does also correlate well with the above-shown assessments. Thus thermal cluster model gives an adequate description of the polymer structure, the quasi-equilibrium state conditions, in particular. In accordance with equations (4), (8) and (7), $D_f^{cr}$ was estimated as 2.85, 2.68 and 2.82, respectively, which also demonstrates conformity between the mentioned techniques [48].

Thereafter, physical aging duration, $t_{age}^{cr}$, required for reaching the quasi-equilibrium state was estimated [48] using the equation (7). For this purpose, parameters in the equation were estimated as follows. Values $\varphi_{cl}^{cr}$ and $\varphi_{cl}^{0}$ are calculated at aging temperature, $T_{age}$, in accordance with equations (4) and (2), respectively. Then at $T_{age}$ parameters $d_f$ and $d_f^{cr}$ are calculated by the equation (7) using appropriate values of $\varphi_{cl}^{0}$ and $\varphi_{cl}^{cr}$. Parameter $t_{age}^{cr}$, calculated by the equation (7) under the mentioned conditions, equals 438 hours. In Figures 16 and 17 it is presented by vertical dashed lines. Clearly $t_{age}^{cr}$ conforms well to $t_{age}$ value corresponded to the transition of $A_p(t_{age})$ and $d_f(t_{age})$ dependencies to the asymptotic branch. This allows determination of duration of transitional (non-stationary) period of the physical aging, $t_{age}^{n}$, during which structure and properties of the polymer are changed ($t_{age}^{n} \le t_{age}^{cr}$).

As mentioned above, $A_p$ increases with molecular mobility, the latter being associated with the mobility of chains in packless zones of the polymer structure [49]. In the framework of the cluster model, such zones are presented by the packless matrix surrounding the clusters. Figure 18 shows the dependence of $A_p$ on the relative part of packless matrix, $\varphi_{p.m.}$. Expectedly [37], linear increase of $A_p$ with $\varphi_{p.m.}$ is observed, and the condition $A_p = 0$is realized at $\varphi_{p.m.} \approx 0.352$ instead of expected $\varphi_{p.m.} = 0$. In accordance with the rule $\varphi_{p.m.} = 1 - \varphi_{cl}$, this value of $\varphi_{p.m.}$ conforms to $\varphi_{cl} \approx 0.648$, which is quite close to $t_{age}^{cr}$ at $T = 293$ K ($\approx$ 0.619). This correlates with the postulate of completely stretched chains in the quasi-equilibrium state limiting $\varphi_{cl}$ increase and thus completely suppressed molecular mobility that leads to $A_p = 0$, e.g. completely friable polymer [48].

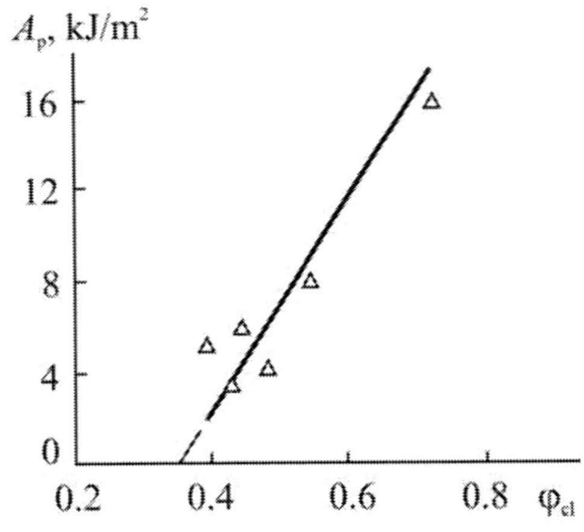


**Figure 18.** Dependence of blow viscosity $A_p$ on relative part of the packless matrix $\varphi_{p.m.}$ for PC polymer [48]

As follows from the equation (8), parameter $D_f^{cr}$ is the function of temperature ($T$) with regard to low logarithmic dependence of it on $f_f$, the value of which hereinafter is accepted constant and equal 0.075 [51]. In its turn, this means that the fractional dimensionality $d_f^{cr}$ of the polymer structure in the quasi-equilibrium state will also be the function of $T$ (refer to equation (4)). Experimentally, this statement can be proved by comparison of $D_f$ and $d_f$ parameters, calculated by the mentioned equations [50].

Thus in the used approximation, $D_f^{cr}$ parameter is the function of $T$ only, which is typical of the thermal cluster [52]. $D_f^{cr}$ increases with temperature that conforms to behavior of the order parameter, described by the equation (3). It is illustrated in Figure 19 by $D_f^{cr}$ ($T$) dependence (curve 1). Values of $D_f^{cr}$ parameter, calculated as the function of $T$ for the quasi-equilibrium state and primary and aged PC film samples by equations (8) and (7), respectively, are also shown in the Figure. The comparison indicates higher $D_f$ values for primary PC samples, rather than for the quasi-equilibrium state. However, heat aging also decreases $D_f$ for aged PC samples, for which $D_f^{cr}$ and $D_f$, calculated by equations (8) and (7), respectively, are in good conformity. All this proves that heat interactions during physical aging of PC induce structural changes in the polymer, and its quasi-equilibrium structure is determined by both configurative and heat interactions that gives an opportunity to consider it as the thermal cluster [52].

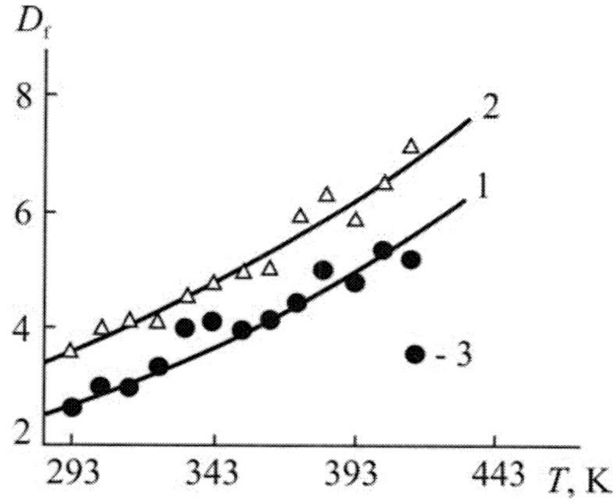


**Figure 19.** Temperature dependencies of dimensionalities of excessive energy localization zones, $D_f$, for PC polymer: 1 – calculated by equation (8); 2, 3 – calculated by equation (6) for primary (2) and aged (3) PC [50]

In: Chemical Physics of Pyrolysis, Combustion and Oxidation ISBN 1-59454-163-9
Editors: A. A. Berlin et al., pp.219-222 

*Chapter 20*

# ATYPICAL BEHAVIOUR TAILORED ELASTOMERIC DURING BURNING

***A. N. Gajdadin, O. O. Tuzhikov,***
***A. N. Garashchenko and V. F. Kablov***
Volgograd State Technical University

The polymeric material collapses under influence of an open flame on the difficult mechanism. Coke is formed, gases of pyrolysis are allocated, goes destruction of internal layers on a surface of a material. The circuit of destruction of polymer is submitted in figure 1. This circuit is enough considered in detail at temperatures of destruction of polymer, formation of gases of pyrolysis, accumulation of a coke layer and influence in the field of a flame. Ways of increase of efficiency of a fire-resistant material are known for due of regulation of processes in each of the named layers.

However the behaviour of elastomer in internal layers is not absolutely clear at temperatures from 623K up to 298K. During a fire elastomer is a subject of influence of mechanical loadings. They arise for due of thermal expansions, influence of weight of designs, etc. The Analysis of a plenty of experimental data shows, that mechanical loadings also operates on behaviour of slow-burning elastomer. During this process elastomer decrepitates and also stratifies at heating, his thermal parameters changes essentially that results in sharp decrease stability of action of a flame. Therefore researches of behaviour of fire-resistant elastomers were carried out at temperature and mechanical loadings simultaneously.

For this purpose behaviour of rubbers was studied on the basis of rubber Dutral-CO-054. The structure of rubbers included usual cure group, filler, some technological and operational additives. Manufacturing of samples carried out in press at temperature 448K. Samples had good durability, they were easily processed. Change of the sizes of a sample (deformation of creep) wes considered as an estimation of parameters at a simultaneous influence of the fixed loading (0.8 mPa) and temperatures from 623K up to 298K. The choice of loadings and temperatures of tests corresponds to conditions of behaviour of internal layers of fire-resistant rubbers at burning. Results of experiment are resulted in figure 2. Apparently from figure, process of creep can be divided into three stages:

- Initial, proceeding during 2 - 3 sec. and connected to initial deformation of a sample;
- A stage of slow deformation;
- A stage of fast deformation with explosive character.

Occurrence of the third stage of deformation is typical for temperatures above 430K.

It is known, that the increase of speed of deformation at influence of loadings for tailored polymeric materials is observed only at temperatures of destruction [1] and corresponds for ethylenepropylene rubbers to temperatures 620 - 640K [2].

However the analysis of a sample after tests shows, that a material "rubber similar", it keeps all external attributes of a porous elastic composition. External attributes of processes of deep pyrolysis (a sticky surface, fragility) are not observed. It is established, that during test structure of a material changes in which matrix big quantity of porouses appear. Scanning on temperature has shown that the beginning of pore-formation corresponds to the beginning of a stage of fast deformation. Besides the termination of influence of temperature and loading on a material at the beginning of a stage of pore-formation results in practically to full restoration of the initial sizes of a sample, which speaks about weak destruction of intermolecular communications in a matrix of elastomer. With achievement of area pore-formation in a material speed of quickly proceeding deformations of explosive character accrues and irreversible changes of structure resulting occurs, resuling to changes of the sizes of a sample. The beginning of a stage of quickly proceeding deformation of explosive character corresponds to rough pore-formation inside a matrix and to changes of structure of a material. It is necessary to note that the temperature of starts of process of pore-formation much below temperature of destruction of a polymeric matrix. It is established, that change of structure of a material and occurrence of a porous phase in a composition reduces straining - strength characteristics and durability (time before destruction under loading) of elastomers. From here follows, that for tailored polymeric substantial is typical growth of speed of deformation in the field of high temperatures help of occurrence of a porous phase and changes of a macrostructure of a material. It is possible to speak about structural current elastomer by analogy with well-known processes of physical and chemical current caused by transformations of the corresponding nature.

Structural current is absent when the temperatures are lower then 423 K. Under such circumstances tests elastomer does not undergo structural transformations at any influence of time. The beginning of process of pore-formation in a matrix is characterized by the minimal temperature, which can be called temperature of the beginning of structural current. Probably the temperature of start of structural current depends on structure and can characterize stability of a material to the influence of temperatures, mechanical pressure and a flame.

At achievement of temperature of the beginning of structural current in a matrix of a material a new porous phase appears. According to the accepted terminology it is possible to assert [1], that the physical transition in a material similar phase is observed. Change of volume and necessity of thermal influence on process show that phase transition of the first sort is observed. As it is known, similar transition is accompanied by changes in internal energy, volume, entropy systems and thermal effect. Processes of crystallization, fusion, condensation and sublimation concern to transitions of the first sort [3]. Parameters of elastomeric material are resulted in table 1 in view of the beginning of pore-formation in internal layers. The obtained experimental datas show that achievement of physical transition results to essential decrease of durability of a composition.

**Table 1. Change of operational parameters of a base material**

| Temperature of tests, K | Time prior to the beginning of a pore-formation, seconds | Durability, seconds |
|---|---|---|
| 423 | Not observed | 804 |
| 473 | 80 | 187 |
| 523 | 60 | 73 |
| 573 | 30 | 44 |
| 623 | 21 | 27 |

The increase of time prior to the beginning of pore-formation promotes the increase of stability of a composition at heating during burning. The increase of serviceability of a material can be only when time of pore-formation in a matrix increases. Probably process of pore-formation in a composition, caused by decomposition of the rests of curing group, evaporation of a residual moisture. During this process the beginning of pore-formation depends on size of the module of high-elasticity of elastomers. We shall lead analogy to the processes proceeding by manufacture of foamed polymeric compositions. It is known, that process of foaming at heating and using special foaming agents proceeds only when pressure in blister of gas is comparable or surpasses value of the module high-elasticity elastomer [4]. Increase of heat stability of matrixes will allow to keep value of the module of high-elasticity in the heated up condition and to increase time before the pore-formation of composition. Therefore the behaviour of rubbers on a basis of ethylene propylene rubbers containing stabilizers of the various nature and mechanisms of action has been considered, during influence of high temperatures and mechanical loadings. Carbamate BNI (dibutylditiocarbamate of nickel), naftam-2, a product 4010NA and systems condensation ability reagents in various concentration were used as stabilizers. It is established, that using of stabilizers allows to change time before the beginning of pore-formation of a material at the specified temperature [5]. Values of time of guaranteed serviceability elastomers with various modifiers are submitted in table 2.

Apparently from the table, using of various systems of stabilizers allows to influence considerably for the time of the beginning of pore-formation and the durability of a material connected to it and fire resistance of a material. As a result of carried out experiments it is shown, that the process of pore-formation is peculiar to the tailored elastic compositions subjected to simultaneous influence of high temperatures, a flame and weak mechanical pressure, which results to fast increase of lengthening of creep and accumulation in a sample of irreversible deformations. The sharp increase of creep is observed at temperatures of considerably lower, than temperature of destruction of a polymeric matrix. At pore-formation there is a process similar phase transition first sort and are considerably reduced mechanical parameters of a composition.

**Table 2. Time of guaranteed serviceability of researched rubbers.**

| Name of the additive | Time prior to the beginning of a pore-formation, sec. (473K) | Durability of an elastomer, sec. |
|---|---|---|
| Elastomer without additives | 80 | 187 |
| Naftam-2 | | |
| 5 mass p. | 102 | 210 |
| 15 mass p. | 70 | 104 |
| 25 mass p. | 22 | 44 |
| Product 4010NA | | |
| 5 mass p. | 50 | 145 |
| 15 mass p. | 42 | 55 |
| 25 mass p. | 25 | 35 |
| Carbamate BNI | | |
| 5 mass p. | 140 | 250 |
| 15 mass p. | 500 | 800 |
| 25 mass p. | 100 | 240 |
| Reactive substances | | |
| 5 mass p. | 110 | 230 |
| 15 mass p. | 65 | 113 |
| 25 mass p. | 42 | 67 |

Time before the beginning of pore-formation in a material can be accepted during guaranteed safe work of a material at the set temperature, action of loading and a flame [6]. Occurrence in a material of a porous phase at temperatures 623K results to decrease of factors of heat diffusivity elastomers. It provides decrease of speed of warm of deep layers at influence on a material of an open flame. Decrease of speed of the heating, connected with change of factor heat diffusivity, provides decrease of time of achievement of temperatures of destraction for each new layer of elastomer. Joint action of all these effects allows to increase fire resistance of a polymeric composition.

## References

[1] Tugov I.I., Kostrikina G.I. *Chemistry and physics of polymers*.-M.: Chemistry, 1989.-432p.

[2] Kuzminskii A.N., Sedov V.V. *Chemical of transformation of elastomers*.-M.: Chemistry,-1984.-230p.

[3] Tagger A.A. Fizikohimija of polymers.-M.: *Chemistry*, 1978,-544p.

[4] Piotrovskij K.B., Tarasova Z.N. Ageing and stabilization of synthetic rubbers and vulcanizats.-M.: *Chemistry*, 1980.-264p

[5] Gajdadin A.N., the *Dissertation on competition of a scientific degree of Cand.Tech.Sci.*, Volgograd: 1995

[6] Features of mechanical behaviour of rubbers on basis SKEPT at temperatures of the beginning destruction., *Rubber and rubber*. 2000. № 3, p. 15

In: Chemical Physics of Pyrolysis, Combustion and Oxidation ISBN 1-59454-163-9
Editors: A. A. Berlin et al., pp.223-241 

*Chapter 21*

# SYNTHESIS AND APPLICATION OF THE NEW PHOSPHORUS ELEMENT FIRE RETARDANTS FOR POLYMERS

***S. N. Bondarenko*, O. I. Tuzhikov**, T. V. Hohlova** and O. O. Tuzhikov*****

*Volzhsky Polytechnical Institute (Branch of the Volgograd State Technical University)
** The Volgograd State Technical University

The direction of combination of components for composite materials is one of perspective ways of development of polymeric technology. Multicomponent polymeric mixes, the fibreglasses, the strengthened elastomers, the polymeric concrete and other kinds of composite polymeric materials have wide application in various technics. Development of the new composites containing an element organic modifiers of the various nature, will allow not only to obtain materials with a unique complex of properties (the increased durability, a thermostability, adhesion to the various substrates, the lowered combustibility), but also to develop scientifically proved principles of their creation.
The certain interest of elementorganic polymers and oligomers as modifier of composites is research. Those materials are not migrating to surface of composite and can interact with polymer matrix.

## 1. SYNTHESIS AND APPLICATION PHOSPHORUS METAL CONTAINING FIRE RETARDANTS

The fire retardants (FR) of polymers may be divided into two groups: inert and reactive. Salts, oxides and hydroxides of metals are inert for polymers. Elementorganic fire retardants can react with polymers and be called reactive for polymers. The phosphorusmetal containing FR includes both fragments obtained by interaction of metal hydroxide and dimethil phosphit can be submitted by scheme (I) [1, 4, 6, 11, 13, 15, 32]:

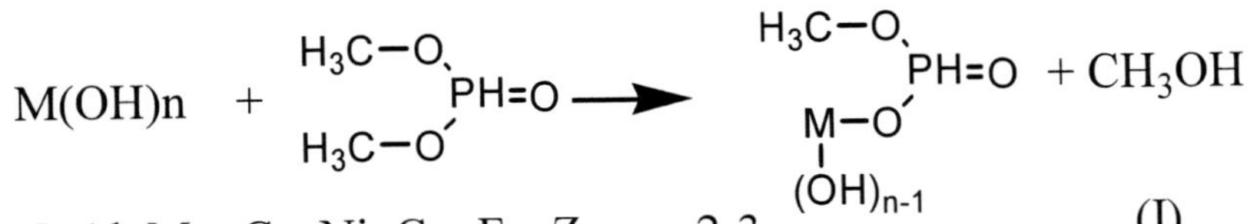

M=Al, Mg, Ca, Ni, Co, Fe, Zn; n=2-3 (I)

**Scheme (I)**

Reaction was carried out at temperature 140-160°C during 2-3 hours with formation of methanol. Kinetics of interactions of aluminium hydroxide and dimethyl phosphite (DMPh) was studied using nitrobenzene as solvent [2, 5, 39].

One of the defects of composition of nonsaturated polyesters is its flammability. This problem may be solved by modification of initials comonomers or using composite fire retardants.

Unsaturated polyesters may be modified by halogen containing dicarboxylic aids (tetrachloro (bromo) phthalic acid, brominated salicylic acid) or glycols. The resins may be modified by halogen- phosphorus containing monomers in at the time of its hardened. The metal hydroxide are also used. They are low-costly and available fire retardants, but they using reduc phisico-chemical property. Composition of unsaturated polyester PN-1 and styrol modified by base salts of aluminium, magnesium, calcium phosphite were hardened in different temperatures by us [22, 29].

Properties of the compositions of unsaturated polyesters and styrene including aluminium, magnesium, calcium hydroxide and their base salts of methyl phosphinic acid in different rates polyester and styrene shown in table 1. We can see that samples containing metal hydroxides haven decrease strength in 1.5 - 8 times (especially samples containing calcium hydroxide). Magnesium and aluminium hydroxides changed their parameters in lower degree. Base salts increase the strength of samples in 1.4-1.5 times when rate PN-1/styrene = 1/1 is used. There is no effect when content of styrene increases.

Interaction of salts with a polyester can be submitted by the scheme (II):

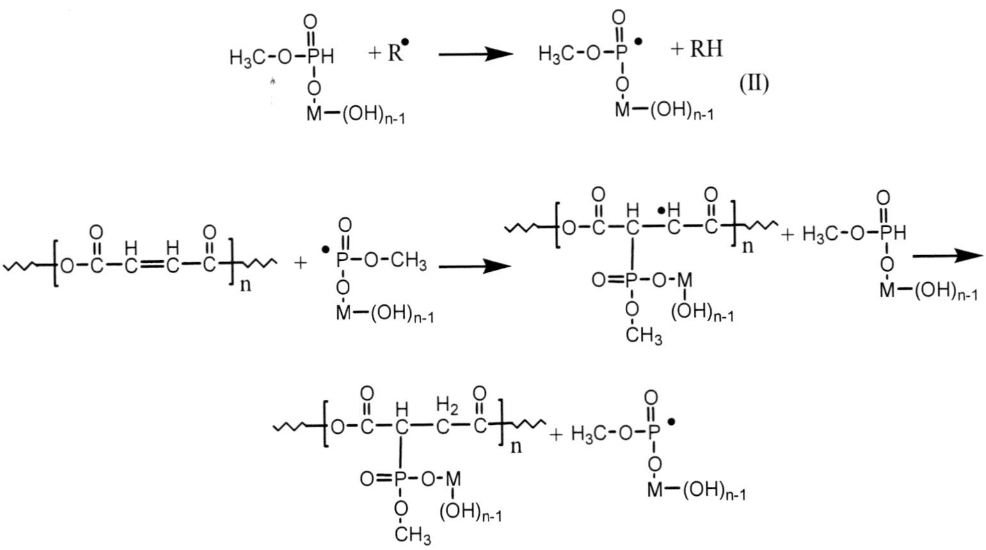

(II)

**Scheme (II):**

**Table 1. Influence of structure of a composition on hardness of the reinforced samples**

| The modifier | The contents of the modifier, % weight | Ratio in a composition of a nonsaturated polyether and styrene; hardness at break of samples | | |
|---|---|---|---|---|
| | | 1 : 1 | 0,4 : 0,6 | 0,3 : 0,7 |
| | | σ, МПа | σ, МПа | σ, МПа |
| - | - | 120,6 | 208,1 | 184,4 |
| Aluminium hydroxide (AH) | 20,0 | 119,8 | 207,5 | 176,8 |
| | 40,0 | 83,3 | 120,3 | 91,8 |
| The basic salt methyl phosphite of aluminium (BSMPA) | 20,0 | 187,2 | 207,8 | 149,7 |
| | 40,0 | 162,9 | 198,1 | 183,7 |
| Oxyhydroxide of magnesium | 20,0 | 98,5 | 158,4 | 144,3 |
| | 40,0 | 58,6 | 14,8 | 58,6 |
| The basic salt methylphosphite of magnesium | 20,0 | 161,3 | 139,2 | 143,1 |
| | 40,0 | 176,4 | 88,5 | 136,8 |
| Oxyhydroxide of calcium | 20,0 | 38,5 | 120,0 | 130,0 |
| | 40,0 | 15,4 | 39,2 | 41,5 |
| The basic salt methylphosphite of calcium | 20,0 | 118,5 | 75,4 | 98,9 |
| | 40,0 | 72,5 | 37,4 | 26,6 |

The results of researches of a thermal degradation developed compositions shown in table 2. The increasing of coke production at introduction phosphorus containing products is observed. Base salts of methyl Aluminium, magnesium, calcium phosphites reduce a thermal degradation and increase the coking after pyrolysis of hardened samples in 3,0 - 3,4 times in compare with unmodified polyester resin and in 1,3 - 1,8 times in compare with corresponding metal hydroxides [34, 35]. Results of flammable investigation shown that samples including base salts have KI by 3-6 units more than samples including metal hydroxides [18, 18, 19, 36, 37].

**Table 2 Influence of the contents of the modifier on a thermal degradation and a carbonization of samples of a composition**

| Modifier | Temperature of loss of weight, $^0C$ | | | | The coking rest, % |
|---|---|---|---|---|---|
| | $T_{20}$ | $T_{40}$ | $T_{60}$ | $T_{80}$ | |
| - | 360 | 370 | 380 | 420 | 10 |
| Aluminium hydroxide (AH) | 320 | 368 | 385 | - | 22 |
| The basic salt methyl phosphite of aluminium (BSMPA) | 350 | 400 | 425 | - | 30 |
| Oxyhydroxide of magnesium | 328 | 358 | 400 | - | 20 |
| The basic salt oxyhydroxide of magnesium | 350 | 400 | 445 | - | 34 |
| Oxyhydroxide of calcium | 360 | 370 | 400 | - | 23 |
| The basic salt of oxyhydroxide of calcium | 360 | 400 | 450 | - | 33 |

**Note.** The contents of the modifier in a composition - 30,0 % mass, ratio PN-I : styrene = 0,4:0,6.

The base metals salts methyl phosphite are effective retardants of burning of compositions on the basis of epoxy resins [9,10].

Dian epoxy resin ED-20 and polythenepolyamine (PEPA) like hardener in amount of 5 mass parts on 100 mass parts of resin were used for preparation of samples. 25, 50, 75 and 100 mass parts BSMPA or AH on 100 mass parts ED-20 were used (table 3).

**Table 3 - Compoundings of the investigated compositions**

| **№ samle** | **1** | **2** | **3** | **4** | **5** | **6** | **7** | **8** | **9** |
|---|---|---|---|---|---|---|---|---|---|
| ED-20 | 100 | 100 | 100 | 100 | 100 | 100 | 100 | 100 | 100 |
| PEPA | 5 | 5 | 5 | 5 | 5 | 5 | 5 | 5 | 5 |
| BSMPA | 0 | 25 | 50 | 75 | 100 | 0 | 0 | 0 | 0 |
| AH | 0 | 0 | 0 | 0 | 0 | 25 | 50 | 75 | 100 |

Compositions were hardened at temperature 100°C within 24 hours

The density, water absorption, factor of inflation, the coking, speed of burning, adhesive properties of the obtained compositions were determined, and also thermomechanical characteristics of samples were investigated.

Research of pyrolysis AH and BSMPA, and also epoxy compositions shown, that the coking rest (CR) of Aluminium hydroxide formed at 500°C was equal theoretically counted when it was decompose up to oxide (table 4).

$$2Al(OH)_3 \longrightarrow Al_2O_3 + 3H_2O \qquad (III)$$

**Scheme (III)**

**Table 4 - Coking rests of samples**

| **Samples** | **1** | **2** | **6** | **4** | **8** | **BSMPA** | **AH** |
|---|---|---|---|---|---|---|---|
| contents | ED-20 | 25 mass parts BSMPA | 25 mass parts AH | 75 mass parts BSMPA | 75 mass parts OhA | – | – |
| CR, % | 6,9 | 21,9 | 13,9 | 38,4 | 27,6 | 76,4 | 67,0 |

Coking rest of BSMPA on absolute value is higher, than for AH. It is probably bonding with formation of polymeric aluminium phosphate during pyrolysis.

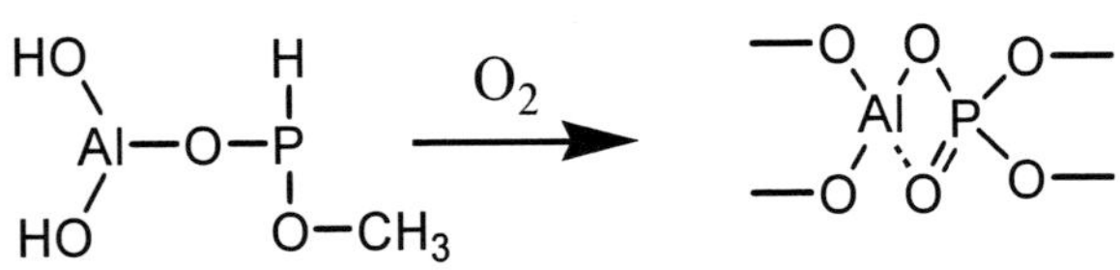


$+ 3H_2O + CO_2$ (IV)

**Scheme IV**

Also it was founded, that a composition containing BSMPA has have the greater coking rest, than the compositions filled AH with identical filling. Thus compositions containing BSMPA has have higher factor of inflation than unfilled compositions of ED-20 resin (fig. 1).

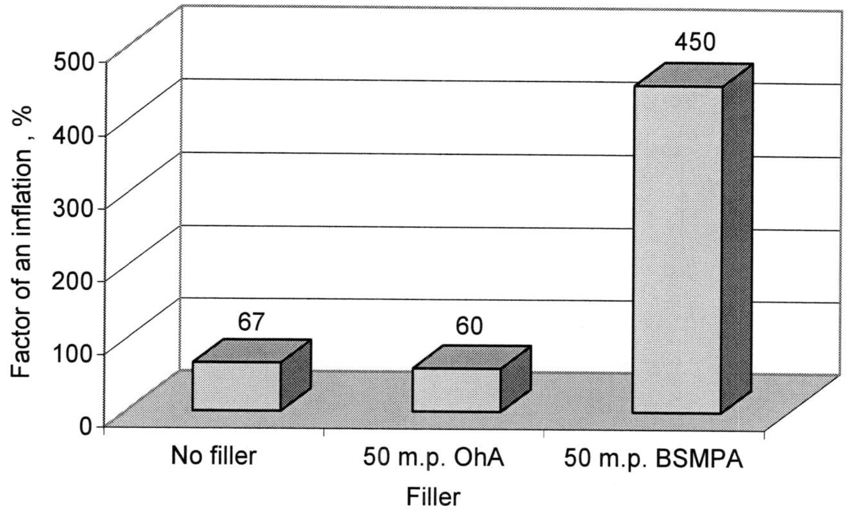


**Figure 1** - Factors of inflation

This fact may be explained by formation of 4 molecules of gaseous products at decomposition of 1 molecule BSMPA, and by formation of only 1,5 molecules at decomposition of Aluminium hydroxide.

Decreases of speed of burning of compositions containing BSMPA are submitted on fig.2.

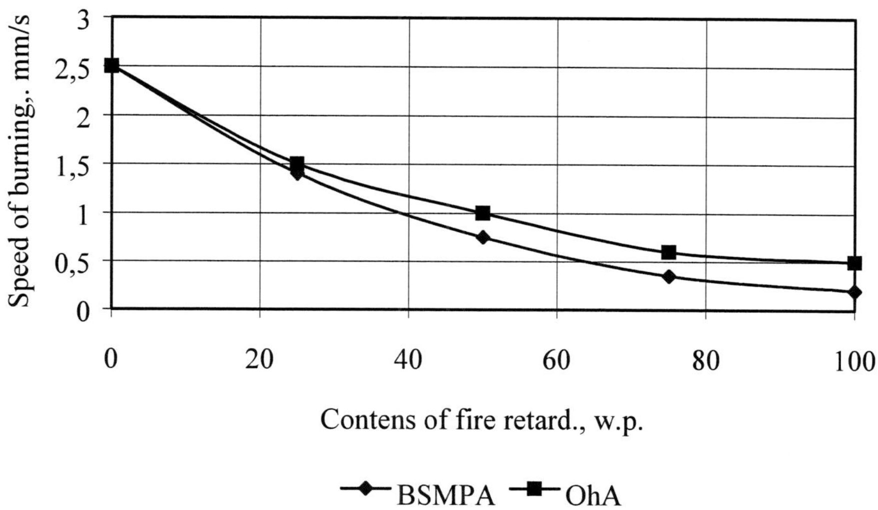


**Figure 2** - The influence of amount of fire retardant on speed of burning of hardened compositions

The influence of quantity of filling on temperature of glass transition is submitted on figure 3.

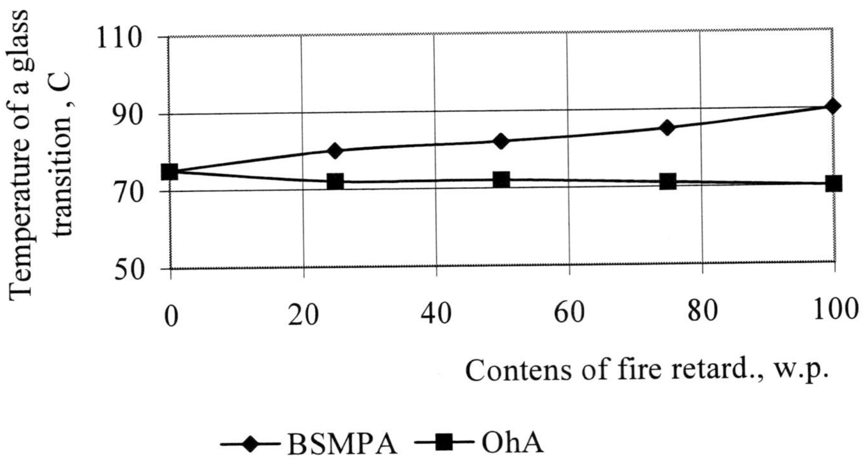


**Figure 3** - The influence of contents of fire retardant on temperature of a glass transition of compositions

The influence of fire retardants on water absorption of hardened compositions for 24 hours is submitted on figure 4.

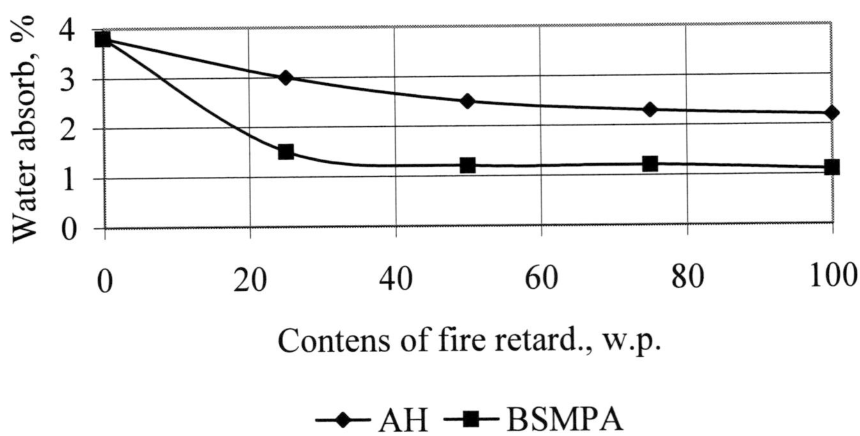


**Figure 4** - Dependence water absorption of compositions by the contents of fire retardant

Adhesive properties of compositions to a metal substrate (change of hardness at shift) are submitted on fig. 5.

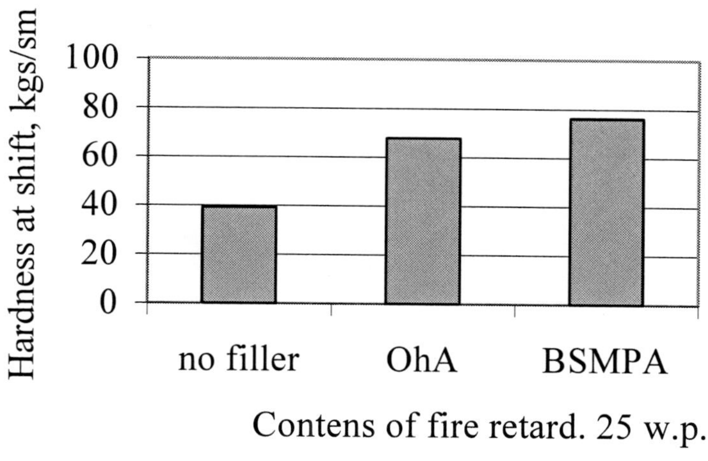


**Figure 5** - The influence of fire retardant on hardness at shift

Samples containing BSMPA had increase hardness at shift in 2 times, containing AH – increase in 1.5 times.

In general it was possible to say that including BSMPA in epoxy resin considerably improves its properties. The obtained epoxy compositions can be used as slow-burning glues, the compounds binding, and fibreglasses - as constructional materials, which have properties surpass the similar compositions filled by Aluminium hydroxide.

## 1.1. Synthesis and Application Boron Phosphorus Fire Retardants

Polymeric substances on the basis of a boric acid such as polyanhydrides and polyesters were known. . Phosphorus-, boron- containing oligomeric and polymeric products and especial the products containing bonds B-O-P-are not so investigated.

Synthesis of phosphorus–boron-containing oligomers by interaction of a boric acid and dimethylphosphite at various temperatures and ratio of reagents may be carried out according by scheme (V):

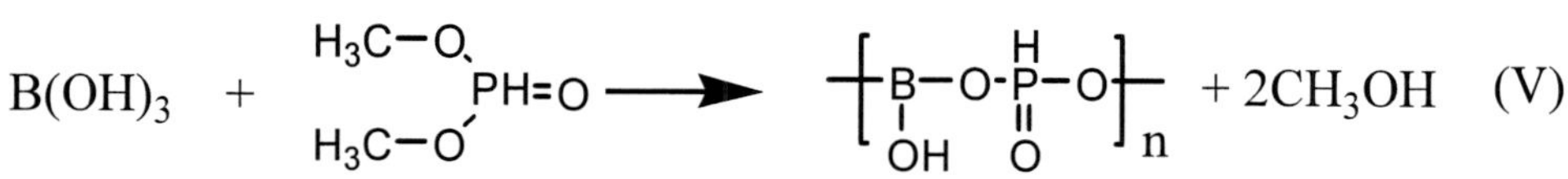


**Scheme (V):**

The influence ratio of initial reagents on physico-chemical properties obtained phosphorus-boron oligomers (PBO) is submitted in table 4.

We have synthesized phosphorus-boron-containing polyols according by following scheme (VI)[23]:

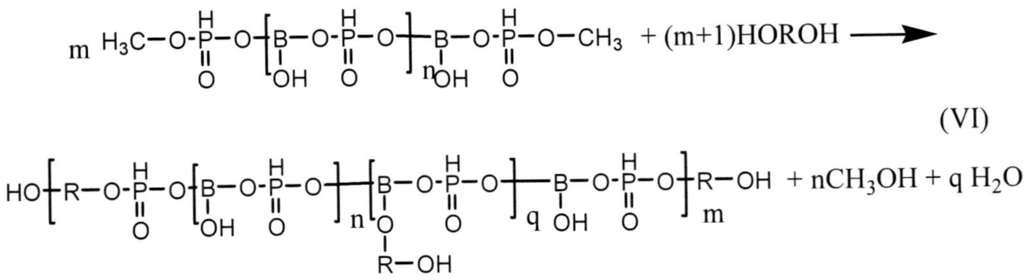

(VI)

where: p + q = n, n = 3 – 4; h + g = m

R = $-CH_2-CH_2-$; $-CH_2-CH_2-O-CH_2-CH_2-$;

$-CH_2-CH_2-CH_2-CH_2-$; $-CH_2-CH(OH)-CH_2-$;

**Scheme (VI)**

PBO are effective phosphorylating agents of polyvinyl alcohol (PVA) [7], and hardeners for epoxy resins [30].

It was shown, when phosphorylation of PVA carried out reactions of cross-linking take place. The products of reaction not solved in water. Therefore the further researches were carried out using preliminary obtained PVA films.

Change of some properties of PVA with change of concentration of PBO is submitted in figures 6-9.

When contents of PBO took place in water increase coking was found.

These data were coordinated to results of research of combustibility of the modified films of PVA. Increase of contents of phosphorus and boron in the modified samples combustibility is sharply reduced. Initial films of PVA burns down completely in open flame. Films of PVA, modified by PBO, fade. Films are not ignited at concentration of PBO for about 20 %.

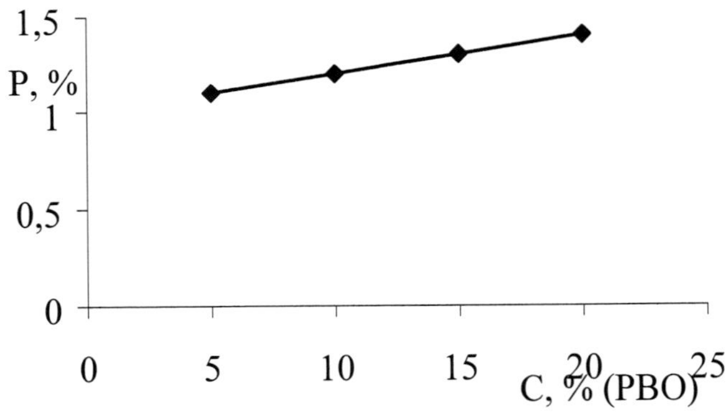


Temperature of reaction - 20 0C, Time - 2 hours

**Figure 6** - The influence of concentration of FBO on contents of phosphorus in films of PVA

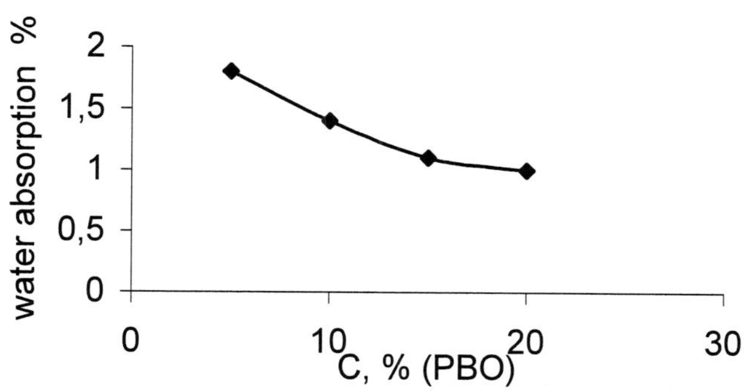


**Figure 7** - The influence of concentration of FBO on water absorption of films of PVA

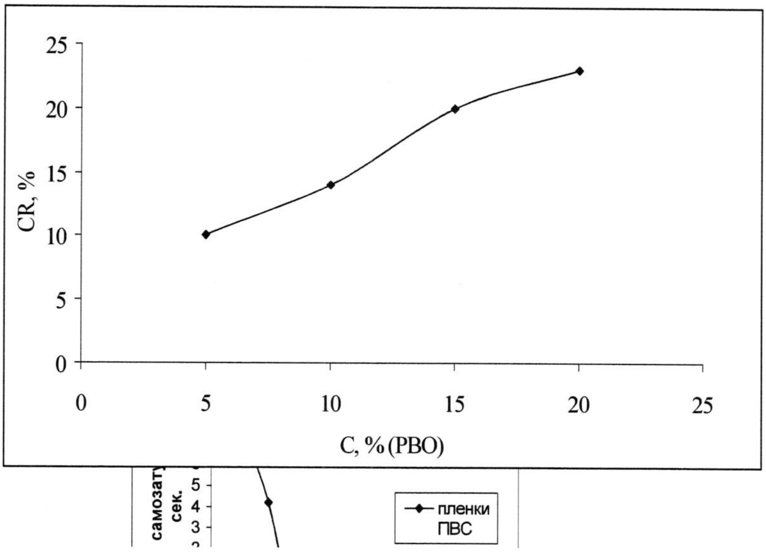


**Figure 8** - The influence of concentration of PBO on the coking rest after pyrolysis (300°C)

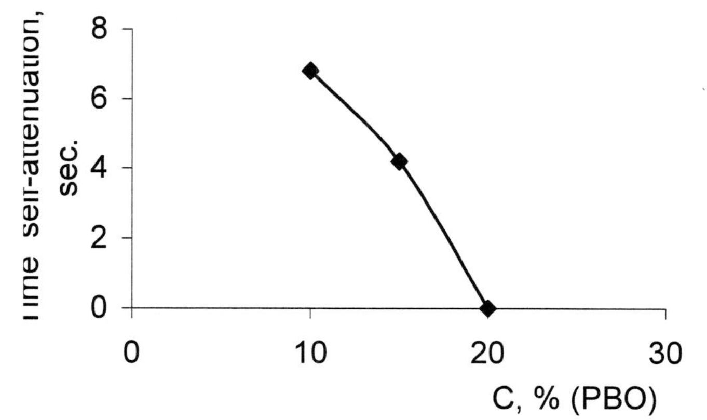


**Figure 9** - The influence of concentration of FBO for the period of self-attenuation of films of PVA

It is necessary to note, that this way of modification of polyvinyl alcohol can be used as for fireproof processing a paper as a cotton fabric by impregnation solutions of modified polyvinyl PVA.[12, 14, 16, 21, 27].

Application of FBO in manufacture of artificial leather increases their fire-resistant property [33]. It was established, that using of PBO in polyurethane compositions in amounts of 5 and 10 mass parts does not cause technological difficulties and practically does not affect the basic physicomechanical properties of obtained materials. The influence of PBO on fire resistance of artificial leather is submitted in table 5.

**Table 5 - Physical - chemical properties of PBO**

| Reagents ratio $B(OH)_3 :(CH_3O)_2P(O)H$, mol/ mol | Molecular weight | $n^{20}_D$ | $d^{20}$, g/sm$^3$ | Viscosity, St | contents of phosphorus % |
|---|---|---|---|---|---|
| 1:1 | 1900 | 1,4509 | 1,7304 | 33,5 | 27,6 |
| 1:2 | 788 | 1, 4599 | 1,6849 | 27,7 | 27,1 |
| 1:3 | 485 | 1, 4538 | 1, 6520 | 3,0 | 26,9 |

Phosphorus-boron-containing polyols can be used as extenders of chain in manufacture of elastic polyurethanes on the basis of a 4,4-diphenylmethane diisocyanate (DMI) and ester of polytetramethylen glycol (PTMEG).

Distinctive feature synthesized phosphorus-boron-containing polyurethanes their increased (in comparison with industrial polymer) thermooxidizing stability (table 6).

**Table 6 - Influence of PBO on fire resistance of samples of artificial leather.**

| Amount of fire retardant, mass parts | Fire resistance after contact of a sample to a flame (sec) | | | | | |
|---|---|---|---|---|---|---|
| | 10 sec | | | 30 sec | | |
| | Residual burning, sec. | Residual decay, sec | Height of a carbonization, mm | Residual burning, sec. | Residual decay, sec | Height of a carbonization, mm |
| Without fire retardant | 31 | 100 | 170 | 3 | 52 | Has burned down for 82 sec. |
| 5,0 | 38 | 3 | 40 | 0 | 0 | 170 |
| 10,0 | 29 | 0 | 20 | 0 | 0 | 50 |

Fire resistance obtained PUR is characterized KI = 28,0-29,0.

## 1.2. Synthesis and Application of Phosphorus Silicon Containing Fire Retardants

Researches of phosphorus containing colloid oxides of silicon (white soot) as fillers, retardants of burning and hardeners of epoxy resins are carried out.

Phosphorus containing white soot was synthesized by their interaction with dimethylphosphite. White soot of marks WS - 50, WS - 100, WS - 120 (WS) were used.

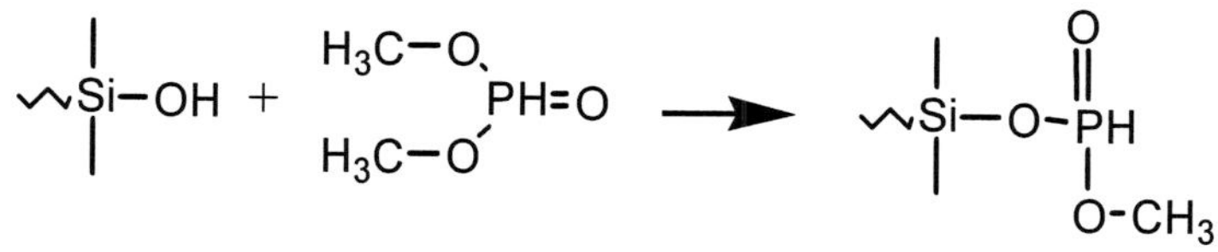

+ $CH_3OH$) (VII)

**Scheme (VII)**

It is established, that with the rise in temperature from 150 up to 180°C the contents of phosphorus grows in products of reaction. The most contents of phosphorus were found after interaction with WS-50 (Fig. 10). Reaction was carried out during 3-4 hours. Products of reaction were separated during 4 hours by acetone in device Soksleta and dried up to constant weight under vacuum.

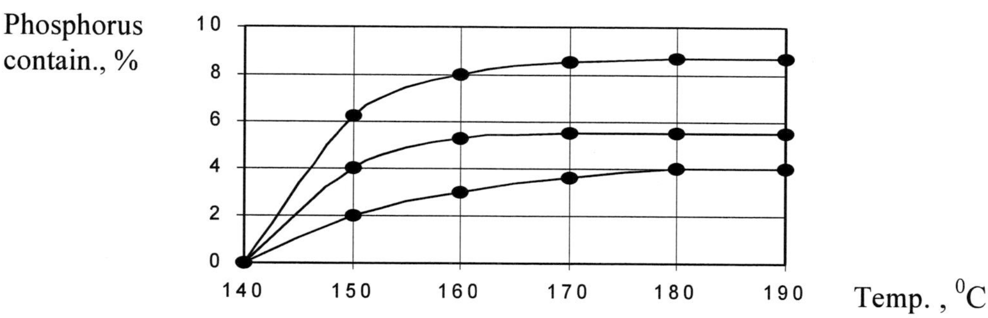


Ratio of reagents: WS:DMP-1:1. Time of reaction - 3 hours
1- WS-120; 2- WS-100; 3- WS-50

**Figure 10 -** The influence of temperature of reaction on the contents of phosphorus in white soots

Some properties phosphorus containing white soots are submitted in table 7.

**Table 7 - Parameters thermooxidizing destruction polyurethanes (PUR)**

| Parameters | PUR | PBPUR-1 | PBPUR-2 | PBPUR-3 | PBPUR-4 |
|---|---|---|---|---|---|
| Contents of phosphorus, % | - | 0,8 | 1,2 | 2,0 | 1,8 |
| Temperature started decomposition, °C | 270 | 280 | 280 | 295 | 290 |
| The coking rest after pyrolysis (600°C) | 2 | 4 | 5 | 8 | 10 |

Fire resistance received PUR is characterized KI = 28,0-29,0.

Obtained products were tested in compositions of ED-20 epoxy resin. Samples were prepared by dispergation phosphorus containing white soots in resin and than they were putted on fiberglass fabric. The compositions were curing during five hours in the temperatures interval 110 – 160°C.

The influence of contents of the modified WS - 50 on hardness of hardened materials, a degree of a curing and combustibility are submitted in table 8.

**Table 8 – Physical-chemical properties phosphorus containing white soots**

| № | Показатель | WS -50 | WS -100 | WS – 120 |
|---|---|---|---|---|
| 1 | Density, | 1,580 | 1,761 | 1,730 |
| 2 | Specific surface, $sm^2/g$ | 27430 | 50210 | 65480 |
| 3 | Contents of phosphorus, % | 10,3 | 4,1 | 3,0 |
| 4 | Temperature of 5% loss of weight, $^0C$ | 200 | 200 | 250 |

Phosphorus containing WS-100 and WS-120 were obtained too. Property of epoxy resins composition including WS-50 shows the best results.

**Table 9 - Influence a phosphorylation of white soot WS-50 on some properties of hardened compositions.**

| Contents of WS-50, % | Hardness at break, MPa | Degree of a curing, % | Time of attenuation of samples after influence of an open flame, sec |
|---|---|---|---|
| 10,0 | 93,5 | 87,6 | 7,0 |
| 20,0 | 76,7 | 90,1 | 4,0 |
| 30,0 | 76,4 | 93,5 | 4,0 |
| 40,0 | 75,9 | 94,2 | 3,8 |

Shown, that obtained element phosphorus containing materials may be used as fire retardants and be surpassed ordinary inorganic and phosphorus containing fire retardants. Synthesized fires retardant does not aggravate property of obtained composition comparison with ordinary inorganic and phosphorus containing fire retardants. Except for that they can be used as retardants of corrosion [40-43].

# 2. Surface Modification of Cured Rubber

## 2.1. Oxidizing Chlorine Phosphorylation of a Crushed Rubber

Surface modification elastomers is one of methods of protection polymer matrix about action of flame. It was shown that engrafting phosphotus element organic functional group changed property of surface significantly. Number of scientific papers informs phosphorylation elastomers. The phosphorylation of carbochain polymers was considered in the review [44], synthesizes of phosphorus containing polymers was considered in monography [45], modification rubber by phosphorus-organic materials was considered in review [46]. Results of phosphorylation crushed cured rubber (car protectors) are introduced as model of modification in this article.

As a reaction of chlorine phosphorilation proceed by radical mechanism [50] that crushed rubber washed off by tetra chlorine ethylene in Sokslet's device for extraction age resistors.

Process phosphorilation of crushed rubber (CR) carried out in reactor with porous bottom where placed phosphorus trichloride and CR. Mass ratio $PCl_3$ : CR was 15:1.

Significant overflow of three-chloride phosphorus was used for mobility of reactionary pulp at barbotage of oxygen. The maximum contain 3,5% of phosphorus was found after 4 hours interaction at 60°C. It was found that contain of phosphorus in product of reaction increased very slowly. For hydrolyzes halogen hydrides groups the product of reaction was treated by hot water. Chlorine contains was changed from 1,5 up 4.0%

As known phosphonats group can be form complexing with salt of metals that it was research of sorption property to ions mercury (II). Results of research are sumitted in table 10.

**Table 10 - The influence of the phosphorus containing on sorption properties of CR Concentration of ions of mercury (II) in a solution 0,1 mg / ml, weight of a sorbent 0,5 g, volume of a solution of 50 ml., time sorption 24 hours, temperature 20-25 °C, pH 2-3**

| Phosphorus contents | pH a solution | Sorption of ions of mercury from a solution % |
|---|---|---|
| 1,5 | 2,3 | 19,4 |
| 2,1 | 2,2 | 49,0 |
| 3,3 | 2,1 | 97,5 |
| 3,5 | 2,1 | 97,5 |

It was shown that at the contents of phosphorus 3,5 % in CR provided almost quantitative absorption ions of mercury.

## 2.2. Modification CR by Dimethyphosphite

Modification of CR by dimethyl phosphit on absents peroxides environment carbon tetrachloride gives phosphorus containing samples with 2,4 %. Reactions on present's peroxides were obtained samples with 7-8% of phosphorus. This phenomenon may be explained by the Todd-Attertons reaction [51]:

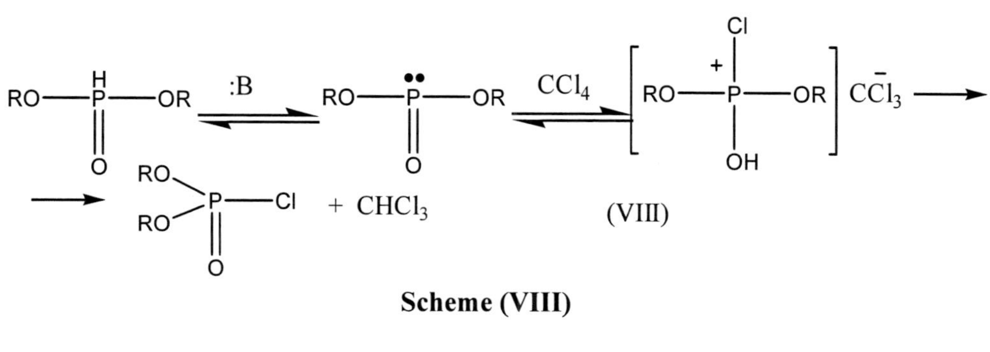


**Scheme (VIII)**

The subsequent reaction of formed chlorine anhydride with double bond [47]:

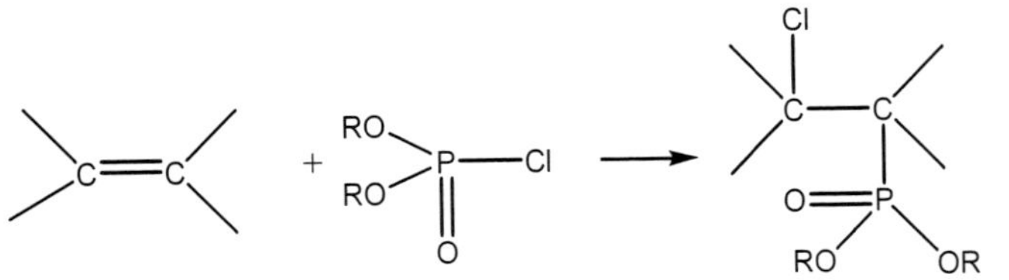


(IX)

where R=CH3; :B - aromatic amine

**Scheme (IX)**

Samples with the various phosphorus containing have been investigated as sorbents of ions $Hg^{+2}$ from water solutions. The obtained results are shown in table 11.

**Table 11 - The influence the contents of phosphorus and pH on a degree of extraction of ions of mercury from water solutions it was used 50 ml of solutions $HgCl_2$ with concentration of 0,1 mg/g and CR 0,5-1,0g.**

| Phosphorus contents | PH solution | Degree of extraction of ions $Hg^{+2}$ |
|---|---|---|
| 2,3 | 2,0 | 90,0 |
| 2,3 | 3,1 | 94,4 |
| 2,3 | 6,7 | 95,0 |
| 4,3 | 2,0 | 96,0 |
| 4,3 | 3,3 | 99,0 |
| 4,3 | 6,7 | 99,0 |
| 6,6 | 2,1 | 97,0 |
| 6,6 | 3,1 | 98,0 |
| 6,6 | 7,1 | 98,0 |
| 7,6 | 2,1 | 99,0 |
| 7,6 | 3,8 | 99,0 |
| 7,6 | 7,2 | 99,0 |

## 2.3. Modification CR by Element - Phosphorus Containing Substrates

The base chemical reaction is of modification [52, 53]:

$P_4O_{10} + 4H_2O_2 + 2H_2O \rightarrow 4H_3PO_5$

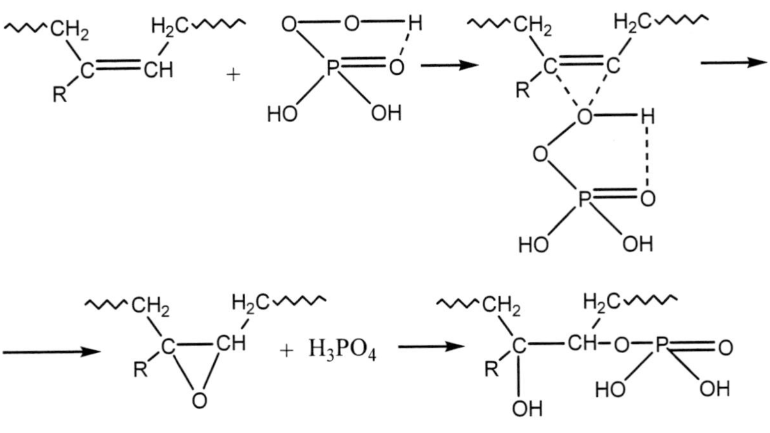

where R = H, CH3

**Scheme (X)**

Reactions of double bonds of cross-linking rubbers on a surface with peroxiphosphoric acid are well described in [54, 55].

Properties obtained CR were submitted in [55] and shown in table 12.

**Table 12 - The influence of conditions of reaction on the contents of phosphorus in rubber and acid number (weight of rubber 5 g).**

| Reagents | | | | Output | | Temperature. | Time of reaction | Phosph. contents | Acid number |
|---|---|---|---|---|---|---|---|---|---|
| $P_2O_5$ | | $H_2O_2$ | | | | | | | |
| g | mol | g | mol | g | mol | °C | hour | % | mgNaOH/g |
| 8,00 | 0,06 | 1,95 | 0,06 | 4,55 | 0,25 | 25-30 | 5,0 | 3,7 | 81,5 |
| 6,00 | 0,045 | 1,46 | 0,45 | 5,41 | 0,31 | 25-30 | 5,2 | 2,2 | 56,7 |
| 8,00 | 0,06 | 1,95 | 0,06 | 9,55 | 0,53 | 20-25 | 5,0 | 1,9 | 48,6 |
| 8,00 | 0,06 | 1,95 | 0,06 | 15,00 | 0,83 | 20-23 | 5,0 | 0,9 | 23,8 |

Shown, that double bonds of cross-linking polymers can react with phosphorus containing reagents and take place to obtaine materials with fire retardant surface.

## REFERENCES


[1] Tuzhikov O.I., Bondarenko S.N., Ljabin M.P., Hohlova T.V. *Research of properties of the composite materials containing the modified hydrooxide of aluminium* / the Collection of proceedings. « Chemistry and technology an element of organic monomers and polymeric materials », Volgograd, 1993, p. 86 - 89.

[2] Ljabin M.P. Development and research of properties of compositions with new phosphorus aluminium containing fire retardants: the Author's abstract of the dissertation on competition of a scientific degree of Cand.Chem.Sci. / VolgGTU. - Volgograd, 1994. - 22 p.

[3] Tuzhikov O.I., Bondarenko S.N., Ljabin M.P., Hohlova T.V. *Kinetic of reaction of oxyhydroxide of magnesium with dimethyl phosphite* / Materials I of the All-Russia conference « Polymeric materials of the lowered combustibility », Volgograd, 1995, p. 28.

[4] Tuzhikov O.I., of Bondarenko S.N., of Ljabin M.P., Hohlova T.B. *Properties of the fibreglasses containing as filler the modified hydrooxide of magnesium.* / In the same place, p. 35 - 37.

[5] Tuzhikov O.I., of Bondarenko S.N. Development new phosphorus-, metallic fire retardants. *Research of a kinetic of reaction of oxyhydroxide of aluminium with dimethylphosphite.* / In the same place, p. 40 - 43.

[6] Bondarenko S.N., of Tuzhikov O.I., of Hohlova T.V. *Modification of a hydrooxide of calcium dimethylphosphite and application of received products as fire retardants.* / In the same place, p. 48 - 50.

[7] Bondarenko S.N., Tuzhikov O.I., Hohlova T.V., Fedotova T.A. *It is limited the combustible PVS, the modified phosphorus-boron-containing fire retardants.* / In the same place, p. 52 - 54.

[8] Bondarenko S.N., Tuzhikov O.I., Hohlova T.V. *Kinetic of reaction of oxyhydroxide of calcium with dimethylphosphite in a nitrobenzene.* - Collection of proceedings «Chemistry and technology an element of organic monomers and polymeric materials», Volgograd, 1996, p. 51 - 56.

[9] Patent RF № 2056444. *A fire-resistant composition.* / Bondarenko S.N., Tuzhikov O.I., Hohlova T.V., Ljabin M.P., Goremykina T.S. - the Bulletin of Inventions, № 8, 1996.

[10] Patent RF № 2056445. *A fire-resistant composition.* / Bondarenko S.N., Tuzhikov O.I., Hohlova T.V., Ljabin M.P., Goremykina T.S. - the Bulletin of Inventions, № 8, 1996.

[11] Bondarenko S.N., Tuzhikov O.I., Hohlova T.V., Koshnareva E.A., Samohina N.V., Lopastejskaja T.I. *Metal-, phosphorus containing fire retardants for fibreglasses.* / Materials of the International symposium «Ecology and safety of ability to live, scientifically - applied aspects, engineering decisions », Volgograd, 1996, p. 61 - 62.

[12] Bondarenko S.N., Tuzhikov O.I., Hohlova T.V., Orlova S.A., Bahtina G.D., Derbisher V.A. *Research of dressing of cellulose of containing fibrous materials phosphorus containing fire retardants a regressive method.* / Collection of proceedings «Chemistry and technology an elementorganic monomers and polymeric materials », Volgograd, 1997, p. 115 - 123.

[13] Bondarenko S.N., Tuzhikov O.I., Hohlova T.V., Avanisjan ZH.G. *Synthesis and research of properties phosphorus of containing additives to polymeric compositions.* / In the same place, p. 101 - 107.

[14] Bondarenko S.N., Tuzhikov O.I., Hohlova T.V., Orlova S.A., Derbisher V.E. *Fireproof modification of fibrous materials by a boratemethyl phosphite.* – Journal of applied chemistry, 1997, v. 70, № 10, p. 1725 – 1728.

[15] Bondarenko S.N., Tuzhikov O.I., Hohlova T.V., Krjachko O.M. *Technological aspects of modification of polymeric composites phosphorus metal containing connections.* / Materials V international scientifically - technical conference «High technology chemical technologies - 98 », Yaroslavl, 1998, p. 345 - 346.

[16] Bondarenko S.N., Tuzhikov O.I., Hohlova T.V., Orlova S.A., Derbisher V.E. *Application phosphorus of containing fire retardants of complex action for dressing textile materials.* / In the same place, p. 366 - 368.

[17] Bondarenko S.N., Orlova S.A., Derbisher V.E. *Decrease in combustibility PUR by introduction phosphorus-boron-containing retardants of burning.* / Materials of III international conference « Polymeric materials of the lowered combustibility », Volgograd, 1998, p. 33.

[18] Bondarenko S.N., Orlova S.A., Derbisher V.E. *Technology of processing PUR with application phosphorus-metal-containing products.* / In the same place, p. 31 - 32.

[19] Bondarenko S.N., Tuzhikov O.I., Hohlova T.V., Dhajbe M.H., Guseva D.A. *Development of polymeric materials of the lowered combustibility on the basis of a nonsaturated polyether.* / In the same place, p. 22 - 23.

[20] Bondarenko S.N., Tuzhikov O.I., Hohlova T.V., Guguev A.V. *Phosphorus silicon containing hardeners and retardants of burning of epoxy resins.* / In the same place, p. 24 - 25.

[21] Bondarenko S.N., Tuzhikov O.I., Hohlova T.V., TSvetkova E.N. *Fireproof modification of a paper phosphorusboron containing oligomers.* / In the same place, p. 25 - 26.

[22] Bondarenko S.N., Tuzhikov O.I., Hohlova T.V., Krekaleva T.V., Derbisher V.E. *Modification of compositions on the basis of a nonsaturated polyether of PN - 1 basic salt methylphosphite of aluminium.* / the Collection of proceedings «Chemistry and technology an element of organic monomers and polymeric materials», Volgograd, 1999, p. 94 - 100.

[23] Bondarenko S.N., Tuzhikov O.I., Hohlova T.V., Orlova S.A., Derbisher V.E., Bahtina G.D. *Modification industrial PUR methacrylic derivatives of acids of phosphorus.* / In the same place, p. 66 - 70.

[24] Bondarenko S.N., Orlova S.A., Derbisher V.E., Vasiljeva V.D. *Modified полиуретаны the lowered combustibility.* / Materials of the international conference « Modern technologies in education and a science », Saratov, 1999, p. 77.

[25] Bondarenko S.N., Derbisher V.E., Vasiljeva V.D. *Phosphorus containing methacrylath as the modifier PUR.* / In the same place, p. 76.

[26] Orlova S.A. *Synthesis and research PUR with use a boron of containing derivatives dimethyl phosphite*: the Author's abstract of the dissertation on competition of a scientific degree of the candidate technical. Sciences. / VolgGTU. - Volgograd, 2000. - 24 p.

[27] Bondarenko S.N., Ljabin M.P., Zheltobrjuhov V.F., Bogacheva L.V., Lavnikova I.V.*modification polycapramid phosphorus a boron containing oligomer.* / Materials of IV international conference «Polymeric materials of the lowered combustibility», Volgograd, 2000, p. 61 – 62

[28] Bondarenko S.N., Tuzhikov O.I., Hohlova T.V., Orlova S.A., Derbisher V.E. *Synthesis and research of properties of oligomers on the basis of a borate methyl of phosphite and glycols.* / In the same place, p. 62 - 63.

[29] Bondarenko S.N., Tuzhikov O.I., Krekaleva T.V., Derbisher V.E. *Fireproof modification of materials on the basis of a polyether of PN - 1 metallic derivative dimethil phosphite.* / In the same place, p. 52 - 53.

[30] Bondarenko S.N., Tuzhikov O.I. *the Covering of the lowered combustibility on the basis of resin ED - 20 and phosphorus-containing oligomer.* / In the same place, p. 55 - 56.

[31] Bondarenko S.N., Tuzhikov O.I., Hohlova T.V., Orlova S.A., Derbisher V.E. *Modification elastic PUR phosphorus a boron containing connections.* / In the same place, p. 106 - 117.

[32] Patent № 2152949. *A way of reception methyl of phosphite of aluminium.* / Bondarenko S.N., Ljabin M.P., Tuzhikov O.I., Zheltobrjuhov V.F., Moskvichev S.M. - Bulletin of Inventions, № 20, 2000.

[33] Patent № 2134737. An artificial leather. / Bondarenko S.N., Tuzhikov O.I., Hohlova T.V., Orlova S.A., Derbisher V.E., Vasiljeva V.D. - *Bulletin of Inventions*, № 23, 2000.

[34] Bondarenko S.N., Tuzhikov O.I., Hohlova T.V., Krekaleva T.V., Derbisher V.E. *Influence metal containing derivatives dimethil phosphite on pyrolysis of some*

*polymers.* / the Collection of proceedings «Chemistry and technology an elementorganic monomers and polymeric materials», Volgograd, 2000, p. 222 - 225.

[35] Bondarenko S.N., Tuzhikov O.I., Hohlova T.V., Krekaleva T.V., Derbisher V.E. *Fireproof modification of polyester resins by the basic salt methylphosphite of magnesium.* / In the same place, p. 218 – 222.

[36] Bondarenko S.N., Tuzhikov O.I., Krekaleva T.V., Derbisher V.E. *Research of pyrolysis of compositions on the basis of the nonsaturated polyether modified by salts of nickel and cobalt.* / Materials All-Russia scientifically - technical conference « Hardin's readings », Volgograd, 2001, p. 122.

[37] Bondarenko S.N., Tuzhikov O.I., Krekaleva T.V., Derbisher V.E. *Influence on parameters of hardness of the reinforced polyester compositions phosphorus metal containing retardants of burning.* / In the same place, p. 123.

[38] Bondarenko S.N., Tuzhikov O.I., Hohlova T.V., Orlova S.A. *Synthesis and research phosphorus containing methacrylaths - perspective retardants of burning of polymeric materials.* / Materials of the international conference «Perspective polymeric composite materials. Alternative technologies. Processing. Application. Ecology », Saratov, 2001, p. 97 - 100.

[39] Bondarenko S.N., Tuzhikov O.I., Hohlova T.V., Ljabin M.P., Moskvichev S.M. Studying of a kinetic of reaction of oxyhydroxide of aluminium with dimethilphosphite in a nitrobenzene. - *Jornal of applied chemistry*, 2001, v. 74, r.12, p. 2059 - 2061.

[40] The patent № 2186877. Retardant of corrosion of steel in neutral water environments / Bondarenko S.N., Ljabin M.P., Moskvichev S.M., Tuzhikov O.I. - *Bulletin of Inventions*, № 22, 2002.

[41] The patent № 2186876. A way of protection of steel from corrosion in neutral water environments. / Bondarenko S.N., Zheltobrjuhov V.F., Ljabin M.P., Moskvichev S.M., Tuzhikov O.I. - *Bulletin of Inventions*, № 22, 2002.

[42] The patent № 2190697. Retardant of corrosion of steel in neutral environments. / Bondarenko S.N., Zheltobrjuhov V.F., Ljabin M.P., Moskvichev S.M., Tuzhikov O.I. - *Bulletin of Inventions*, № 28, 2002.

[43] Bondarenko S.N., Tuzhikov O.I., Ljabin M.P., Hohlova T.V., Moskvichev S.M. Studying of a kinetic of reaction of oxyhydroxide of magnesium with dimethil phosphite. - *Jornal of applied chemistry*, 2003, v.76, r.8, c. 1251-1253.

[44] Hardin A.P.,Bondarenko S.M., Tuzhikov O.I. Phosphorylation of carbon-chain polymers / *Successes of chemistry*, v.52, exp.7, 1983, 1173-1195.

[45] Allavediniv A.B. *Synthesis both research of properties and application of phosphorus of containing materials*

[46] Ratnikova T.V. Modification rubber by phosphorus containing substances. *Rubber and rubber*, 1979, №12, page 18-23

[47] Drozdovskii V.F., Razgon D.R. Processing and use of the worn out trunks. *Rubber and rubber*, 1995, №2, page 2-8

[48] Makarov V.M., Drozdovskii V.F. Use of left from the use of trunks and wastage of rubber products. *Л., Chemistry*, 1981, 249p.

[49] Drozdovskii V.F. Application of the modified and unmodified crushed cured rubber. *Rubber and rubber*, 1997, №2, p. 48-50

[50] Rahimov A.I. Synthesis phosphorus organic substans. *M., Science*, 1983г. 248 p.

[51] Nifantev E.E. Chemistry of hidrophosphorils substans / *M.: Science*, 1983-264 p.

[52] T.V.Hohlova, O.O.Tuzhikov, G.R.Gadzhiev, T.V.Ivanov, O.I.Tuzhikov. *Sorbent on the basis of the crushed automobile tire covers for extraction of organic compounds from waste waters.* // Science chemical technologies - 2002: materials [the Theses of the reports] of the Eighth International scientific and technical conference., Ufa, October 7-10 2002 /Ufa state petroleum technical universities - Ufa, 2002, - p. 230-231

[53] O.I.Tuzhikov, T.V.Hohlova, O.O.Tuzhikov, L.G.Lazebnaia, A.U.Samoilenko. *Research of phosphorylation and ozonization of the crushed rubber* // of a Slip of mountain institute [St.-Petersburg] Exp. Ecology and rational nature use, 2003 V.154, page 212.

# INDEX

## A



## B



## C



## D



## E

## F



## G



## H



## I



## L



## M



## N



## O

## P



## Q



## R



## S



## T



## V



## W